NANOTHERMODYNAMICS

Theory and Applications

NANOTHERMODYNAMICS

Theory and Applications

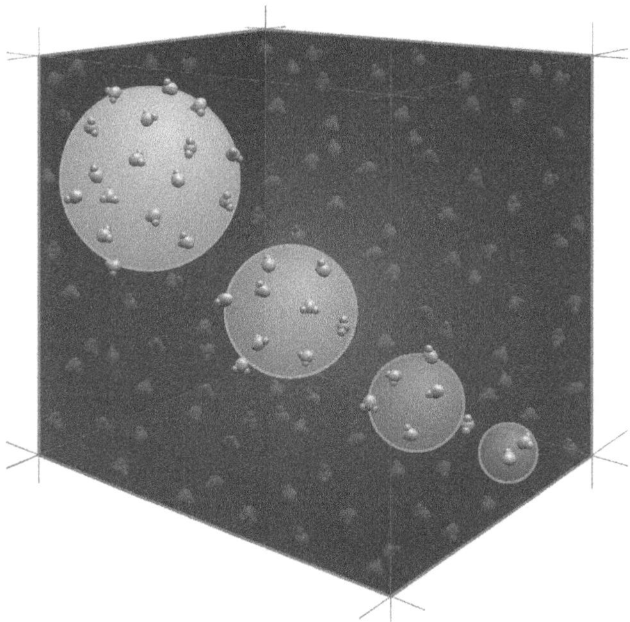

Dick Bedeaux
Signe Kjelstrup
Sondre K Schnell
Norwegian University of Science and Technology, Norway

World Scientific

NEW JERSEY • LONDON • SINGAPORE • BEIJING • SHANGHAI • HONG KONG • TAIPEI • CHENNAI • TOKYO

Published by

World Scientific Publishing Co. Pte. Ltd.
5 Toh Tuck Link, Singapore 596224
USA office: 27 Warren Street, Suite 401-402, Hackensack, NJ 07601
UK office: 57 Shelton Street, Covent Garden, London WC2H 9HE

Library of Congress Control Number: 2023941927

British Library Cataloguing-in-Publication Data
A catalogue record for this book is available from the British Library.

NANOTHERMODYNAMICS
Theory and Applications

ISBN 978-981-127-499-2 (hardcover)
ISBN 978-981-127-500-5 (ebook for institutions)
ISBN 978-981-127-501-2 (ebook for individuals)

For any available supplementary material, please visit
https://www.worldscientific.com/worldscibooks/10.1142/13372#t=suppl

Desk Editor: Nur Syarfeena Binte Mohd Fauzi

Typeset by Stallion Press
Email: enquiries@stallionpress.com

Terrell Hill passed away in 2014 at the age of 96. We like to honor his memory with this book.

Preface

This book grew out of an idea to study properties of small subsystems of a large reservoir [1, 2]. Observations were at the time not explainable with standard thermodynamics. But the theory of Hill [3, 4] on thermodynamics of small systems provided the systematic procedure needed to address the problem. It was in 1962 that Hill proposed that thermodynamics can be formulated for the nanoscale! Since then, numerous examples have been studied, see [5–9] and references therein.

The purpose of this book is to expand — and demonstrate — Hill's theory. The theory adds a new term to the fundamental Gibbs equation, that is specific for systems at the nanoscale. The properties that follow may be counterintuitive. The equation of state for a small system, for instance, is not given once and for all. We shall see that it changes with the environmental variables that control the small system. The statistical mechanical machinery remains as before, however.

Within small systems' thermodynamics, we can now deal with single molecules, unlike in classical thermodynamics. Equilibrium properties will change upon confinement and depend on system shape. Such features make nanothermodynamics very different from classical thermodynamics. It extends the classical theory down to the nanometer-scale, and can affect systems up to the micrometer scale. As soon as the system's surface energy is sizable compared to the bulk energy, like when the Young–Laplace equation applies, we may benefit from the small system analysis.

The world of small systems challenges the standard knowledge; that the number of particles in a system must be very large for thermodynamic equations to apply. We shall see that thermodynamic equations apply perfectly well also for small particle numbers, provided that small-system effects are accounted for correctly. In the world where size and shape are central, we shall find that equations of state can be used down to one particle in a box! There are scaling laws, which help us determine and understand the large system limit better!

Readers of the book will need knowledge of classical and statistical thermodynamics, such as the books of Callen [10], Smith and van Ness [11], Widom [12] and McQuarrie [13]. We are building on these well accepted books, as well as on the books of Hill [3, 4, 14] on equilibrium thermodynamics of small systems. A preliminary version of the first 11 chapters were printed on April 2020 [15]. Since then, a special issue has appeared in *Nanomaterials* counting ten applications [16]. We have also spelled out the relevance of nanothermodynamics to porous media science [17].

The authors would like to thank all colleagues for fruitful discussions during the work on the book, in particular Jean-Marc Simon and Thijs J.H. Vlugt, who were central at the start of the work. In addition, we want to thank in alphabetical order other co-authors on original publications which have provided much of the basis for new chapters; Eivind Bering, Vilde Bråten, Máté Erdös, Alex Hansen, Jianying He, Peter Krüger, Ivan Latella, Othonas A. Moultos, Olav Galteland, Bjørn Hafskjold, Michael T. Rauter, J. Miguel Rubi, Bjørn A. Strøm, Thuat T. Trinh, Astrid de Wijn and Øivind Wilhelmsen. We are grateful to PoreLab, with its scientific director Alex Hansen and administrative leader Marie-Laure Olivier, for generous support, discussions and for providing us with a creative, supportive atmosphere.

The authors are grateful to the Norwegian Research Council for its Center of Excellence Funding Scheme, project no 262644, Porelab, and SKS is grateful for the Norwegian Research Council funding through its Young Researcher Talents program, project no 275754. The Sigma2 project no. NN9414K is thanked for CPU-time used to generate data for figures. Michel Uiterwijk is thanked for the cover design and for figures to Chapter 3.

Dick Bedeaux, Signe Kjelstrup, and Sondre K. Schnell
PoreLab Center of Excellence,
Norwegian University of Science and Technology
dick.bedeaux@ntnu.no
signe.kjelstrup@ntnu.no
sondre.k.schnell@ntnu.no

Trondheim, January 2023

Contents

Preface vii

1. Scope **1**

1.1 The Simulation Tool 2
1.2 Small Systems at Equilibrium 2
1.3 Expansion to Systems Away from Equilibrium . . . 4
1.4 Limitations 5
1.5 Book Outline 5
1.6 Perspectives 8

2. Large Versus Small Thermodynamic Systems **11**

2.1 Small Systems and Their Environment 13
2.2 Hill's Postulate: Small System Definition 16
2.3 Legendre Transforms for Ensembles of Replicas 19
2.4 Concluding Remarks 21

3. Nanothermodynamics **23**

3.1 Environmental Variables U, V, $\mathbf{N}$ 26
3.2 Environmental Variables T, V, $\mathbf{N}$ 29
3.3 Environmental Variables T, p, $\mathbf{N}$ 33
3.4 Environmental Variables T, V, $\boldsymbol{\mu}$ 37
3.5 Environmental Variables T, p, N_1, $\boldsymbol{\mu}_1$ 42
3.6 Environmental Variables T, p, $\boldsymbol{\mu}$ 46
3.7 The Subdivision Potential 49

3.8 Equilibrium Between Small Systems 50
3.9 Concluding Remarks 52

4. Statistical Mechanics 55

4.1 The Hamiltonian of Small Systems 56
4.2 Microcanonical Ensemble: U, V, $\mathbf{N}$ 57
4.3 Canonical Ensemble: T, V, $\mathbf{N}$ 59
4.4 Isobaric Ensemble: T, p, $\mathbf{N}$ 62
4.5 Grand Canonical Ensemble: T, V, $\boldsymbol{\mu}$ 67
4.6 Osmotic Ensemble: T, p, N_1, $\boldsymbol{\mu}_1$ 69
4.7 Nanocanonical Ensemble: T, p, $\boldsymbol{\mu}$ 73
4.8 Legendre Transforms for Partition Functions 76
4.9 Legendre–Fenchel Transforms 78
4.10 The Liouville–von Neumann and the Liouville equation 80
4.11 Concluding Remarks 82

5. The Small System Method 83

5.1 Size-Dependent Scaling: A Hypothesis 84
5.2 Fluctuation Formulas 87
5.3 A System of Repulsive Particles 88
5.4 The Small System in a Large Reservoir 89
5.5 The Sampling Routine: A Prescription 95
5.6 Examples 96
5.7 Concluding Remarks 100

6. Ideal Gas Mixture with Surface Energy 101

6.1 A Model for Surface Energy 103
6.2 Canonical Ensemble: T, V, $\mathbf{N}$ 105
6.3 Grand Canonical Ensemble: T, V, $\boldsymbol{\mu}$ 109
6.4 Nanocanonical Ensemble: T, p, $\boldsymbol{\mu}$ 112
6.5 Concluding Remarks 115

7. Internal Energy **117**

7.1 Partial Internal Energies in a Small System 118
7.2 An Integral Relation for the Internal Energy 119
7.3 Concluding Remarks 122

8. Small System Correlation Functions **123**

8.1 The Maxwell Velocity Distribution 125
8.2 One- and Two-Particle Densities 126
8.3 Pair Correlation Functions 128
8.4 Concluding Remarks 130

9. Fluctuations **133**

9.1 Microcanonical Ensemble: U, V, $\mathbf{N}$ 134
9.2 Canonical Ensemble: T, V, $\mathbf{N}$ 135
9.3 Isobaric Ensemble: T, p, $\mathbf{N}$ 137
9.4 Grand Canonical Ensemble: T, V, $\boldsymbol{\mu}$ 139
9.5 Osmotic Ensemble: T, p, N_1, $\boldsymbol{\mu}_1$ 143
9.6 Nanocanonical Ensemble: T, p, $\boldsymbol{\mu}$ 145
9.7 Concluding Remarks 147

10. Kirkwood–Buff Integrals **149**

10.1 Finite-Volume KB-Integrals 150
10.2 Finite-Volume KB-Integrals from Fluctuations . . . 151
10.3 The Small System Method 155
10.4 The Traditional Method 156
10.5 The Thermodynamic Limit Value 160
10.6 Concluding Remarks 161

11. Pressures and Partial Molar Enthalpies **163**

11.1 The Integral Pressure and the Integral Enthalpy Contributions 164
11.2 Integral Relation for the Differential Pressure . . . 166

11.3 Partial Molar Enthalpies 168
11.4 Transformation in the Thermodynamic Limit . . . 170
11.5 Determining Partial Molar Enthalpies 172
11.6 Concluding Remarks 175

12. Hill's versus Gibbs' Thermodynamics 177

12.1 Hill's Description of Small, Curved Systems 179
12.2 Volume- and Surface-Contributions 180
12.3 Comparing Hill's and Gibbs' Expressions 182
12.4 Concluding Remarks 184

13. Maxwell–Stefan's and Fick's Diffusivities 185

13.1 Fick's and Maxwell–Stefan's Equations 187
13.2 Transformation Relations 189
13.3 Diffusion Models 190
13.4 Maxwell–Stefan Diffusivities 191
13.5 The Thermodynamic Factor at Work 192
13.6 Concluding Remarks 194

14. Chemical Reactions 195

14.1 The Hydrogen Dissociation Reaction 196
14.2 Fluctuations, Scaling Laws 199
14.3 Thermodynamic Limit Values 203
14.4 Numerical Examples 208
14.5 Concluding Remarks 217

15. Adsorption on Flat, Large Surfaces 219

15.1 System 220
15.2 Surface Densities 222
15.3 Thermodynamic Factors 223
15.4 Adsorption Thermodynamics 227
15.5 Data Reduction Procedure 231
15.6 Concluding Remarks 232

16. Adsorption on Small, Curved Surfaces 233

16.1 An Osmotic-like Ensemble 234
16.2 The Reference State 238
16.3 Size-Dependent Thermodynamic Properties 239
16.4 Single-Component Adsorption 242
16.5 Simulating Adsorption on a Small Sphere 245
16.6 Procedure for Data Reduction 248
16.7 Films on Small Surfaces 250
16.8 Concluding Remarks 252

17. A Representative Elementary Volume of a Porous Medium 255

17.1 Additivity of Variables: Densities 258
17.2 Estimate of the Configurational Contribution to the Entropy 266
17.3 Additivity of Variables: Statistical Mechanical Basis 268
17.4 Gibbs' Equation 271
17.5 Concluding Remarks 274

18. Integral and Differential Pressures in a Slit-Pore 275

18.1 Thermodynamic Relations for Fluids in Small Pores 277
18.2 Representative Elementary Volumes 279
18.3 The Integral Pressure 281
18.4 Pressure Components 283
18.5 Molecular Simulations 285
18.6 Pressure Variations Perpendicular and Parallel to the Pore Wall 289
18.7 Pressure Differences Across the Liquid–Vapor Surface 292
18.8 A Scaling Law for Pressures 294
18.9 Concluding Remarks 295

19. Entropy Production in a Porous Medium **297**

19.1 The Entropy Production 298
19.2 Constitutive Equations 302
19.3 Fluctuation–Dissipation Theorems 303
19.4 Non-isothermal Transport of Two Immiscible Fluids 305
19.5 Concluding Remarks 309

20. Transport of a Single Fluid Through a Porous Medium **311**

20.1 The Variables of the Representative Elementary Volume 313
20.2 Simulation Procedures 318
20.3 Integral and Differential Pressures 321
20.4 Transport at Varying Porosities 329
20.5 Concluding Remarks 332

21. Stretching of Small Polymers **333**

21.1 Theory 335
21.2 Model 340
21.3 The Stretching Regimes 341
21.4 Friction Coefficients 349
21.5 Entropy Production 352
21.6 Concluding Remarks 354

A. Edge and Corner Effects: T, V, N **357**

B. Edge and Corner Effects: T, V, μ **361**

C. Transformations in the Thermodynamic Limit **365**

D. Interaction Potentials in Atomistic Simulations **371**

References 375

Index 393

Chapter 1

Scope

We present the foundation of nanothermodynamics; the method of Hill. We explain how the method extends Gibbs' thermodynamics, when dealing with small systems, and discuss why the extension is important.

The laws that govern energy conversion are the laws of thermodynamics. These laws are normally formulated for the so-called thermodynamic limit, where system properties are average properties of large collections of molecules. Conversion of energy from one form to another can, however, often take place on a scale where only a few molecules are involved, say when a chemical reaction takes place catalyzed by an enzyme in a membrane. Such systems are abundant in nature, but also in modern technology: nanotechnology. One immediate question therefore arises: Does normal thermodynamic theory apply when the system becomes small? In this book, we explain that the short answer to this question is no, but that normal thermodynamics can be extended to deal with small systems.

The purpose of this book is to explain and apply the systematic thermodynamic theory of small systems; proposed by Hill in a paper from 1962 [3]. We shall use and extend his method. For more details, see the books *Thermodynamics of Small Systems, Part 1 and 2* [3, 4, 14]. In the first part of this book we highlight the basic idea of the theory. In the last part, we demonstrate the power of the theory in various applications in and away from equilibrium; see also [5–9].

Reading this book requires knowledge in classical and statistical thermodynamics. Callen [10] emphasizes the internal structure and formal aspects of thermodynamics, suitable as background for the present text. Chemists may find useful the classical book by Smith and van Ness, see, e.g., the 8th edition [11]. Hill's theory gives a systematic approach to small systems, which goes beyond these books. The connection between the molecular and the system level is provided by statistical mechanics, and we refer to McQuarrie [13] or Widom [12]. These texts apply for large as well as small systems, and give a good background for the present text.

1.1 The Simulation Tool

Throughout the book we shall use molecular dynamics simulations as a microscope into the world of small systems, see for instance Frenkel and Smit [18] and Appendix D. Molecular Dynamics simulations are invaluable when it comes to produce new information on small systems properties, but also to test assumptions made in the theory development. This is how we will use them here.

1.2 Small Systems at Equilibrium

Clausius [19] was first to write the measurable change in a system's internal energy, U, in terms of the changes in entropy, S, and volume, V:

$$dU = TdS - pdV \tag{1.1}$$

where the temperature is T, and the pressure is p.

Gibbs [20] extended this equation by a term most important to chemists; the chemical potential μ_j of component j, in a mixture of n components. He introduced μ_j to measure the change in the internal energy due to the change in the amount N_j of component j, all other quantities kept constant. The specific internal energy, or chemical potential, is then obtainable from the equation that now bears Gibbs' name:

$$dU = TdS - pdV + \sum_{j=1}^{n} \mu_j dN_j \tag{1.2}$$

The basis set of variables in Eq. (1.2) are the extensive variables, S, V, N_j. They are all proportional to the system's volume. The internal energy U is therefore an Euler homogeneous function of degree one in S, V and N_j [10], which means that Eq. (1.2) is a total differential in these variables.

But what will happen when we move from a macroscopic number down to a few molecules? The normal thermodynamic relations cease to apply. Can we modify them? In particular, U is no longer an Euler homogeneous function of S, V and N_j for a small system. Hill proposed already in the 1960s to add a term to Gibbs' equation, which measures the degree of *smallness* of the system. Doing this, he could restore extensivity and deal with thermodynamic systems down to one molecule.

In order to accomplish the task, he took advantage of the classical and statistical thermodynamic machinery and constructed an *ensemble* of $\mathcal{N}$ replicas of the small system. The total internal energy, U_t, is then the energy of all replicas in the ensemble, and the differential change in the total internal energy is

$$dU_t = TdS_t - pdV_t + \sum_{j=1}^{n} \mu_j dN_{j,t} + \varepsilon d\mathcal{N} \tag{1.3}$$

Subscript t indicates from now on an ensemble property, and we refer to this equation as the *Hill–Gibbs equation.* The subdivision potential, ε, is the increase in the total internal energy when we add one replica, keeping the total entropy S_t, the total volume V_t and the total amounts of $N_{j,t}$ constant. Because ε gives the internal energy needed to subdivide S_t, V_t and $N_{j,t}$ over one more replica, Hill gave it the telling name: *subdivision potential.* For a large number of replicas, the ensemble is a large system and has the usual thermodynamic properties. In particular, U_t is an Euler homogeneous function of degree one in $S_t, V_t, N_{j,t}$ and $\mathcal{N}$. We shall examine the meaning of the subdivision potential and the connected property, the replica energy X, in Chapter 3. Both explain the energy change upon a change in subdivision.

The present book builds on Hill's idea, and uses Eq. (1.3) as a starting point for all elaborations. Hill called his theory *Thermodynamics of Small Systems*, but he and Chamberlin [21, 22] also used the word *Nanothermodynamics* to describe the field, a name which we shall adopt. The scale relevant to energy conversion, where ε is finite, concerns systems where the surface energy becomes relevant. Their typical size is measured in nanometer, but systems on larger scales may also benefit from the method. The name Thermodynamics of Small Systems is therefore more general.

Applications of Hill's theory have so far been scattered, but we believe that his theory deserves to be further explored and used. This book is a step in that direction. We give an extended theoretical basis followed by applications. To the best of our knowledge, this book presents a first attempt to address this issue after Hill's books. A preliminary version of the theoretical part of the present book was published in 2020 [15].

1.3 Expansion to Systems Away from Equilibrium

Classical thermodynamics describes the internal energy of a system in equilibrium. Two states, each in equilibrium, can be compared. According to the second law of thermodynamics, the entropy change in the system and the surroundings will determine which of the two states is the stable one. This can be computed with classical thermodynamics. In order to describe the *rate* of the process, how the energy conversion proceeds between the states, we need nonequilibrium thermodynamics. This theory is well understood for homogeneous systems [23, 24]. It has been extended to heterogeneous systems [25], but is hardly used for energy conversion in small systems, see however [26, 27].

We are aiming to broaden and expand the theoretical basis of Hill's equilibrium theory, and take steps in the direction of formulating the entropy production for small systems. Doing this, we also hope to set a basis for the field of nonequilibrium nanothermodynamics [23, 25]. In this theory, the entropy production is found by introducing the balance laws into the Gibbs equation. For a small system, it follows that the balance laws must be introduced in the

Hill–Gibbs equation (Eq. (1.3)):

$$dS_t = \frac{1}{T}dU_t + \frac{p}{T}dV_t - \frac{\boldsymbol{\mu}}{T} \cdot d\mathbf{N}_t - \frac{\varepsilon}{T}d\mathcal{N} \tag{1.4}$$

where the shorthand notations $\boldsymbol{\mu} \equiv (\mu_1, \ldots, \mu_n)$ and $\mathbf{N}_t \equiv (N_{1,t}, \ldots, N_{n,t})$ have been introduced. The centered dot $\cdot$ indicates a contraction of two vectors.

In order to describe transport processes, we need to first split the change in entropy of the ensemble of replicas dS_t into an exchange contribution $dS_{t,e}$ and a production term $dS_{t,i}$. This can be done using balance equations for internal energy U_t, momentum, volume V_t, particle numbers $\mathbf{N}_t$, and number of replicas $\mathcal{N}$. Examples are given in Chapters 19–21 [27].

1.4 Limitations

The small systems dealt with in this book are topologically speaking *simply connected*, meaning that shapes like, e.g., donuts are excluded from the analysis. We also do not consider complicated surface structures like those of a fractal nature. We furthermore restrict ourselves to short-range potentials. Long-range (Coulombic) potentials were discussed by Latella *et al.* [28]. A connection to quantum mechanics is made, but not applied, in Chapter 4.

Well-known size-dependent properties in statistical mechanics contain contributions which are logarithmic in the particle number. The logarithmic dependence arises from the number of ways to interchange N particles, which is $N!$ The logarithm of $N!$ gives, in addition to other terms, a contribution proportional to $\ln N$. This term can dominate in molecular simulations with periodic boundary conditions, see, e.g., Landau and Binder [29], and Siepmann *et al.* [30] for a discussion. This book considers systems where logarithmic contributions are negligible, cf. Chapter 6.

1.5 Book Outline

The book is dedicated to the development of a theoretical foundation of nanothermodynamics. Examples are added to illustrate various aspects of the theory and of possibilities (needs) for further use.

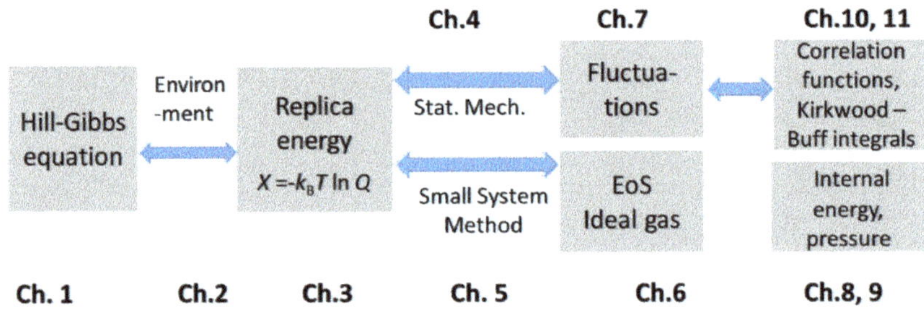

Figure 1.1. The structure of the book's first 11 chapters. The relation between the chapters is explained in the text.

We present the book structure and outline in what follows, before we end with perspectives on future work.

The relations between the first eleven chapters are illustrated in Fig. 1.1. The present Chapter 1 gives the scope of the book and presents the Hill–Gibbs equation. In Chapter 2, we explain the importance of the environment that controls the small system, and how the environmental variables have a different status in small and large systems. We derive the replica energy X in Chapter 3. The statistical mechanical machinery and the partition function Q in Chapter 4 are the same for small as for large systems.

A practical scaling method (the *Small System Method*) is presented in Chapter 5 [1, 2, 31, 32]. It is used to give equations of state for an ideal gas with surface energy in Chapter 6. Chapters 7 and 11 explain the consequences of Hill's description for the internal energy, the pressure and the enthalpy. The bridge to molecular properties is made for all sets of environmental variables in Chapter 8. In Chapter 9, we discuss the fluctuations in energy, volume and particle numbers for all ensembles, and apply the relation between the replica energy and the partition function Q, $X = -k_B\, T \ln Q$, where k_B is the Boltzmann constant. Chapter 10 *defines* and describes finite-volume Kirkwood–Buff integrals [33–35]. Thermodynamic properties of solutions can be found from these. They will appear on the nanoscale in a different way than on the macroscale. This is also demonstrated in Chapter 11.

Hill's thermodynamics can be reduced to Gibbs' for spherical and cylindrical small systems, see Chapter 12 [36]. Once we have the Small System Method to compute thermodynamic factors of mixtures, we can compare experimental values of Fick's diffusion coefficients to values obtained from Maxwell–Stefan diffusion coefficients as calculated with molecular dynamics simulations [1, 2, 31, 32], see Chapter 13. This opens up a way to simulate coefficients which are difficult to access by experiments. The Small System Method can be extended to deal with partial molar energies and enthalpies for chemical reactions, see Chapter 14, and find activity coefficients for layers adsorbed to large, flat (Chapter 15, [37]) or small, curved surfaces (Chapter 16, [38, 39]).

An interesting result of fluid confinement to pores is that two, and not only one, pressures become necessary in the description of the representative volume element of the system. We see this in Chapter 17 [40, 41], which deals with the thermodynamics of two fluid phases in a porous medium. In Chapter 18, we discuss two fluid phases in a slit pore. A new equilibrium condition is found; the so-called integral pressure is constant across phase boundaries. A thorough discussion is given of the so-called representative elementary volume. We obtain scaling laws that relate properties at the large and small scale, similar to the Small System Method for determination of the thermodynamic factor.

The last three chapters concern systems away from global equilibrium. The simple one force–one flux relationships depend on the manner in which we experimentally stretch a polymer, see Chapter 21. Transport equations follow, as usual, from the system's entropy production [27]. In order to deal with transport in porous media, we need equilibrium properties of a representative volume element, see Chapter 19 [42]. The entropy production, linear flux–force relations and fluctuation–dissipation theorems can then be found. In a face centered cubic-type lattice with confined fluids, the driving force is minus the gradient in the integral pressure, see Chapter 20.

1.6 Perspectives

This book is intended to present and expand the nanothermodynamic theory of Hill, and to illustrate how it can be applied. Hill gave numerous examples, but not many have been further applied. Small system thermodynamic properties are as of yet scarcely systematically investigated and tabulated, like large systems are. The same applies to transport properties on the small scale. The hope is to inspire new simulations or experiments, and contribute to the further development of nanothermodynamics.

The book grew out of an idea some years ago to study equilibrium properties of small systems controlled by a reservoir [1, 2, 31, 32]. Some of the first results from these studies are already in use, most prominently by Vlugt and coworkers [1, 2, 31, 32], van der Vegt and coworkers [43, 44] and Cortes-Huerto and coworkers [8]. These groups have taken advantage of the small system scaling law to compute the thermodynamic factor in the thermodynamic limit. Such factors describe the non-ideality of mixtures, see Chapters 12 and 15. From the scaling law, Krüger and coworkers [33–35, 45] improved significantly the precision in the determination of finite volume Kirkwood–Buff integrals, see Chapter 10. Skorpa *et al.* [46, 47] found a scaling law for partial molar enthalpies of various ensembles, and Strøm *et al.* [48] gave a scaling law for the compressibility of small volume elements of water. He and coauthors have also shown how thermodynamic properties of adsorbed CO_2 are changing with the area curvature of the adsorbent [39], see Chapter 16. The scaling laws seem useful for films [38], particles [39], colloids and more.

A series of new applications may follow from these early developments. We mention in particular the effort to explore small system properties of confined fluids in porous materials. The so-called integral pressure of a representative volume element of a nanoporous medium, $\widehat{p}$, enters the driving force of transport [40, 41, 49]. The need to use the integral pressure in confined systems was recently underlined [50]. We expect this to be important for descriptions of porous media transport.

Once thermodynamic equations have been given for the state of equilibrium, the next step is to find the laws of transport. By using

Hill–Gibbs' equation, the assumption of local equilibrium, and the balance laws, we can construct the entropy production. Bering *et al.* [27] gave the entropy production for simple stretching of a single molecule in different environments. The stretching behaved in accordance with nanothermodynamics. This may be regarded a first of numerous possible examples, which could be described by this theory.

Chapter 2

Large Versus Small Thermodynamic Systems

We explain what the concept "small" means in terms of Hill's subdivision potential. Thermodynamic theories of large and small systems are analogous, but have different ways to find system properties. What is special about small systems, is their control by environmental variables.

Energy conversion cannot be described without the laws of thermodynamics. Classical thermodynamics is set up to give these laws for a system with a large number of particles, an ensemble [10, 11, 19, 20]. A system is *large in the thermodynamic sense* when the basis set of thermodynamic variables (U, S and $\mathbf{N}$) are proportional to the system's volume. The total energy of two large systems, combined into one system, is then equal to the sum of the separate energies. A system is small in the thermodynamic sense when the total energy of two small systems combined is *not* equal to the sum of the separate energies. The variables of the basis set are then not extensive. In this sense, the small and large systems differ. The method of Hill proposes a way to deal with the *non-extensive* character of small systems, keeping the internal structure of classical thermodynamics.

How can we take advantage of the systematic structure of thermodynamics and describe the energy conversion with non-extensive variables? It was Hill who first saw the possibility to do so [3, 4].

Rather than looking at one small system, he introduced an *ensemble of replicas* of the small system. This allowed him to use normal thermodynamic relations for the ensemble. Afterwards, the single system property could be computed as an average ensemble property.

Remark 2.1. Small systems are no longer extensive in V. As a system becomes smaller, a substantial part of U, S and $\mathbf{N}$ becomes proportional to the surface area, or to $\sim V^{2/3}$. When the surface tension depends on the surface curvature, there are also contributions proportional to $V^{1/3}$, as well as constant contributions. When the surface has edges and nooks, like we see in a cube, pyramid or triangular prism, there are also contributions proportional to $V^{1/3}$ as well as constant contributions.

The number of replicas $\mathcal{N}$ was added as a new extensive variable in Gibbs' equation, cf. Eq. (1.3). The subdivision potential ε expressed the change in internal energy per replica added at constant $S_t, V_t, N_{j,t}$. The so-called replica energy, X, is equal to the subdivision potential for the basis set of variables, but the expression for the replica energy changes when we change the set of variables that control the system, see Chapter 3. The properties of the ensemble of small systems are, however, mathematically speaking, always Euler homogeneous functions of degree one in the number of replicas. In this way, extensivity is restored.

We shall determine expressions for the replica energy of six different ensembles in Chapter 3. This list of ensembles is not exhaustive, but provides typical examples and can be used to demonstrate the method. The replica energy depends not only on the size, but also on the shape of the small system. Hill's nanothermodynamics is suited to deal with size- and shape-dependence of thermodynamic properties. The construction has the big advantage that we can keep on using the statistical thermodynamic machinery as we do in classical thermodynamics. The results may be surprising. In Chapter 6, we shall derive equations of state for an ideal gas with surface absorption particular for the nanoscale. We calculate the subdivision potential, ε, for three sets of control variables, and show that it depends on the choice. The equation of state of a small system depends on the set of environmental control variables. In all cases, however, ε disappears in the thermodynamic limit. The variables of the *single* small system remain, as we shall see, *non-extensive.* Legendre transforms can be set up as usual for the ensemble of replicas, but not for single replicas, see Section 2.3 and Chapter 21. Legendre–Fenchel transforms are more general.

Before we start to describe the small system, let us recapitulate what a macroscopic thermodynamic system is [10, 11]. This familiar system we shall simply call "large". The basis set of variables was presented already in Eq. (1.2); S, V and N_j. The internal energy is a total differential in these variables:

$$dU = \left(\frac{\partial U}{\partial S}\right)_{V,N_j} dS + \left(\frac{\partial U}{\partial V}\right)_{S,N_j} dV + \sum_{j=1}^{n} \left(\frac{\partial U}{\partial N_j}\right)_{S,V,N_l} dN_j \tag{2.1}$$

The partial derivatives of the internal energy define the temperature, T, minus the pressure, $-p$, and the chemical potentials, μ_j, respectively. The system has n components, cf. also Eq. (1.2). For a large system, the internal energy can be described by these extensive variables. The description can also be transformed to alternative variable sets, like T, V, μ_j.

The strength of classical thermodynamics is its systematic structure [10, 11] leading to Maxwell's and other relations. The fact that we can transform equations of state from one set of control variables to another by Legendre transforms (see Section 2.3) is characteristic of large thermodynamic systems, but also of whole ensembles of small systems.

We start this introductory chapter presenting six of the ensembles, as they are used in all of classical thermodynamics. The Hill–Gibbs equation will be further explained, and the structural differences in the thermodynamic theory of small and large systems will be pointed out.

2.1 Small Systems and Their Environment

Via the environment, we can control selected system variables. We call such variables *environmental variables* [3, 4]. An alternative name is *control variables.* Heat- and particle-baths can be used to control, for instance, the temperature and the chemical potentials in a system. Some properties of large systems depend on the variables that we control during their determination. The heat capacity at constant system pressure is, for instance, different from the heat capacity at constant system volume. But, the equation of state for a large system remains the same, independent of how the system is controlled. In a large system, the intensive variables, like T, p, μ_j, are *independent* of the size and shape of the system. This is not so for small systems.

The size and shape of a small system can also be controlled by the environment. When we speak of variables controlled by the environment, we shall include shape and size- variables. Table 2.1 presents six sets of environmental variables. The system size is accounted for by the volume, or the particle number. In addition to the usual variables, we have listed the shape of the small system. These environmental variables are used in Chapters 3 and 4 to define the replica energies of six ensembles.

Table 2.1 (line 1) tells that $U, V, \mathbf{N} \equiv (N_1, \ldots, N_n)$ and shape are controlled in the microcanonical ensemble. In the canonical ensemble (line 2), T, V, $\mathbf{N}$ and shape are controlled; in the isobaric ensemble (line 3), p, T, $\mathbf{N}$ and shape are controlled; in the grand canonical ensemble (line 4), T, V, $\boldsymbol{\mu} \equiv (\mu_1, \ldots, \mu_n)$ and shape are controlled, and in the osmotic ensemble (line 5), T, p, N_1, $\boldsymbol{\mu}_1 \equiv (\mu_2, \ldots, \mu_n)$ and shape are controlled. In the nanocanonical ensemble (last line), T, p, $\boldsymbol{\mu}$ are controlled. This ensemble is completely open and can only be

Table 2.1. Terminology for six sets of environmental variables.

Ensemble name	Abbreviation	Environmental variables
Microcanonical	MC	U, V, $\mathbf{N}$, shape
Canonical	C	T, V, $\mathbf{N}$, shape
Isobaric	P	T, p, $\mathbf{N}$, shape
Grand canonical	GC	T, V, $\boldsymbol{\mu}$, shape
Osmotic	O	T, p, N_1, $\boldsymbol{\mu}_1$, shape
Nanocanonical	NC	T, p, $\boldsymbol{\mu}$, shape

used for a system of finite size. In this ensemble, we compute $U, V, \mathbf{N}$ from T, p, $\boldsymbol{\mu}$; see Section 3.6.

The top five sets are common in statistical mechanics. The first is most relevant to Molecular Dynamics simulations; the next four to experiments. The last set is the nanocanonical ensemble, defined and discussed by Hill [3, 4, 14].

The system shape can be cubic, spherical or any other symmetric or asymmetric shape. The surface can be differentiable, like for a sphere, or more generally, a spheroid. It can also have edges and corners as, e.g., in a cube or a platonic solid. A small system can be two-dimensional. The volume is then replaced by the surface area as a variable. A circle or an ellipse has a differentiable boundary line, while a square or a parallelogram has edges and corners. We consider only simply connected systems, see Limitations in Chapter 1. The equation of state of a small system will depend on how the system is controlled. Therefore, there is in principle, a different equation of state for each way the system can be controlled. We give examples in Chapter 6.

A small system is naturally composed of a small number of particles. Clearly, the smallest number of particles in any system is one. It is more difficult to decide how many particles are needed for a system to be large. In Chapter 6, we estimate that 27 particles are needed for the surface energy to be comparable to the bulk energy of an ideal gas in a cube with surface adsorption. The system is thus still small at this number of particles. Given that the minimum distance between hard spheres is equal to the diameter of the hard spheres, 27 particles will fit in a cube with a size of 3 particle diameters.

Single molecules can be regarded as small systems; e.g., colloidal particles, RNA molecules, or polymers [26, 27]. Asymmetric shapes, like spheroids or cylinders, may introduce orientational order in a small system; which is not present in the bigger system. Such order may also have an impact on the equation of state. In the derivation of the equations of states, we shall compare large and small systems with the same shape.

The problem is now: how can we create an overarching framework for a description of all these properties, dependencies and peculiarities? This is where Hill's postulate enters.

2.2 Hill's Postulate: Small System Definition

Hill's idea [3, 4] was to introduce an ensemble of replicas of the small system. The ensemble of replicas has a total internal energy U_t, a total entropy S_t, a total volume V_t, a total number of particles $N_{j,t}$ of component j and a number of replicas, $\mathcal{N}$. By construction, the ensemble variables U_t, S_t, V_t, $\mathbf{N}_t$, $\mathcal{N}$ are Euler homogeneous functions of the first order. Figure 2.1 illustrates three cases. In the first, $\mathcal{N}$ is large, in the central one, $\mathcal{N} = 26$, and in the one to the right, $\mathcal{N} = 8$. The total differential of U_t follows the Hill–Gibbs relation, cf. Eq. (1.3):

$$dU_t = TdS_t - pdV_t + \boldsymbol{\mu} \cdot d\mathbf{N}_t + \varepsilon d\mathcal{N} \tag{2.2}$$

The partial derivatives of U_t define the temperature, pressure and chemical potentials as before:

$$\left(\frac{\partial U_t}{\partial S_t}\right)_{V_t,\mathbf{N}_t,\mathcal{N}} = T, \quad \left(\frac{\partial U_t}{\partial V_t}\right)_{S_t,\mathbf{N}_t,\mathcal{N}} = -p, \quad \left(\frac{\partial U_t}{\partial \mathbf{N}_t}\right)_{S_t,V_t,\mathcal{N}} = \boldsymbol{\mu} \tag{2.3}$$

The new thermodynamic derivative, ε, is defined by

$$\left(\frac{\partial U_t}{\partial \mathcal{N}}\right)_{S_t,V_t,\mathbf{N}_t} = \varepsilon \tag{2.4}$$

with the ensemble variables $S_t, V_t, \mathbf{N}_t$. The *subdivision potential* ε was introduced by Hill, cf. Eq. (1.3). It measures the increase in the total internal energy as the number of replicas increases, keeping S_t, V_t, $\mathbf{N}_t$ constant as we go to the left in the figure. But S_t, V_t, $\mathbf{N}_t$ are

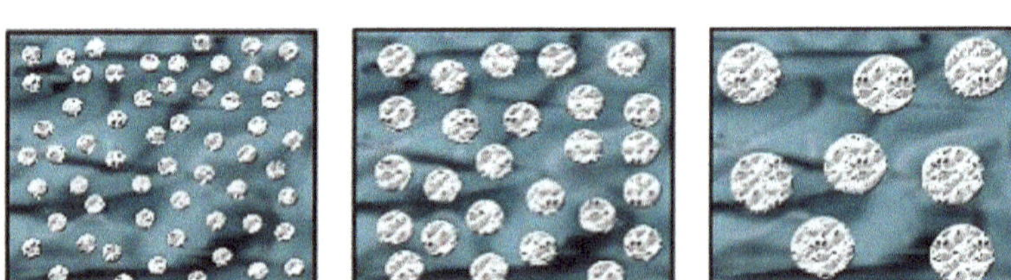

Figure 2.1. Three ensembles of small systems. The total particle number is the same in all ensembles, but the particles are distributed between different numbers of clusters (small systems).

subdivided over more replicas, which explains the name subdivision potential. The subdivision potential ε may be positive or negative, depending on whether reversible work is needed or gained when we introduce a new replica, thereby changing the ratio of the surface area to the volume of the ensemble.

We are interested in the properties of one small system (one replica) and compute the average internal energy U, entropy S, volume V and number of particles $\mathbf{N}$ per replica

$$U \equiv \frac{U_t}{\mathcal{N}}, \quad S \equiv \frac{S_t}{\mathcal{N}}, \quad V \equiv \frac{V_t}{\mathcal{N}}, \quad \mathbf{N} \equiv \frac{\mathbf{N}_t}{\mathcal{N}} \tag{2.5}$$

To increase the number of replicas, keeping S_t, V_t, $\mathbf{N}_t$ constant, means to decrease S, V, N, because S_t, V_t, $\mathbf{N}_t$ are subdivided over a larger number of replicas.

Equations (2.2)–(2.5) are true for all choices of environmental variables. For large thermodynamic systems, U, S and $\mathbf{N}$ are proportional to the volume. The variables U, S, V and $\mathbf{N}$ are then independent of the number of replicas $\mathcal{N}$, and the subdivision potential ε is zero.

Remark 2.2. The entropy S of a small system is determined by the probability distribution of U_t, $\mathbf{N}_t$ and V_t over the separate replicas. This probability distribution is *the same* for each replica of the system. The entropy of all separate replicas is therefore also the same and equal to S [3, 4].

The ensemble variables $U_t, S_t, V_t, \mathbf{N}_t$ are all proportional to the number of replicas $\mathcal{N}$ (Euler homogeneous functions of degree one). We integrate the differential of U_t at constant $T, p, \boldsymbol{\mu}, \varepsilon$, and obtain the Euler equation for the ensemble of replicas as follows:

$$U_t = TS_t - pV_t + \boldsymbol{\mu} \cdot \mathbf{N}_t + \varepsilon\mathcal{N} \tag{2.6}$$

It follows that $T, p, \boldsymbol{\mu}, \varepsilon$ are Euler homogeneous functions of degree zero: *they do not depend on the number of replicas.* With Eq. (2.5) we obtain from Eq. (2.6)

$$U = TS - pV + \boldsymbol{\mu} \cdot \mathbf{N} + \varepsilon \tag{2.7}$$

When we substitute Eq. (2.5) into Eq. (2.2) and use Eq. (2.7), we find the Gibbs equation for one small system

$$dU = TdS - pdV + \boldsymbol{\mu} \cdot d\mathbf{N} \tag{2.8}$$

Remark 2.3. Equation (2.8) looks like the usual equation for a large thermodynamic system, which is the reason why we stick to the name: *The Gibbs equation.* But there is an important difference: The variables U, S, $\mathbf{N}$ are *not* Euler homogeneous of the first order in the volume V. If they were, it would follow that $U = TS - pV + \boldsymbol{\mu} \cdot \mathbf{N}$ so that the subdivision potential in Eq. (2.7) would be zero. The small system properties $(T, p, \boldsymbol{\mu})$ are *not* Euler homogeneous of the zeroth order in the volume V. This implies that $(T, p, \boldsymbol{\mu})$ for small systems, in general, depend on the volume.

When we differentiate Eq. (2.6) and use Eq. (2.2), we obtain a Gibbs–Duhem-like equation for the whole system of replicas

$$0 = S_t dT - V_t dp + \mathbf{N}_t \cdot d\boldsymbol{\mu} + \mathcal{N} d\varepsilon \tag{2.9}$$

After division by $\mathcal{N}$, we obtain

$$d\varepsilon = -SdT + Vdp - \mathbf{N} \cdot d\boldsymbol{\mu} \tag{2.10}$$

We refer to this equation as the *Hill–Gibbs–Duhem equation*, to signify that it is dedicated to small systems. It reduces to the usual Gibbs–Duhem equation for large systems, for which the subdivision potential is zero.

We define a system as small in Hill's sense when the subdivision potential is finite.

On the other hand, we cannot conclude that any small system has a finite subdivision potential. We can, e.g., introduce the surface area as an explicit variable, and find a small system description with zero subdivision potential [51]. A small system is characterized by a smaller number of microscopic states (see Chapter 4).

Remark 2.4. Following Gibbs [20], it is possible to introduce the area as an additional variable in the Gibbs equation. By introducing also curvature variables [52, 53], an alternative description can be obtained without making use of the ensemble idea of Hill. Gibbs,

Tolman and Helfrich chose to add terms of significance when needed. Hill's method includes all effects at the outset.

We have explained above how Hill gave the small system a thermodynamic basis, by introducing the ensemble of replicas. The ensemble of $\mathcal{N}$ replicas of the small system can be described by the ordinary thermodynamic framework. While the single system is small, the ensemble of system replicas is large.

The replicas are independent of one another, and their individual properties appear as averages of the ensemble. The introduction of replicas gives an ensemble-distribution in phase space. This made it natural for Hill to adopt the *ergodic hypothesis*. Each small system will in time go through all possible microscopic states and the time average is the same as the average over the ensemble of replicas.

The replicas of the small system have microscopic states that depend on the environmental variables, see for instance Chapter 21. Also, the state's probability distribution in phase space will depend on the environmental variables, as we shall see (Chapter 4). It is thus appropriate to name an ensemble of replicas by the name of the ensemble in statistical mechanics that controls it, cf. Table 2.1.

2.3 Legendre Transforms for Ensembles of Replicas

Equation (2.2) can be written as

$$dU_t\left(S_t, V_t, \mathbf{N}_t, \mathcal{N}\right) = TdS_t - pdV_t + \boldsymbol{\mu} \cdot d\mathbf{N}_t + \varepsilon d\mathcal{N} \tag{2.11}$$

Three Legendre transforms for the whole ensemble of replicas of the small system are as follows:

$$\begin{aligned} d(U_t - TS_t) &= dF_t(T, V_t, \mathbf{N}_t, \mathcal{N}) \\ &= -S_t dT - pdV_t + \boldsymbol{\mu} \cdot d\mathbf{N}_t + \varepsilon d\mathcal{N} \end{aligned} \tag{2.12}$$

$$\begin{aligned} d(U_t + pV_t) &= dH_t(S_t, p, \mathbf{N}_t, \mathcal{N}) \\ &= -TdS_t + V_t dp + \boldsymbol{\mu} \cdot d\mathbf{N}_t + \varepsilon d\mathcal{N} \end{aligned} \tag{2.13}$$

$$\begin{aligned} d(U_t - TS_t + pV_t) &= dG_t(T, p, \mathbf{N}_t, \mathcal{N}) \\ &= -S_t dT + V_t dp + \boldsymbol{\mu} \cdot d\mathbf{N}_t + \varepsilon d\mathcal{N} \end{aligned} \tag{2.14}$$

We gave the transformations from the $S_t, V_t, \mathbf{N}_t, \mathcal{N}$ to the $T, V_t, \mathbf{N}_t, \mathcal{N}$ ensemble; from the $S_t, V, \mathbf{N}_t, \mathcal{N}$ to the $S_t, p, \mathbf{N}_t, \mathcal{N}$ ensemble; and from the $S_t, V_t, \mathbf{N}_t, \mathcal{N}$ to the $T, p, \mathbf{N}_t, \mathcal{N}$ ensemble. Equations (2.12), (2.13) and (2.14) give a simple route to the Helmholtz energy, $F_t(T, V_t, \mathbf{N}_t, \mathcal{N})$, the enthalpy, $H_t(S_t, p, \mathbf{N}_t, \mathcal{N})$ and the Gibbs energy $G_t(T, p, \mathbf{N}_t, \mathcal{N})$; all properties of the ensemble of replicas. The total energy of the *ensemble* of replicas is an Euler homogeneous function of degree one in the number of replicas. The variables $S_t, V_t, \mathbf{N}_t, \mathcal{N}$ as well as F_t, H_t and G_t are all Euler homogeneous functions of degree one. The variables T, p, $\boldsymbol{\mu}$ and ε are Euler homogeneous functions of degree zero in the number of replicas.

Legendre transforms for the whole ensemble of replicas, where $\mathcal{N}$ is taken as an additional variable, work the same way as they work for a large system, where $\mathcal{N}$ is not an additional variable. For the whole ensemble of replicas one can transform the equation of state from one ensemble to another, just as this can be done in the thermodynamic limit for the properties per replica. We have

$$
\begin{aligned}
F_t(T, V_t, \mathbf{N}_t, \mathcal{N}) &= -p(T, V, \mathbf{N})V_t + \boldsymbol{\mu}(T, V, \mathbf{N}) \cdot \mathbf{N}_t + \varepsilon(T, V, \mathbf{N})\mathcal{N} \\
H_t(S_t, p, \mathbf{N}_t, \mathcal{N}) &= T(S, p, \mathbf{N})S_t + \boldsymbol{\mu}(S, p, \mathbf{N}) \cdot \mathbf{N}_t + \varepsilon(S, p, \mathbf{N})\mathcal{N} \\
G_t(T, p, \mathbf{N}_t, \mathcal{N}) &= \boldsymbol{\mu}(T, p, \mathbf{N}) \cdot \mathbf{N}_t + \varepsilon(T, p, \mathbf{N})\mathcal{N}
\end{aligned}
\tag{2.15}
$$

where we used that T, p, $\boldsymbol{\mu}$ and ε are Euler homogeneous functions of degree zero in the number of replicas. The expressions reduce to their usual form when the subdivision potential ε is zero (the system is large).

Maxwell relations apply to the ensemble of replicas. From Eq. (2.12) it follows for instance that

$$
\begin{aligned}
\left(\frac{\partial S_t}{\partial V_t}\right)_{T,\mathbf{N}_t,\mathcal{N}} &= \left(\frac{\partial p}{\partial T}\right)_{V_t,\mathbf{N}_t,\mathcal{N}} \\
\left(\frac{\partial \varepsilon}{\partial T}\right)_{V_t,\mathbf{N}_t,\mathcal{N}} &= -\left(\frac{\partial S_t}{\partial \mathcal{N}}\right)_{T,V_t,\mathbf{N}_t}
\end{aligned}
\tag{2.16}
$$

Many more of such relations can be given. Since the variables T, p and ε are Euler homogeneous functions of degree zero in the number of replicas, it follows from these relations that

$$\begin{aligned}\left(\frac{\partial S(V,T,\mathbf{N})}{\partial V}\right)_{T,\mathbf{N}} &= \left(\frac{\partial p(V,T,\mathbf{N})}{\partial T}\right)_{V,\mathbf{N}} \\ \left(\frac{\partial \varepsilon(V,T,\mathbf{N})}{\partial T}\right)_{V,\mathbf{N}} &= -\left(\frac{\partial S_t}{\partial \mathcal{N}}\right)_{T,V_t,\mathbf{N}_t} = -S(V,T,\mathbf{N})\end{aligned} \tag{2.17}$$

Per replica, Eq. (2.15) gives

$$\begin{aligned}F &= U - TS = -pV + \boldsymbol{\mu}\cdot\mathbf{N} + \varepsilon \\ H &= U + pV = TS + \boldsymbol{\mu}\cdot\mathbf{N} + \varepsilon \\ G &= U - TS + pV = \boldsymbol{\mu}\cdot\mathbf{N} + \varepsilon\end{aligned} \tag{2.18}$$

In small systems $S, V, \mathbf{N}$ as well as F, H and G, are not Euler homogeneous functions of degree one in the size of the system, V. All the equations of state for different ensembles are therefore different. Only in the large system limit, when ε approaches zero, they become the same. We shall see in Chapter 21 that Legendre–Fenchel transforms apply to a small system. This type of transform is further described in Section 4.9.

The entropy of the ensemble is proportional to $\mathcal{N}$. This is yet a characteristic of small systems:

$$\left(\frac{\partial S_t}{\partial \mathcal{N}}\right)_{T,V_t,\mathbf{N}_t} = S(T,V,\mathbf{N}) \tag{2.19}$$

which we already used above. A similar relation is true for V_t and $\mathbf{N}_t$.

2.4 Concluding Remarks

Figure 2.2 gives a summary of this chapter. Characteristic properties of small and large systems are listed, allowing us to compare similarities and differences.

A large thermodynamic system is extensive in the basis set of variables. A small thermodynamic system is non-extensive in this set. But an ensemble of system replicas is extensive in the sense that the energy is an Euler homogeneous function of degree one in the number of replicas. The thermodynamic description of the

Characteristic properties of thermodynamic systems

Large	Small
• Variables U, S, V, N_j are Euler homogeneous in V • Legendre transforms apply • Maxwell relations exist • The large system equation of state does not depend on shape • Zero subdivision potential	• Variables $U_t, S_t, V_t, N_{t,j}, \mathcal{N}$ are Euler homogeneous in $\mathcal{N}$ • Variables $U, S, V, N_{,j}$ are not Euler homogeneous in V • Legendre transforms apply to the ensemble of replicas only • Maxwell relations exist • The small system equation of state depends on shape • Finite subdivision potential

Figure 2.2. Large and small system properties compared.

ensemble of replicas is therefore formally similar to the description of large thermodynamic systems. The new feature is caused by the new variable; the number of replicas $\mathcal{N}$.

Legendre transforms can be used in the usual way for the whole ensemble of replicas, but not for small systems. It is therefore interesting that the Legendre–Fenchel transform may apply. A first example of this, concerning stretching of small polymers [27, 54], is described in Chapter 21.

For small as well as large thermodynamic systems, we can use standard statistical mechanics. Also, small systems are typically ergodic. The subdivision potential, ε, represents the internal energy change by adding one replica to the ensemble with variables $S_t, V_t, \mathbf{N}_t, \mathcal{N}$ keeping $S_t, V_t, \mathbf{N}_t$ constant. This quantity is negligible for large systems. Only experiments or simulations can tell when this happens.

The replica energy, X, of the different ensembles will be defined and related to the subdivision potential in the next chapter. It depends on the environmental variables which control the system. Each small system has a replica energy that depends on the set of controlled variables.

Chapter 3

Nanothermodynamics

Thermodynamic properties of small systems are size- and shape-dependent. We develop thermodynamic expressions for the replica energy of a small system, given its shape, and the environmental variables which are used to control it. Conditions for equilibrium between two small systems are formulated.

All thermodynamic properties can be found from a system's equation of state. For a large system, the equation of state does not depend on the environmental variables. This is no longer true for a small system. Such equations will be derived for an ideal gas with surface adsorption in Chapter 6. In this chapter, we take a first step in the direction of finding the equation of state and defining the replica energy for six common ensembles presented in Chapter 2.

We discussed in the previous chapter that the small system internal energy is not proportional to its volume, when contributions from, e.g., *surface energies* or *line energies* are comparable to the energy of the volume. Surface and line energies determine the subdivision potential. Properties of small systems depend on the volume, but they depend also on the shape. For instance, the surface area of a sphere differs from that of a dodecahedron, for the same volume, V, see Fig. 3.1. The sphere and the dodecahedron have therefore different internal energies. This is the reason why "shape" is included in the set of environmental variables of a small system in Chapter 2, see Table 2.1. The shape can be controlled, e.g., by confinement. A film thickness, for example, can be controlled by the positioning of adjacent walls. When we want to replace V by the

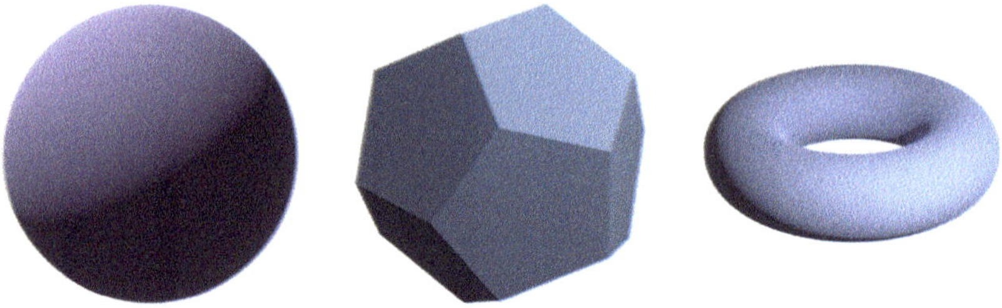

(a) Sphere (b) Dodecahedron (c) Torus

Figure 3.1. Two simple connected and one complex connected geometry.

pressure p as an environmental variable in the Gibbs equation, we must keep the shape of the small system the same, in order to have a well-defined operation. By environmental variables we mean variables that are used to control the system. These variables are needed for a complete definition of the small system. Small systems with different volumes can only be directly compared when their shape is the same. We need to specify the small system shape; spherical, cubic, tetrahedral, etc.

Remark 3.1. Environmental variables need be specified in order to have a small system that is well defined. When the system is small, the pressure, for instance, will change if we switch environmental variables. To see this, consider first U, V and $\mathbf{N}$ as environmental variables. We can calculate the temperature for this microcanonical ensemble and obtain $T^{\mathrm{MC}}(U, V, \mathbf{N})$, see Table 3.1, Section 3.1. If we next choose this temperature $T = T^{\mathrm{MC}}$, V and $\mathbf{N}$ to compute the pressure, we find that $p^{\mathrm{MC}}(U, V, \mathbf{N}) \neq p^{\mathrm{C}}(T = T^{\mathrm{MC}}(V, U, \mathbf{N}), V, \mathbf{N})$. The same is true for other thermodynamic quantities. Only in the thermodynamic limit, we obtain the same pressure.

The Hill–Gibbs equation (see Eq. (2.2)) gives the starting point for any computation of thermodynamic properties of small systems. It was formulated in Chapters 1 and 2 for an ensemble of replicas with the basis set of variables $U_t, V_t, \mathbf{N}_t$. We shall now reformulate this equation for applications with the sets of environmental variables from Chapter 2. The subdivision potential, introduced in Eq. (2.2),

Table 3.1. Ensembles, corresponding environmental variables and replica energy, X.

Ensemble name	Environmental variables	Replica energy, X	Sec.
MC	U, V, N_j	$T^{\mathrm{MC}}(U,V,N_j)S^{\mathrm{MC}}(U,V,N_j)$	3.1
C	T, V, N_j	$F^{\mathrm{C}}(T,V,N_j)$	3.2
P	T, p, N_j	$G^{\mathrm{P}}(T,p,N_j)$	3.3
GC	T, V, μ_j	$-\widehat{p}(T,V,\mu_j)V$	3.4
O	$T,p,N_1,\mu_2,\ldots,\mu_n$	$\widehat{\mu}_1^{\mathrm{O}}(T,p,N_1,\mu_2,\ldots,\mu_n)N_1$	3.5
NC	T, p, μ_j	$X^{\mathrm{NC}}(T,p,\mu_j)=\varepsilon(T,p,\mu_j)$	3.6

Notes: The shape of the small system is always controlled. The section number refers to the section where the property is defined.

is relevant for all ensembles. Another central property is the replica energy X [3, 4], which equals the subdivision potential only for the nanocanonical ensemble. We shall see that the expressions both for the replica energy and the subdivision potential differ substantially between the sets of environmental variables. We want to bring out how the replica energy can be determined in each case; and find the conditions that make the case unique. The outcome of all determinations is summarized in Table 3.1. In the analysis we refer to U, S, V and N_j as additive variables. The method is general and can be used for other ensembles.

The procedure to find the replica energy X can be summarized as follows:

- Choose the set of environmental variables used to control the small system of interest.
- Introduce the additive variables of the set into the Hill–Gibbs equation for the ensemble (Eq. (2.2)).
- Determine the replica energy, X, as the prefactor of $d\mathcal{N}$.

Remark 3.2. To illustrate the procedure of finding X, consider for simplicity the case of one additive, environmental variable, V. The total volume is $V_t = \mathcal{N}V$. We therefore have $pdV_t = pd(\mathcal{N}V) = p\mathcal{N}dV + pVd\mathcal{N}$. In the Hill–Gibbs equation, we replace

$p\mathrm{d}V_t$ by $p\mathcal{N}\mathrm{d}V$ and combine $pVd\mathcal{N}$ with the $\varepsilon d\mathcal{N}$ term. This gives: $(\varepsilon - pV)\, d\mathcal{N} \equiv X d\mathcal{N}$, which identifies X.

After we have specified the environmental variables, we replace dU_t, TdS_t and $\boldsymbol{\mu}\cdot d\mathbf{N}_t$ in the Hill–Gibbs equation (2.2) following this recipe. One outcome of the systematic approach is six expressions for the subdivision potential, one for each ensemble, cf. Table 3.2 in Section 3.7.

3.1 Environmental Variables U, V, N

Consider a small system where the variables U, V, $\mathbf{N}$ and shape are controlled. This is accomplished by completely isolating the small system. Figure 3.2 gives a sphere (disk) as example. The ensemble of such a system is a *microcanonical* ensemble (superscript MC). There is no exchange of volume, internal energy or particles of the separate replicas with the environment. Therefore, we cannot do experiments on the system, but simulations are feasible. The ensemble is important in molecular dynamics simulations [55].

Following the prescription above, we substitute $U_t = \mathcal{N}U$, $V_t = \mathcal{N}V$, and $\mathbf{N}_t = \mathcal{N}\mathbf{N}$ into Eq. (2.2), and find

$$\begin{aligned} dS_t^{\mathrm{MC}} &= \frac{1}{T^{\mathrm{MC}}}\mathcal{N}dU + \frac{p^{\mathrm{MC}}}{T^{\mathrm{MC}}}\mathcal{N}dV - \frac{\boldsymbol{\mu}^{\mathrm{MC}}}{T^{\mathrm{MC}}}\mathcal{N}\cdot d\mathbf{N} \\ &\quad - \frac{1}{T^{\mathrm{MC}}}\left(\varepsilon^{\mathrm{MC}} - U - p^{\mathrm{MC}}V + \boldsymbol{\mu}^{\mathrm{MC}}\cdot\mathbf{N}\right)d\mathcal{N} \\ &= \frac{1}{T^{\mathrm{MC}}}\mathcal{N}dU + \frac{p^{\mathrm{MC}}}{T^{\mathrm{MC}}}\mathcal{N}dV - \frac{\boldsymbol{\mu}^{\mathrm{MC}}}{T^{\mathrm{MC}}}\mathcal{N}\cdot d\mathbf{N} - \frac{X^{\mathrm{MC}}}{T^{\mathrm{MC}}}d\mathcal{N} \end{aligned} \tag{3.1}$$

The replica energy, X^{MC}, the prefactor of $d\mathcal{N}$, is thus defined by

$$X^{\mathrm{MC}}(U, V, \mathbf{N}) \equiv \varepsilon^{\mathrm{MC}} - U - p^{\mathrm{MC}}V + \boldsymbol{\mu}^{\mathrm{MC}}\cdot\mathbf{N} \tag{3.2}$$

The replica energy is also equal to minus the temperature T^{MC} times the specific replica entropy, cf. Eq. (3.1):

$$X^{\mathrm{MC}}(U, V, \mathbf{N}) = -T^{\mathrm{MC}}\left(\frac{\partial S_t^{\mathrm{MC}}}{\partial\mathcal{N}}\right)_{V,U,\mathbf{N}} \tag{3.3}$$

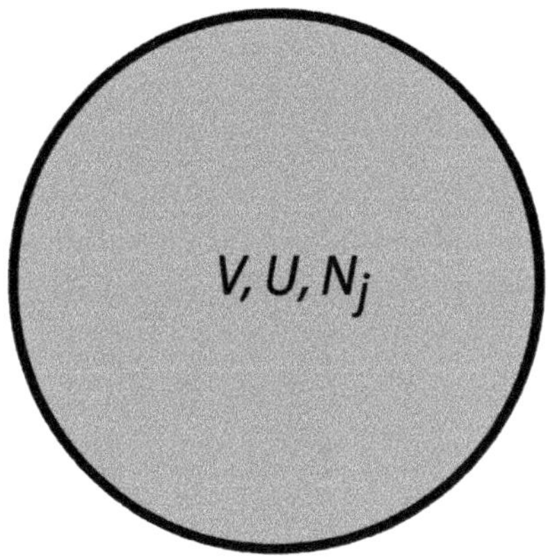

Figure 3.2. A small spherical system where the variables V, U, N_j are controlled. The system is isolated from the environment.

The specific replica entropy $\left(\partial S_t^{\mathrm{MC}}/\partial \mathcal{N}\right)_{U_t,V,\mathbf{N}}$ is the increase in entropy S_t^{MC} by adding one replica to the ensemble of replicas at constant U, V and $\mathbf{N}$.

By integrating Eq. (3.1) at constant U, V, $\mathbf{N}$ and X^{MC}, we obtain

$$0 = T^{\mathrm{MC}} S_t^{\mathrm{MC}} + X^{\mathrm{MC}} \mathcal{N} \tag{3.4}$$

The entropy of all replicas is the same [3, 4], see Remark 2.2, so that

$$S_t^{\mathrm{MC}}(U, V, \mathbf{N}, \mathcal{N}) = \mathcal{N} S^{\mathrm{MC}}(U, V, \mathbf{N}) \tag{3.5}$$

Equations (3.5) and (3.4) imply that X^{MC} equals minus the replica entropy times the temperature, cf. Table 3.1

$$X^{\mathrm{MC}}(U, V, \mathbf{N}) = -T^{\mathrm{MC}}(U, V, \mathbf{N}) S^{\mathrm{MC}}(U, V, \mathbf{N}) \tag{3.6}$$

Equation (3.6) also follows by introducing Eq. (2.7) into Eq. (3.2). By introducing Eqs. (3.4) and (3.5) into Eq. (3.1), we obtain the Gibbs equation for the small system:

$$dS^{\mathrm{MC}} = \frac{1}{T^{\mathrm{MC}}} dU + \frac{p^{\mathrm{MC}}}{T^{\mathrm{MC}}} dV - \frac{\boldsymbol{\mu}^{\mathrm{MC}}}{T^{\mathrm{MC}}} \cdot d\mathbf{N} \tag{3.7}$$

This equation has exactly the same form as for a large system. The important distinction is that the properties of the small system

$S^{\mathrm{MC}}, U, V, \mathbf{N}$ are not Euler homogeneous functions of degree one. Large systems have this property, giving $T^{\mathrm{MC}}S^{\mathrm{MC}} = U + p^{\mathrm{MC}}V - \boldsymbol{\mu}^{\mathrm{MC}} \cdot \mathbf{N}$, so that $\varepsilon^{\mathrm{MC}} = 0$ in that case. In small systems, T^{MC}, p^{MC}, $\boldsymbol{\mu}^{\mathrm{MC}}$ depend on the volume V. In order to make this distinction more visible, we define the hat-variable, entropy-hat with symbol $\widehat{S}^{\mathrm{MC}}$, in analogy with other hat-variables explained in the other sections of this chapter;

$$T^{\mathrm{MC}}\widehat{S}^{\mathrm{MC}} \equiv U + p^{\mathrm{MC}}V - \boldsymbol{\mu}^{\mathrm{MC}} \cdot \mathbf{N} \tag{3.8}$$

Together with Eqs. (3.2) and (3.6), this implies that

$$T^{\mathrm{MC}}(S^{\mathrm{MC}} - \widehat{S}^{\mathrm{MC}}) = -\varepsilon^{\mathrm{MC}} \tag{3.9}$$

The difference between the entropy and the entropy-hat, times the temperature, is equal to minus the subdivision potential. This difference is characteristic for small systems. Hill did not introduce $\widehat{S}^{\mathrm{MC}}$.

Figure 3.3 pictures ensemble examples with, respectively, $\mathcal{N} = 3, 4$ and 5 spherical replicas of a one-component system. Not only the size but also the shape of the replicas are the same. All replicas have three particles, the same volume V and internal energy U. For U, this is symbolized by use of the same shading. The addition of one replica implies that the total volume $\mathcal{N}V$, the total internal energy $\mathcal{N}U$, and the total number of particles $\mathcal{N}N = \mathcal{N}3$ increase with V, U and 3, respectively. It follows from Eq. (3.7) that $1/T^{\mathrm{MC}}$, $p^{\mathrm{MC}}/T^{\mathrm{MC}}$ and $\mu_k^{\mathrm{MC}}/T^{\mathrm{MC}}$ are given by

$$\begin{aligned}
\frac{1}{T^{\mathrm{MC}}(U, V, \mathbf{N})} &= \left(\frac{\partial S^{\mathrm{MC}}}{\partial U}\right)_{V,\mathbf{N}} \\
\frac{p^{\mathrm{MC}}(U, V, \mathbf{N})}{T^{\mathrm{MC}}(U, V, \mathbf{N})} &= \left(\frac{\partial S^{\mathrm{MC}}}{\partial V}\right)_{U,\mathbf{N}} \\
\frac{\mu_k^{\mathrm{MC}}(U, V, \mathbf{N})}{T^{\mathrm{MC}}(U, V, \mathbf{N})} &= -\left(\frac{\partial S^{\mathrm{MC}}}{\partial N_k}\right)_{U,V,\mathbf{N}_k}
\end{aligned} \tag{3.10}$$

where $\mathbf{N}_k \equiv \{N_1, \ldots, N_{k-1}, N_{k+1}, \ldots, N_n\}$.

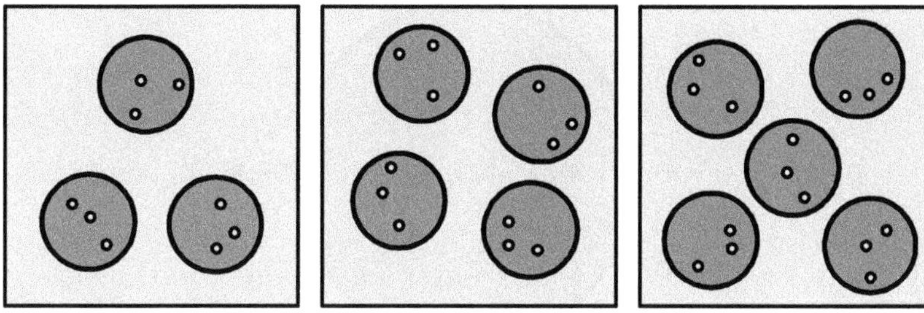

Figure 3.3. Three ensembles of replicas, all with environmental variables U, V and $N = 3$. All replicas are isolated from the environment. The ensembles have 3, 4 or 5 replicas, respectively.

Remark 3.3. Equation (3.7) forms a starting point for a non-equilibrium thermodynamic analysis of a small system, with the additive variables U, V, $\mathbf{N}$, $\mathcal{N}$ and shape as environmental variables.

3.2 Environmental Variables T, V, N

Consider next a small system where the variables T, V, $\mathbf{N}$ and the shape are controlled. The system can now exchange energy with the surroundings. Figure 3.4 illustrates a system with spherical shape, controlled by the variables T, V, $\mathbf{N}$. The exchange of internal energy with the environment is indicated by arrows. There is no exchange of volume or particles with the environment. This means that energy exchange takes place through heat exchange. We now consider $\mathcal{N}$ replicas of the small system, and construct an ensemble (the total system), which is large enough to follow the laws of classical thermodynamics. The ensemble of systems is *canonical*, and this is denoted by superscript C. In all replicas, the volume and the number of particles are the same.

Following the prescription of the three bullet points in the introduction to this chapter, we introduce $V_t = \mathcal{N}V$ and $\mathbf{N}_t = \mathcal{N}\mathbf{N}$ into Eq. (2.2). This gives

$$\begin{aligned} dU_t^{\mathrm{C}} &= TdS_t^{\mathrm{C}} - p^{\mathrm{C}}\mathcal{N}dV + \boldsymbol{\mu}^{\mathrm{C}}\mathcal{N} \cdot d\mathbf{N} + \left(\varepsilon^{\mathrm{C}} - p^{\mathrm{C}}V + \boldsymbol{\mu}^{\mathrm{C}} \cdot \mathbf{N}\right) d\mathcal{N} \\ &= TdS_t^{\mathrm{C}} - p^{\mathrm{C}}\mathcal{N}dV + \boldsymbol{\mu}^{\mathrm{C}}\mathcal{N} \cdot d\mathbf{N} + X^{\mathrm{C}}d\mathcal{N} \end{aligned} \tag{3.11}$$

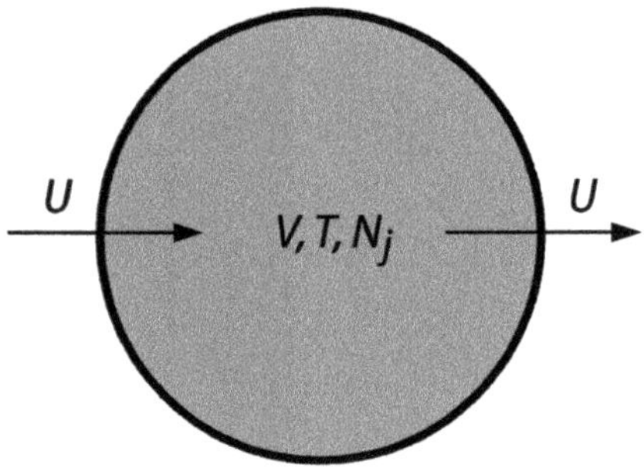

Figure 3.4. A small spherical system with environmental variables V, T, N_j, exchanging internal energy with the environment.

The replica energy, the prefactor of $d\mathcal{N}$, becomes

$$X^{\mathrm{C}}(T,V,\mathbf{N}) \equiv \varepsilon^{\mathrm{C}} - p^{\mathrm{C}}V + \boldsymbol{\mu}^{\mathrm{C}} \cdot \mathbf{N} \tag{3.12}$$

The replica energy is also equal to the specific internal energy of the ensemble of replicas, cf. Eq. (3.11)

$$X^{\mathrm{C}}(T,V,\mathbf{N}) = \left(\frac{\partial U_t^{\mathrm{C}}}{\partial \mathcal{N}}\right)_{S_t,V,\mathbf{N}} \tag{3.13}$$

The replica energy can be seen as an expression for the *reversible* work needed to add one replica of the small system at constant S_t^{C}, V, $\mathbf{N}$ and shape to the ensemble. The addition of a replica at constant S_t^{C}, V and $\mathbf{N}$ implies that S_t^{C} must be redistributed over one more replica, while the total volume $\mathcal{N}V$ and the total number of particles of each component $\mathcal{N}\mathbf{N}$ increase.

By integrating Eq. (3.11) at constant T, V, $\mathbf{N}$ and X^{C}, we obtain

$$U_t^{\mathrm{C}}(T,V,\mathbf{N},\mathcal{N}) = TS_t^{\mathrm{C}}(T,V,\mathbf{N},\mathcal{N}) + X^{\mathrm{C}}(T,V,\mathbf{N})\mathcal{N} \tag{3.14}$$

The *average* variables of the small system are obtained from the variables of the total system from

$$\begin{aligned} U_t^{\mathrm{C}}(T,V,\mathbf{N},\mathcal{N}) &\equiv \mathcal{N}U^{\mathrm{C}}(T,V,\mathbf{N}) \\ S_t^{\mathrm{C}}(T,V,\mathbf{N},\mathcal{N}) &\equiv \mathcal{N}S^{\mathrm{C}}(T,V,\mathbf{N}) \end{aligned} \tag{3.15}$$

By construction, these are averages in the canonical ensemble. The entropy per replica is due to the distribution of the internal energy, which is the same for all replicas. By introducing the variables of the small system into Eq. (3.14), we identify the replica energy as the Helmholtz energy:

$$X^{\mathrm{C}}(T,V,\mathbf{N}) = U^{\mathrm{C}}(T,V,\mathbf{N}) - TS^{\mathrm{C}}(T,V,\mathbf{N}) = F^{\mathrm{C}}(T,V,\mathbf{N}) \quad (3.16)$$

Equation (3.16) also follows when we substitute Eq. (2.7) for ε^{C} into Eq. (3.12). The replica energy X^{C} is equal to the Helmholtz energy F^{C} for controlled variables T, V, $\mathbf{N}$ and shape, see also Table 3.1. By introducing Eqs. (3.15) and (3.16) into Eq. (3.11), we obtain the Gibbs equation for the small system

$$dU^{\mathrm{C}} = TdS^{\mathrm{C}} - p^{\mathrm{C}}dV + \boldsymbol{\mu}^{\mathrm{C}} \cdot d\mathbf{N} \quad (3.17)$$

This equation has exactly the same form as for a large system. The important distinction is that the properties of the small system U^{C}, $S^{\mathrm{C}}, V, \mathbf{N}$ are not Euler homogeneous functions of degree one in the volume. Large systems have this property, giving $F^{\mathrm{C}} = -p^{\mathrm{C}}V + \boldsymbol{\mu}^{\mathrm{C}} \cdot \mathbf{N}$, so that $\varepsilon^{\mathrm{C}} = 0$ in that case. For small systems p^{C}, $\boldsymbol{\mu}^{\mathrm{C}}$ depend on the volume V. In order to make this distinction more visible, we define

$$\widehat{F}^{\mathrm{C}} \equiv -p^{\mathrm{C}}V + \boldsymbol{\mu}^{\mathrm{C}} \cdot \mathbf{N} \quad (3.18)$$

Together with Eqs. (3.12) and (3.16), this implies that

$$F^{\mathrm{C}} - \widehat{F}^{\mathrm{C}} = \varepsilon^{\mathrm{C}} \quad (3.19)$$

The subdivision potential is the difference between the Helmholtz energy and the Helmholtz energy-hat. The hat-variable is yet another way to describe the system smallness.

We have pictured three ensembles, for $\mathcal{N} = 3, 4$ and 5 spherical replicas, in the one-component case in Fig. 3.5. All replicas have four particles and the same volume V. The internal energies of the replicas differ, which is symbolized by different colors. The addition of one more replica implies that the total internal energy $\mathcal{N}U$, the total volume $\mathcal{N}V$, and the total number of particles $\mathcal{N}N = \mathcal{N}4$ increase with U, V and $N = 4$, respectively. In terms of the Helmholtz energy,

$$dF^{\mathrm{C}} = -S^{\mathrm{C}}dT - p^{\mathrm{C}}dV + \boldsymbol{\mu}^{\mathrm{C}} \cdot d\mathbf{N} \quad (3.20)$$

This equation contains the changes due to changes in the environmental variables T, V, $\mathbf{N}$ and it is therefore more useful in the laboratory

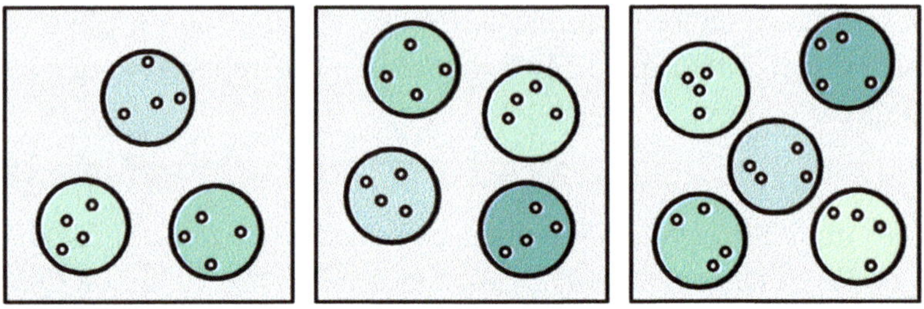

Figure 3.5. Three ensembles of replicas with environmental variables T, V, $N = 4$. All replicas can exchange internal energy with the environment. Their difference in energy is indicated by different color shades.

than Eq. (3.17). The Maxwell relations are

$$\left(\frac{\partial S^{\mathrm{C}}}{\partial V}\right)_{T,\mathbf{N}} = \left(\frac{\partial p^{\mathrm{C}}}{\partial T}\right)_{V,\mathbf{N}} \tag{3.21}$$

$$\left(\frac{\partial S^{\mathrm{C}}}{\partial N_j}\right)_{T,V,\mathbf{N}_j} = -\left(\frac{\partial \mu_j^{\mathrm{C}}}{\partial T}\right)_{V,\mathbf{N}} \tag{3.22}$$

$$\left(\frac{\partial p^{\mathrm{C}}}{\partial N_j}\right)_{T,V,\mathbf{N}_j} = -\left(\frac{\partial \mu_j^{\mathrm{C}}}{\partial V}\right)_{T,\mathbf{N}} \tag{3.23}$$

Remark 3.4. It is possible to obtain Maxwell relations from Eq. (3.17). An example is $(\partial T/\partial V)_{S^{\mathrm{C}},\mathbf{N}} = -(\partial p^{\mathrm{C}}/\partial S^{\mathrm{C}})_{V,\mathbf{N}}$. To calculate a derivative, keeping S^{C} constant, is difficult, however. For practical reasons, we prefer to use the Maxwell relations which can be derived from Eq. (3.20).

From Eq. (3.20), we identify the pressure, entropy and chemical potentials, p^{C}, S^{C}, μ_j^{C}:

$$\begin{aligned} p^{\mathrm{C}}(T,V,\mathbf{N}) &= -\left(\frac{\partial F^{\mathrm{C}}}{\partial V}\right)_{T,\mathbf{N}} \\ S^{\mathrm{C}}(T,V,\mathbf{N}) &= -\left(\frac{\partial F^{\mathrm{C}}}{\partial T}\right)_{V,\mathbf{N}} \\ \mu_j^{\mathrm{C}}(T,V,\mathbf{N}) &= \left(\frac{\partial F^{\mathrm{C}}}{\partial N_j}\right)_{T,V,\mathbf{N}_j} \end{aligned} \tag{3.24}$$

The change in the total entropy of the replicas follows from the Hill–Gibbs equation (Eq. (3.11)) and is given by

$$dS_t^{\mathrm{C}} = \frac{1}{T}dU_t^{\mathrm{C}} + \frac{p^{\mathrm{C}}}{T}\mathcal{N}dV - \frac{\boldsymbol{\mu}^{\mathrm{C}}}{T}\mathcal{N}\cdot d\mathbf{N} - \frac{X^{\mathrm{C}}}{T}d\mathcal{N} \tag{3.25}$$

The change in the entropy per replica follows from the Gibbs equation (Eq. (3.17)):

$$dS^{\mathrm{C}} = \frac{1}{T}dU^{\mathrm{C}} + \frac{p^{\mathrm{C}}}{T}dV - \frac{\boldsymbol{\mu}^{\mathrm{C}}}{T}\cdot d\mathbf{N} \tag{3.26}$$

These two equations can be a starting point for a non-equilibrium thermodynamic analysis of a small system with V, T, $\mathbf{N}$, $\mathcal{N}$ and the shape as environmental variables, i.e., the entropy production.

3.3 Environmental Variables T, p, N

This small system can exchange volume and internal energy with the environment. It is an isobaric, isothermal system, where T, p, $\mathbf{N}$ and shape are controlled. The system is illustrated in Figs. 3.6 and 3.7. The exchange of energy is indicated by arrows through the wall. This keeps the temperature constant. The pressure is controlled by adjusting the position of the piston. This may change the volume somewhat, keeping the shape (approximately) the same. Superscript P indicates that the small system is isobaric. Properties are then calculated as a function of the environmental variables T, p, $\mathbf{N}$ and shape.

The only additive variable is $\mathbf{N}$. By introducing $\mathbf{N}_t = \mathcal{N}\mathbf{N}$ into Eq. (2.2), the Hill–Gibbs equation for this ensemble can be written as

$$\begin{aligned} dU_t^{\mathrm{P}} &= TdS_t^{\mathrm{P}} - pdV_t^{\mathrm{P}} + \boldsymbol{\mu}^{\mathrm{P}}\mathcal{N}\cdot d\mathbf{N} + \left(\varepsilon^{\mathrm{P}} + \boldsymbol{\mu}^{\mathrm{P}}\cdot\mathbf{N}\right)d\mathcal{N} \\ &= TdS_t^{\mathrm{P}} - pdV_t^{\mathrm{P}} + \boldsymbol{\mu}^{\mathrm{P}}\mathcal{N}\cdot d\mathbf{N} + X^{\mathrm{P}}d\mathcal{N} \end{aligned} \tag{3.27}$$

The replica energy becomes

$$X^{\mathrm{P}}(T,p,\mathbf{N}) \equiv \varepsilon^{\mathrm{P}} + \boldsymbol{\mu}^{\mathrm{P}}\cdot\mathbf{N} \tag{3.28}$$

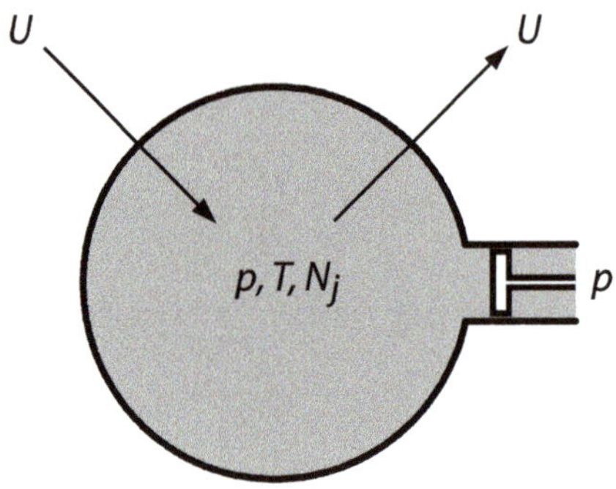

Figure 3.6. A small system with environmental variables p, T, N_j exchanges volume and internal energy with the environment.

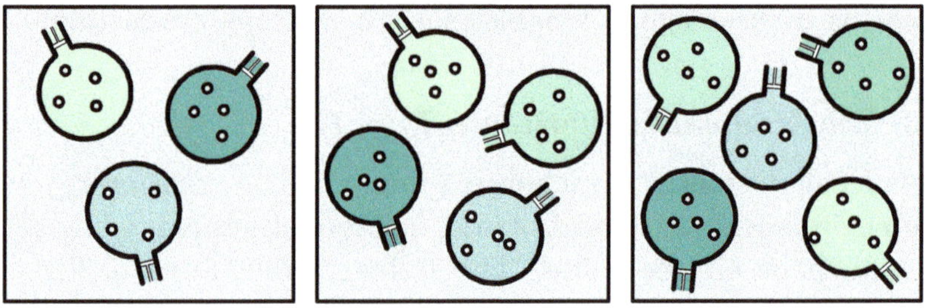

Figure 3.7. Three ensembles of replicas with environmental variables T, p, N_j. All exchange volume and internal energy with the environment. Different color shades indicate different internal energies.

The replica energy is also equal to the partial specific internal energy of the ensemble of replicas, cf. Eq. (3.27)

$$X^{\mathrm{P}}(T, p, \mathbf{N}) = \left(\frac{\partial U_t^{\mathrm{P}}}{\partial \mathcal{N}}\right)_{S_t, V_t, \mathbf{N}} \tag{3.29}$$

The values for the ensembles, U_t^{P}, S_t^{P} and V_t^{P}, are also functions of $\mathcal{N}$. The replica energy can be understood as the reversible work needed to add one replica of the small system, now at constant S_t^{P}, V_t^{P}, $\mathbf{N}$ and shape. The addition of a replica at constant S_t^{P}, V_t^{P} and $\mathbf{N}$ implies that S_t^{P} and V_t^{P} have to be redistributed over one more replica, while the total number of particles of each component $\mathcal{N}\mathbf{N}$ increases.

There is a shape-dependence of the thermodynamic properties in all small systems for all choices of environmental variables. When we replace the size of the volume V by the pressure p as a control variable, the dependence on the shape of the volume remains.

By integrating Eq. (3.27) at constant T, p, $\mathbf{N}$ and X^{P}, we obtain

$$U_t^{\mathrm{P}}(T,p,\mathbf{N},\mathcal{N}) = TS_t^{\mathrm{P}}(T,p,\mathbf{N},\mathcal{N}) - pV_t^{\mathrm{P}}(T,p,\mathbf{N},\mathcal{N}) + X^{\mathrm{P}}(T,p,\mathbf{N})\mathcal{N} \tag{3.30}$$

T and p are given by the values of these quantities in the environment. The *average* variables of the small system are related to the variables of the total system by

$$\begin{aligned} U_t^{\mathrm{P}}(T,p,\mathbf{N},\mathcal{N}) &\equiv \mathcal{N}U^{\mathrm{P}}(T,p,\mathbf{N}) \\ V_t^{\mathrm{P}}(T,p,\mathbf{N},\mathcal{N}) &\equiv \mathcal{N}V^{\mathrm{P}}(T,p,\mathbf{N}) \\ S_t^{\mathrm{P}}(T,p,\mathbf{N},\mathcal{N}) &\equiv \mathcal{N}S^{\mathrm{P}}(T,p,\mathbf{N}) \end{aligned} \tag{3.31}$$

The entropy per replica is due to the distribution of internal energy and volume, which is the same for all replicas. By introducing the variables for the small system into Eq. (3.27), we obtain

$$\begin{aligned} X^{\mathrm{P}}(T,p,\mathbf{N}) &= U^{\mathrm{P}}(T,p,\mathbf{N}) - TS^{\mathrm{P}}(T,p,\mathbf{N}) + pV^{\mathrm{P}}(T,p,\mathbf{N}) \\ &= G^{\mathrm{P}}(T,p,\mathbf{N}) \end{aligned} \tag{3.32}$$

X^{P} is therefore equal to the Gibbs energy of a single replica — a confirmation of the interpretation of X^{P} in terms of reversible work at constant pressure. Equation (3.32) follows by introducing Eq. (2.7) for ε^{P} into Eq. (3.28).

By introducing Eqs. (3.31) and (3.32) into Eq. (3.27), we obtain the Gibbs relation for the small isobaric system,

$$dU^{\mathrm{P}} = TdS^{\mathrm{P}} - pdV^{\mathrm{P}} + \boldsymbol{\mu}^{\mathrm{P}} \cdot d\mathbf{N} \tag{3.33}$$

This equation has exactly the same form as for a large system. The important distinction is that the properties of the small system U^{P}, S^{P}, V^{P}, $\mathbf{N}$ are not Euler homogeneous functions of degree one in the volume. Large systems have this property, giving $G^{\mathrm{P}} = \boldsymbol{\mu}^{\mathrm{P}} \cdot \mathbf{N} = U^{\mathrm{P}} - TS^{\mathrm{P}} + pV^{\mathrm{P}}$, so that $\varepsilon^{\mathrm{P}} = 0$ in that case. For small systems, $\boldsymbol{\mu}^{\mathrm{P}}$ depends on $\mathbf{N}$. In order to make this distinction more visible, we define the Gibbs energy-hat as follows:

$$\widehat{G}^{\mathrm{P}} \equiv \boldsymbol{\mu}^{\mathrm{P}} \cdot \mathbf{N} \quad \text{and} \quad G^{\mathrm{P}} = X^{\mathrm{P}} \equiv \widehat{\boldsymbol{\mu}}^{\mathrm{P}} \cdot \mathbf{N} \tag{3.34}$$

where we have introduced the chemical potential-hat. Together with Eq. (3.28), this implies that

$$G^{\mathrm{P}} - \widehat{G}^{\mathrm{P}} = \left(\widehat{\boldsymbol{\mu}}^{\mathrm{P}} - \boldsymbol{\mu}^{\mathrm{P}}\right) \cdot \mathbf{N} = \varepsilon^{\mathrm{P}} \tag{3.35}$$

The difference between the Gibbs energy and the Gibbs energy-hat is the subdivision potential.

The corresponding Hill–Gibbs–Duhem equation is

$$dG^{\mathrm{P}} = -S^{\mathrm{P}} dT + V^{\mathrm{P}} dp + \boldsymbol{\mu}^{\mathrm{P}} \cdot d\mathbf{N} \tag{3.36}$$

This equation contains the changes due to changes in the environmental variables T, p, $\mathbf{N}$, and is more useful in the laboratory than Eq. (3.33). It is the appropriate equation from which we obtain the following Maxwell relations:

$$\begin{aligned}
\left(\frac{\partial S^{\mathrm{P}}}{\partial p}\right)_{T,\mathbf{N}} &= -\left(\frac{\partial V^{\mathrm{P}}}{\partial T}\right)_{p,\mathbf{N}} \\
\left(\frac{\partial S^{\mathrm{P}}}{\partial N_j}\right)_{T,p,\mathbf{N}_j} &= -\left(\frac{\partial \mu_j^{\mathrm{P}}}{\partial T}\right)_{p,\mathbf{N}} \\
\left(\frac{\partial V^{\mathrm{P}}}{\partial N_j}\right)_{T,p,\mathbf{N}_j} &= \left(\frac{\partial \mu_j^{\mathrm{P}}}{\partial p}\right)_{T,\mathbf{N}}
\end{aligned} \tag{3.37}$$

Furthermore, we derive S^{P}, V^{P}, μ_j^{P} from

$$\begin{aligned}
S^{\mathrm{P}}(T,p,\mathbf{N}) &= -\left(\frac{\partial G^{\mathrm{P}}}{\partial T}\right)_{p,\mathbf{N}} \\
V^{\mathrm{P}}(T,p,\mathbf{N}) &= \left(\frac{\partial G^{\mathrm{P}}}{\partial p}\right)_{T,\mathbf{N}} \\
\mu_j^{\mathrm{P}}(T,p,\mathbf{N}) &= \left(\frac{\partial G^{\mathrm{P}}}{\partial N_j}\right)_{T,p,\mathbf{N}_j}
\end{aligned} \tag{3.38}$$

Remark 3.5. Observe that G^{P} is *not* equal to $\boldsymbol{\mu}^{\mathrm{P}} \cdot \mathbf{N}$, as it is for large systems. This is because the chemical potentials depend on $\mathbf{N}$ for small systems.

The change in the total entropy of the ensemble of replicas follows from the Hill–Gibbs equation (3.27) and is given by

$$dS_t^{\mathrm{P}} = \frac{1}{T}dU_t^{\mathrm{P}} + \frac{p}{T}dV_t^{\mathrm{P}} - \frac{\boldsymbol{\mu}^{\mathrm{P}}}{T}\mathcal{N} \cdot d\mathbf{N} - \frac{X^{\mathrm{P}}}{T}d\mathcal{N} \tag{3.39}$$

The change in the entropy per replica is given by the Gibbs equation (Eq. (3.33)):

$$dS^{\mathrm{P}} = \frac{1}{T}dU^{\mathrm{P}} + \frac{p}{T}dV^{\mathrm{P}} - \frac{\boldsymbol{\mu}^{\mathrm{P}}}{T} \cdot d\mathbf{N} \tag{3.40}$$

These two equations can be used as starting point of a non-equilibrium thermodynamic analysis of a small system with p, T, $\mathbf{N}$ and $\mathcal{N}$ and the shape as environmental variables. Equation (3.40) was used to find the entropy production for isotensional stretching of small polymers in Chapter 21 [27].

3.4 Environmental Variables T, V, $\boldsymbol{\mu}$

This small system can exchange energy and particles with the environment to keep T, V and $\boldsymbol{\mu}$ constant. The system is illustrated in Figs. 3.8 and 3.9. The possible exchange is indicated by arrows. The variables are calculated as functions of the environmental variables T, V, $\boldsymbol{\mu}$ and the shape, using a grand canonical ensemble (indicated by superscript GC).

The additive variable is V. By introducing $V_t = \mathcal{N}V$ in Eq. (2.2), the Hill–Gibbs equation obtained is

$$\begin{aligned} dU_t^{\mathrm{GC}} &= TdS_t^{\mathrm{GC}} - p^{\mathrm{GC}}\mathcal{N}dV + \boldsymbol{\mu} \cdot d\mathbf{N}_t^{\mathrm{GC}} + \left(\varepsilon^{\mathrm{GC}} - p^{\mathrm{GC}}V\right)d\mathcal{N} \\ &= TdS_t^{\mathrm{GC}} - p^{\mathrm{GC}}\mathcal{N}dV + \boldsymbol{\mu} \cdot d\mathbf{N}_t^{\mathrm{GC}} + X^{\mathrm{GC}}d\mathcal{N} \end{aligned} \tag{3.41}$$

The replica energy (the prefactor of $d\mathcal{N}$) becomes

$$X^{\mathrm{GC}}(T, V, \boldsymbol{\mu}) \equiv \varepsilon^{\mathrm{GC}} - p^{\mathrm{GC}}V \tag{3.42}$$

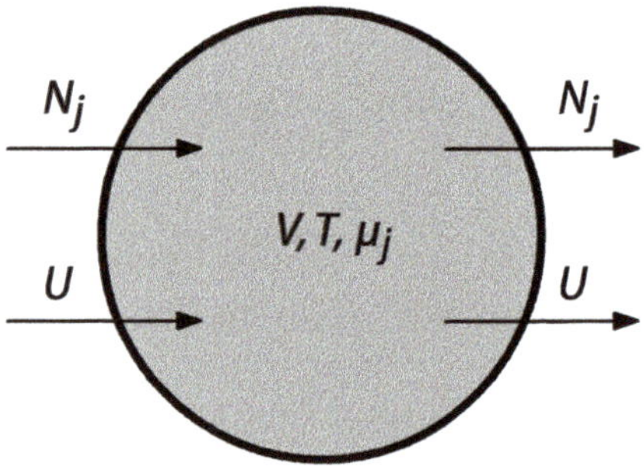

Figure 3.8. A small system with environmental variables V, T, μ_j exchanges internal energy and particles with the environment.

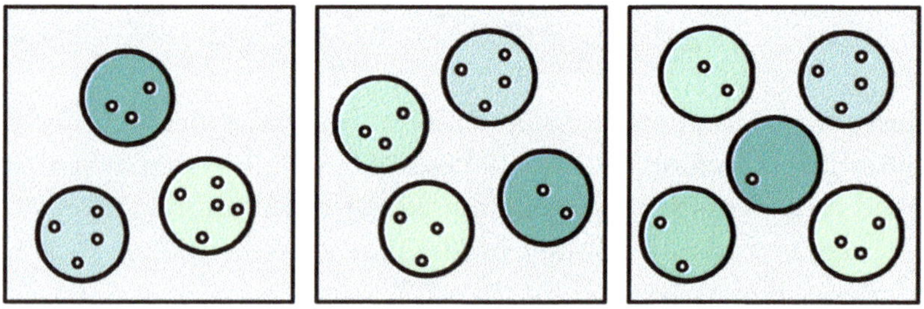

Figure 3.9. Three ensembles of replicas with environmental variables V, T, μ_j. All replicas exchange internal energy and particles with the environment. Differnt color shades indicate different internal energies.

It is equal to the specific internal energy of the ensemble of replicas, cf. Eq. (3.41):

$$X^{\mathrm{GC}}(T, V, \boldsymbol{\mu}) = \left(\frac{\partial U_t^{\mathrm{GC}}}{\partial \mathcal{N}} \right)_{S_t, V, \mathbf{N}_t} \tag{3.43}$$

The replica energy gives again the reversible work needed to add one replica of the small system, but now at constant S_t^{GC}, V, $\mathbf{N}_t^{\mathrm{GC}}$ and shape. The addition of a replica at constant S_t^{GC}, V and $\mathbf{N}_t^{\mathrm{GC}}$ implies that S_t^{GC} and $\mathbf{N}_t^{\mathrm{GC}}$ have to be redistributed over one more replica, while the total volume $\mathcal{N}V$ increases. The replica energy is compressional energy and was denoted $-\widehat{p}V$ by Hill [3, 4]:

$$\widehat{p}(T, V, \boldsymbol{\mu})V \equiv -X^{\mathrm{GC}}(T, V, \boldsymbol{\mu}) = -\left(\frac{\partial U_t^{\mathrm{GC}}}{\partial \mathcal{N}} \right)_{S_t, V, \mathbf{N}_t} \tag{3.44}$$

Remark 3.6. The reason to use the symbol $\widehat{p}V$ for X^{GC} is that the quantity can be associated with work. Mechanical work is normally done by extending the volume, V. Work can also be done when the number of systems with volume V increases by adding replicas. The last operation is additive and gave Hill the reason to introduce the name *integral pressure*, or p-hat, for $\widehat{p}$ [3, 4]. It has also been called the effective pressure. The quantity $\widehat{p}V$ obtains its meaning from Eq. (3.44). Once we refer to $\widehat{p}$ as the integral pressure, p^{GC} becomes the *differential pressure.*

By integrating Eq. (3.41) at constant T, V and $\boldsymbol{\mu}$, we obtain

$$\begin{aligned} U_t^{\mathrm{GC}}(T, V, \boldsymbol{\mu}, \mathcal{N}) = TS_t^{\mathrm{GC}}(T, V, \boldsymbol{\mu}, \mathcal{N}) + \boldsymbol{\mu} \cdot \mathbf{N}_t^{\mathrm{GC}}(T, V, \boldsymbol{\mu}, \mathcal{N}) \\ - \widehat{p}(T, V, \boldsymbol{\mu}) V \mathcal{N} \end{aligned} \tag{3.45}$$

Here, T and $\boldsymbol{\mu}$ are given by the values that these quantities have in the environment. The average variables of the small system are related to the variables of the total system by

$$\begin{aligned} U_t^{\mathrm{GC}}(T, V, \boldsymbol{\mu}, \mathcal{N}) &\equiv \mathcal{N} U^{\mathrm{GC}}(T, V, \boldsymbol{\mu}) \\ S_t^{\mathrm{GC}}(T, V, \boldsymbol{\mu}, \mathcal{N}) &\equiv \mathcal{N} S^{\mathrm{GC}}(T, V, \boldsymbol{\mu}) \\ N_{j,t}^{\mathrm{GC}}(T, V, \boldsymbol{\mu}, \mathcal{N}) &\equiv \mathcal{N} N_j^{\mathrm{GC}}(T, V, \boldsymbol{\mu}) \end{aligned} \tag{3.46}$$

The entropy S^{GC} is determined by the probability distribution over $\mathbf{N}$ and U, which is the same for each replica.

By introducing the variables of the small system into Eq. (3.41), we obtain

$$\begin{aligned} X^{\mathrm{GC}}(T, V, \boldsymbol{\mu}) = -\widehat{p}(T, V, \boldsymbol{\mu}) V = U^{\mathrm{GC}}(T, V, \boldsymbol{\mu}) \\ - TS^{\mathrm{GC}}(T, V, \boldsymbol{\mu}) - \boldsymbol{\mu} \cdot \mathbf{N}^{\mathrm{GC}}(T, V, \boldsymbol{\mu}) \end{aligned} \tag{3.47}$$

Equation (3.47) follows by introducing Eq. (2.7) for $\varepsilon^{\mathrm{GC}}$ into Eq. (3.42). It implies that the replica energy is equal to the grand potential. This equation furthermore pinpoints that U^{GC} is not an Euler homogeneous function of degree one in S^{GC}, V and $\mathbf{N}^{\mathrm{GC}}$. The small system follows classical thermodynamics only if $\widehat{p} = p^{\mathrm{GC}}$.

In general, it follows from Eqs. (3.42) and (3.47) that

$$\left(p^{\mathrm{GC}} - \widehat{p}\right) V = \varepsilon^{\mathrm{GC}} \tag{3.48}$$

The difference between these two pressures times the volume is the subdivision potential. Rauter *et al.* [40] gave this difference for the slit pore as a scaling law with the pore wall distance. The corresponding law for a cylindrical pore could be predicted [49]. The system is small when $\widehat{p} \neq p^{\mathrm{GC}}$. The value of $\widehat{p}$ can deviate significantly from p^{GC} due to surface energy [41].

By introducing Eq. (3.46) into Eq. (3.41), using Eq. (3.47), we obtain the Gibbs relation for the small system

$$dU^{\mathrm{GC}} = TdS^{\mathrm{GC}} - p^{\mathrm{GC}}dV + \boldsymbol{\mu} \cdot d\mathbf{N}^{\mathrm{GC}} \tag{3.49}$$

The corresponding Hill–Gibbs–Duhem equation is

$$d\left(\widehat{p}V\right) = S^{\mathrm{GC}}dT + p^{\mathrm{GC}}dV + \mathbf{N}^{\mathrm{GC}} \cdot d\boldsymbol{\mu} \tag{3.50}$$

This equation contains changes of $\widehat{p}V$ due to changes in the environmental variables T, V, $\boldsymbol{\mu}$ and is more useful than Eq. (3.49). We use it to obtain the following Maxwell relations:

$$\begin{aligned}
\left(\frac{\partial S^{\mathrm{GC}}}{\partial V}\right)_{T,\boldsymbol{\mu}} &= \left(\frac{\partial p^{\mathrm{GC}}}{\partial T}\right)_{V,\boldsymbol{\mu}} \\
\left(\frac{\partial S^{\mathrm{GC}}}{\partial \mu_j}\right)_{T,V,\boldsymbol{\mu}_j} &= \left(\frac{\partial N_j^{\mathrm{GC}}}{\partial T}\right)_{V,\boldsymbol{\mu}} \\
\left(\frac{\partial p^{\mathrm{GC}}}{\partial \mu_j}\right)_{T,V,\boldsymbol{\mu}_j} &= \left(\frac{\partial N_j^{\mathrm{GC}}}{\partial V}\right)_{T,\boldsymbol{\mu}}
\end{aligned} \tag{3.51}$$

Furthermore, we derive p^{GC}, S^{GC}, N_j^{GC}:

$$\begin{aligned}
p^{\mathrm{GC}}(T,V,\boldsymbol{\mu}) &= \left(\frac{\partial \widehat{p}V}{\partial V}\right)_{T,\boldsymbol{\mu}} = \widehat{p} + V\left(\frac{\partial \widehat{p}}{\partial V}\right)_{T,\boldsymbol{\mu}} \\
S^{\mathrm{GC}}(T,V,\boldsymbol{\mu}) &= \left(\frac{\partial \widehat{p}V}{\partial T}\right)_{V,\boldsymbol{\mu}} \\
N_j^{\mathrm{GC}}(T,V,\boldsymbol{\mu}) &= \left(\frac{\partial \widehat{p}V}{\partial \mu_j}\right)_{T,V,\boldsymbol{\mu}_j}
\end{aligned} \tag{3.52}$$

These equations are all special for small systems, as they imply non-zero partial derivatives of $\widehat{p}$. We see from Eq. (3.52) that p^{GC} differs from $\widehat{p}$ when $\widehat{p}$ depends on V. This dependence is typical for small GC-systems.

Remark 3.7. Equation (3.52) served as an inspiration for the names integral pressure for $\widehat{p}$ and differential pressure for p^{GC}, see Hill [3, 4], cf. Remark 3.6.

The existence of $\widehat{p}$ and p^{GC} imply that two definitions are possible for the enthalpy of the small systems [46, 47]:

$$\begin{aligned}\widehat{H}(T,V,\boldsymbol{\mu}) &\equiv U^{\mathrm{GC}}(T,V,\boldsymbol{\mu}) + \widehat{p}(T,V,\boldsymbol{\mu})V \\ &= TS^{\mathrm{GC}}(T,V,\boldsymbol{\mu}) + \boldsymbol{\mu}\cdot\mathbf{N}^{\mathrm{GC}}(T,V,\boldsymbol{\mu})\end{aligned} \tag{3.53}$$

and

$$H^{\mathrm{GC}}(T,V,\boldsymbol{\mu}) \equiv U^{\mathrm{GC}}(T,V,\boldsymbol{\mu}) + p^{\mathrm{GC}}(T,V,\boldsymbol{\mu})V \tag{3.54}$$

Remark 3.8. In large systems, $\widehat{p}$ and p^{GC}, and consequently $\widehat{H}$ and H^{GC}, are equal. In analogy to $\widehat{p}$ and p^{GC} we name $\widehat{H}$ and H^{GC} the integral and the differential enthalpy. We understand H^{GC} as the common enthalpy, which is equal to the internal energy plus the compressional energy $p^{\mathrm{GC}}V$. The quantity $\widehat{H}$ contains the compressional energy $\widehat{p}V$ of replicas. The differences are given by the subdivision potential: $(p^{\mathrm{GC}} - \widehat{p})V = H^{\mathrm{GC}} - \widehat{H} = \varepsilon^{\mathrm{GC}}$.

The change in the total entropy of the replicas follows from Hill–Gibbs' equation (Eq. (3.41)) and is given by

$$dS_t^{\mathrm{GC}} = \frac{1}{T}dU_t^{\mathrm{GC}} + \frac{p^{\mathrm{GC}}}{T}\mathcal{N}dV - \frac{\boldsymbol{\mu}}{T}\cdot d\mathbf{N}_t^{\mathrm{GC}} - \frac{X^{\mathrm{GC}}}{T}d\mathcal{N} \tag{3.55}$$

The change in the entropy per replica has the form of the Gibbs equation (Eq. (3.49))

$$dS^{\mathrm{GC}} = \frac{1}{T}dU^{\mathrm{GC}} + \frac{p^{\mathrm{GC}}}{T}dV - \frac{\boldsymbol{\mu}}{T}\cdot d\mathbf{N}^{\mathrm{GC}} \tag{3.56}$$

These two equations can serve as starting points for a non-equilibrium thermodynamic analysis of a small system with T, V, $\boldsymbol{\mu}$, $\mathcal{N}$ and the shape as environmental variables.

3.5 Environmental Variables T, p, N_1, $\boldsymbol{\mu}_1$

This system can exchange volume, energy and particles of components $2, \ldots, n$ with the environment. It is an isobaric and isothermal system, with constant T, p, N_1, $\boldsymbol{\mu}_1 \equiv (\mu_2, \ldots, \mu_n)$ and shape. The ensemble of small systems is called osmotic (indicated by superscript O), since one of the components is controlled. Superscript O means that all variables are functions of the environmental variables T, p, N_1, $\boldsymbol{\mu}_1$, shape, and U_t^{O}, S_t^{O}, V_t^{O}, $\mathbf{N}_{1,t}^{\mathrm{O}}$ are also functions of $\mathcal{N}$. The system is illustrated in Figs. 3.10 and 3.11.

By introducing $N_{1,t} = \mathcal{N} N_1$ into Eq. (1.3), the Hill–Gibbs equation for this ensemble becomes

$$\begin{aligned} dU_t^{\mathrm{O}} &= TdS_t^{\mathrm{O}} - pdV_t^{\mathrm{O}} + \mu_1^{\mathrm{O}} \mathcal{N} dN_1 + \boldsymbol{\mu}_1 \cdot d\mathbf{N}_{1,t}^{\mathrm{O}} + \left(\varepsilon^{\mathrm{O}} + \mu_1^{\mathrm{O}} N_1\right) d\mathcal{N} \\ &= TdS_t^{\mathrm{O}} - pdV_t^{\mathrm{O}} + \mu_1^{\mathrm{O}} \mathcal{N} dN_1 + \boldsymbol{\mu}_1 \cdot d\mathbf{N}_{1,t}^{\mathrm{O}} + X^{\mathrm{O}} d\mathcal{N} \end{aligned} \tag{3.57}$$

The replica energy, the prefactor of $d\mathcal{N}$, is

$$X^{\mathrm{O}} \left(T, p, N_1, \boldsymbol{\mu}_1\right) \equiv \varepsilon^{\mathrm{O}} + \mu_1^{\mathrm{O}} N_1 \tag{3.58}$$

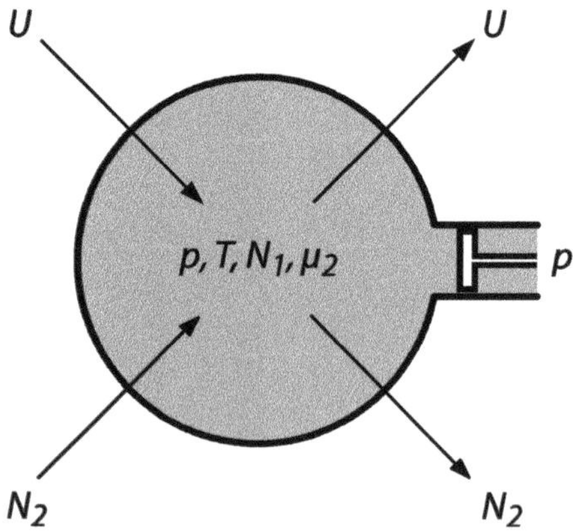

Figure 3.10. The two-component system with environmental variables T, p, N_1, $\mu_2 = \boldsymbol{\mu}_1$ exchanges internal energy, volume and particles of the second component with the environment.

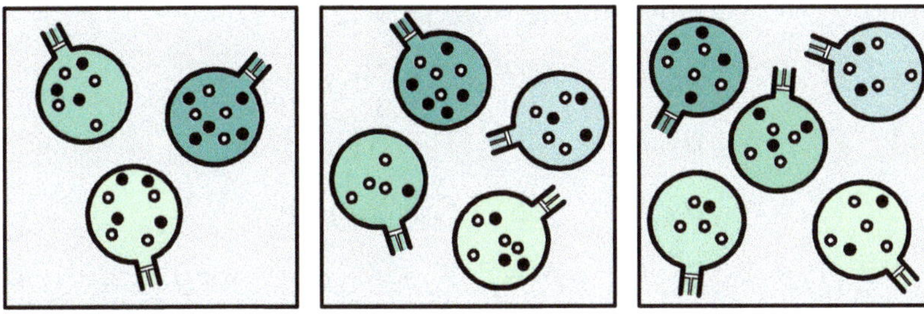

Figure 3.11. Three ensembles of replicas with environmental variables p, T, N_1, μ_2. All replicas exchange volume, internal energy and particles of the second component of two with the environment. Different color shades indicate different internal energies.

The replica energy is the partial specific internal energy of the ensemble of replicas, cf. Eq. (3.57)

$$X^{\mathrm{O}}\left(T,p,N_1,\boldsymbol{\mu}_1\right)=\left(\frac{\partial U_t^{\mathrm{O}}}{\partial \mathcal{N}}\right)_{S_t^{\mathrm{O}},V_t^{\mathrm{O}},N_1,\mathbf{N}_{1,t}^{\mathrm{O}}} \tag{3.59}$$

The replica energy can, as before, be interpreted as the reversible work needed to add one replica of the small system now at constant S_t^{O}, V_t^{O}, N_1, $\mathbf{N}_{1,t}^{\mathrm{O}}$. The addition of a replica at constant S_t^{O}, V_t^{O}, N_1, $\mathbf{N}_{1,t}^{\mathrm{O}}$ implies that S_t^{O}, V_t^{O}, $\mathbf{N}_{1,t}^{\mathrm{O}}$ have to be redistributed over one more replica, while the total additive number of particles of the first component $\mathcal{N}N_1$ increases.

The shape-dependence is present for all choices of environmental variables when the system is small. This is so also here. When we replace the size of volume V by the pressure p as an environmental variable, the dependence on the shape of the volume remains.

By integrating Eq. (3.57) at constant T, p, N_1, $\boldsymbol{\mu}_1$ and X^{O}, we obtain

$$\begin{aligned} &U_t^{\mathrm{O}}\left(T,p,N_1,\boldsymbol{\mu}_1,\mathcal{N}\right)\\ &\quad=TS_t^{\mathrm{O}}\left(T,p,N_1,\boldsymbol{\mu}_1,\mathcal{N}\right)-pV_t^{\mathrm{O}}\left(T,p,N_1,\boldsymbol{\mu}_1,\mathcal{N}\right)\\ &\qquad+\boldsymbol{\mu}_1\cdot\mathbf{N}_{1,t}^{\mathrm{O}}\left(T,p,N_1,\boldsymbol{\mu}_1,\mathcal{N}\right)+X^{\mathrm{O}}\left(T,p,N_1,\boldsymbol{\mu}_1\right)\mathcal{N} \end{aligned} \tag{3.60}$$

where T, p, $\boldsymbol{\mu}_1$ are given by the values that these quantities have in the environment. The average variables of the small system are related to the variables of the total system by

$$\begin{aligned}
U_t^{\mathrm{O}}(T,p,N_1,\boldsymbol{\mu}_1,\mathcal{N}) &\equiv \mathcal{N}U^{\mathrm{O}}(T,p,N_1,\boldsymbol{\mu}_1) \\
V_t^{\mathrm{O}}(T,p,N_1,\boldsymbol{\mu}_1,\mathcal{N}) &\equiv \mathcal{N}V^{\mathrm{O}}(T,p,N_1,\boldsymbol{\mu}_1) \\
S_t^{\mathrm{O}}(T,p,N_1,\boldsymbol{\mu}_1,\mathcal{N}) &\equiv \mathcal{N}S^{\mathrm{O}}(T,p,N_1,\boldsymbol{\mu}_1) \\
N_{j,t}^{\mathrm{O}}(T,p,N_1,\boldsymbol{\mu}_1,\mathcal{N}) &= \mathcal{N}N_j^{\mathrm{O}}(T,p,N_1,\boldsymbol{\mu}_1) \quad \text{for } j=2,\ldots,n
\end{aligned} \tag{3.61}$$

By construction, these are averages in the osmotic ensemble. The entropy per replica is due to the distribution of internal energy, volume and the number of partcles of components 2-n, which is the same for all replicas. By introducing the variables of the small system into Eq. (3.57), we obtain

$$\begin{aligned}
X^{\mathrm{O}}(T,p,N_1,\boldsymbol{\mu}_1) &= U^{\mathrm{O}}(T,p,N_1,\boldsymbol{\mu}_1) - TS^{\mathrm{O}}(T,p,N_1,\boldsymbol{\mu}_1) \\
&\quad + pV^{\mathrm{O}}(T,p,N_1,\boldsymbol{\mu}_1) - \boldsymbol{\mu}_1 \cdot \mathbf{N}_1^{\mathrm{O}}(T,p,N_1,\boldsymbol{\mu}_1) \\
&\equiv \widehat{\mu}_1^{\mathrm{O}}(T,p,N_1,\boldsymbol{\mu}_1)\,N_1
\end{aligned} \tag{3.62}$$

Equation (3.62) also follows when we substitute Eq. (2.7) for ε^{O} into Eq. (3.59). In the last equality, we defined the *integral* chemical potential of the first component.

Remark 3.9. In large systems, $\widehat{\mu}_1^{\mathrm{O}}$ and μ_1^{O} are equal. In analogy to $\widehat{p}$ and p^{GC}, we name $\widehat{\mu}_1^{\mathrm{O}}$ and μ_1^{O} the integral and the differential chemical potential, respectively. We understand μ_1^{O} as the common chemical potential. It follows from Eq. (3.58) that the difference is given by the subdivision potential

$$(\mu_1^{\mathrm{O}} - \widehat{\mu}_1^{\mathrm{O}})N_1 = -\varepsilon^{\mathrm{O}} \tag{3.63}$$

By introducing Eqs. (3.60) and (3.61) into Eq. (3.57), we obtain the Gibbs equation of the small system

$$dU^{\mathrm{O}} = TdS^{\mathrm{O}} - pdV^{\mathrm{O}} + \mu_1^{\mathrm{O}}dN_1 + \boldsymbol{\mu}_1 \cdot d\mathbf{N}_1^{\mathrm{O}} \tag{3.64}$$

A two-component system of white and black particles is illustrated in Fig. 3.11. The figure shows that the replicas can have different internal energy. The number of N_1, chemical potential of the other components (here only μ_2), and the pressure is kept constant. The corresponding Hill–Gibbs–Duhem equation is

$$dX^{\mathrm{O}} = -S^{\mathrm{O}}dT + V^{\mathrm{O}}dp + \mu_1^{\mathrm{O}}dN_1 - \mathbf{N}_1^{\mathrm{O}} \cdot d\boldsymbol{\mu}_1 \tag{3.65}$$

or alternatively

$$d\left(\widehat{\mu}_1^{\mathrm{O}}N_1\right) = -S^{\mathrm{O}}dT + V^{\mathrm{O}}dp + \mu_1^{\mathrm{O}}dN_1 - \mathbf{N}_1^{\mathrm{O}} \cdot d\boldsymbol{\mu}_1 \tag{3.66}$$

In the thermodynamic limit $\mu_1^{\mathrm{O}} = \widehat{\mu}_1^{\mathrm{O}}$. The equation contains changes due to changes in environmental variables $T, p, N_1, \boldsymbol{\mu}_1$, making it more useful than Eq. (3.64). We obtain the Maxwell relations from Eq. (3.66).

By using Eq. (3.65), we can derive S^{O}, V^{O}, μ_1^{O}, $\mathbf{N}_1^{\mathrm{O}}$ as follows:

$$\begin{aligned}
S^{\mathrm{O}}(T,p,N_1,\boldsymbol{\mu}_1) &= -\left(\frac{\partial X^{\mathrm{O}}}{\partial T}\right)_{p,N_1,\boldsymbol{\mu}_1} = -\left(\frac{\partial \widehat{\mu}_1^{\mathrm{O}}N_1}{\partial T}\right)_{p,N_1,\boldsymbol{\mu}_1}\\
\overline{V}^{\mathrm{O}}(T,p,N_1,\boldsymbol{\mu}_1) &= \left(\frac{\partial X^{\mathrm{O}}}{\partial p}\right)_{T,N_1,\boldsymbol{\mu}_1} = \left(\frac{\partial \widehat{\mu}_1^{\mathrm{O}}N_1}{\partial p}\right)_{T,N_1,\boldsymbol{\mu}_1}\\
\mu_1^{\mathrm{O}}(T,p,N_1,\boldsymbol{\mu}_1) &= \left(\frac{\partial X^{\mathrm{O}}}{\partial N_1}\right)_{T,p,\boldsymbol{\mu}_1} = \left(\frac{\partial \widehat{\mu}_1^{\mathrm{O}}N_1}{\partial N_1}\right)_{T,p,\boldsymbol{\mu}_1}\\
&= \widehat{\mu}_1^{\mathrm{O}} + N_1\left(\frac{\partial \widehat{\mu}_1^{\mathrm{O}}}{\partial N_1}\right)_{T,p,\boldsymbol{\mu}_1}\\
N_j^{\mathrm{O}}(T,p,N_1,\boldsymbol{\mu}_1) &= -\left(\frac{\partial X^{\mathrm{O}}}{\partial \mu_j}\right)_{T,p,N_1,\mu_{l\neq 1,j}}\\
&= -\left(\frac{\partial \widehat{\mu}_1^{\mathrm{O}}N_1}{\partial \mu_j}\right)_{T,p,N_1,\mu_{l\neq 1,j}} \qquad \text{for } j = 2,\ldots,n
\end{aligned} \tag{3.67}$$

The replica energy is equal to the chemical potential hat times N_1, $\widehat{\mu}_1^{\mathrm{O}}N_1$.

Remark 3.10. Equations (3.67) motivate the two names; integral ($\widehat{\mu}_1^{\mathrm{O}}$) and differential ($\mu_1^{\mathrm{O}}$) chemical potential, in analogy to the naming of $\widehat{p}$ and p, cf. Remark 3.6. We cannot use the name small-system chemical potential for the quantity $\widehat{\mu}_1^{\mathrm{O}}$ only. *Both* chemical potentials are particular for the small system.

The change in the total entropy of the replicas follows from the Hill–Gibbs equation (3.57) and is given by

$$dS_t^{\mathrm{O}} = \frac{1}{T}dU_t^{\mathrm{O}} + \frac{p}{T}dV_t^{\mathrm{O}} - \frac{\mu_1^{\mathrm{O}}}{T}\mathcal{N}dN_1 - \frac{\boldsymbol{\mu}_1}{T}\cdot d\mathbf{N}_{1,t}^{\mathrm{O}} - \frac{X^{\mathrm{O}}}{T}d\mathcal{N} \quad (3.68)$$

The change in the entropy per replica is given by the Gibbs equation (3.64)

$$dS^{\mathrm{O}} = \frac{1}{T}dU^{\mathrm{O}} + \frac{p}{T}dV^{\mathrm{O}} - \frac{\mu_1^{\mathrm{O}}}{T}dN_1 - \frac{\boldsymbol{\mu}_1}{T}\cdot d\mathbf{N}_1^{\mathrm{O}} \quad (3.69)$$

These two equations can serve as starting point of a non-equilibrium thermodynamic analysis of a small system with T, p, N_1, $\boldsymbol{\mu}_1$, $\mathcal{N}$ and the shape as environmental variables. We use Eq. (3.69) to find the entropy production for isometric stretching of small polymers in Chapter 21 [27].

3.6 Environmental Variables T, p, μ

Consider next a small system that can exchange volume, energy and particles of all components with the environment. This small system is completely open. We call it nanocanonical following Hill [3, 4]. This is indicated by superscript NC. The properties must be calculated as functions of the environmental variables, and U_t^{NC}, S_t^{NC}, V_t^{NC}, $\mathbf{N}_t^{\mathrm{NC}}$, and also as functions of $\mathcal{N}$.

The Gibbs equation for this ensemble is, cf. Eq. (2.2),

$$\begin{aligned} dU_t^{\mathrm{NC}} &= TdS_t^{\mathrm{NC}} - pdV_t^{\mathrm{NC}} + \boldsymbol{\mu}\cdot d\mathbf{N}_t^{\mathrm{NC}} + \varepsilon^{\mathrm{NC}}d\mathcal{N} \\ &= TdS_t^{\mathrm{NC}} - pdV_t^{\mathrm{NC}} + \boldsymbol{\mu}\cdot d\mathbf{N}_t^{\mathrm{NC}} + X^{\mathrm{NC}}d\mathcal{N} \end{aligned} \quad (3.70)$$

The replica energy is equal to the subdivision potential and is given by

$$X^{\mathrm{NC}}(T, p, \boldsymbol{\mu}) \equiv \varepsilon^{\mathrm{NC}}(T, p, \boldsymbol{\mu}) \quad (3.71)$$

The replica energy is the partial internal energy of the ensemble of replicas, cf. Eq. (3.70),

$$X^{\mathrm{NC}}(T, p, \boldsymbol{\mu}) = \left(\frac{\partial U_t^{\mathrm{NC}}}{\partial \mathcal{N}}\right)_{S_t^{\mathrm{NC}}, V_t^{\mathrm{NC}}, \mathbf{N}_t^{\mathrm{NC}}} \quad (3.72)$$

The replica energy can again be interpreted as the reversible work needed to add one replica of the small system now at constant S_t^{NC}, V_t^{NC}, $\mathbf{N}_t^{\mathrm{NC}}$. The addition of a replica at constant S_t^{NC}, V_t^{NC}, $\mathbf{N}_t^{\mathrm{NC}}$ implies that S_t^{NC}, V_t^{NC}, $\mathbf{N}_t^{\mathrm{NC}}$ have to be redistributed over one more replica. The shape dependence is also present here. We can replace the size of the volume V by the pressure p as an environmental variable, but the dependence on the shape of the volume remains.

By integrating Eq. (3.70) at constant T, p, $\boldsymbol{\mu}$ and X^{NC}, we obtain

$$\begin{aligned} U_t^{\mathrm{NC}}(T,p,\boldsymbol{\mu},\mathcal{N}) &= TS_t^{\mathrm{NC}}(T,p,\boldsymbol{\mu},\mathcal{N}) - pV_t^{\mathrm{NC}}(T,p,\boldsymbol{\mu},\mathcal{N}) \\ &\quad + \sum_{j=1}^{n} \mu_j N_{j,t}^{\mathrm{NC}}(T,p,\boldsymbol{\mu},\mathcal{N}) + X^{\mathrm{NC}}(T,p,\boldsymbol{\mu})\mathcal{N} \end{aligned} \tag{3.73}$$

The values of T, p, $\boldsymbol{\mu}$ are the same as in the environment. The average variables of the small system are given by the variables of the total system by

$$\begin{aligned} U_t^{\mathrm{NC}}(T,p,\boldsymbol{\mu},\mathcal{N}) &\equiv \mathcal{N}U^{\mathrm{NC}}(T,p,\boldsymbol{\mu}) \\ V_t^{\mathrm{NC}}(T,p,\boldsymbol{\mu},\mathcal{N}) &\equiv \mathcal{N}V^{\mathrm{NC}}(T,p,\boldsymbol{\mu}) \\ S_t^{\mathrm{NC}}(T,p,\boldsymbol{\mu},\mathcal{N}) &\equiv \mathcal{N}S^{\mathrm{NC}}(T,p,\boldsymbol{\mu}) \\ N_{j,t}^{\mathrm{NC}}(T,p,\boldsymbol{\mu},\mathcal{N}) &= \mathcal{N}N_j^{\mathrm{NC}}(T,p,\boldsymbol{\mu}) \end{aligned} \tag{3.74}$$

By construction, these are averages of the nanocanonical ensemble. The entropy per replica is due to the distribution of the internal energy, the volume and the number of particles of the components, which is the same for all replicas. By introducing the variables of the small system into Eq. (3.70), we obtain

$$\begin{aligned} X^{\mathrm{NC}}(T,p,\boldsymbol{\mu}) &= U^{\mathrm{NC}}(T,p,\boldsymbol{\mu}) - TS^{\mathrm{NC}}(T,p,\boldsymbol{\mu}) \\ &\quad + pV^{\mathrm{NC}}(T,p,\boldsymbol{\mu}) - \boldsymbol{\mu}\cdot\mathbf{N}^{\mathrm{NC}}(T,p,\boldsymbol{\mu}) \end{aligned} \tag{3.75}$$

This equation is identical to Eq. (2.7) for $\varepsilon^{\mathrm{NC}} = X^{\mathrm{NC}}$. It follows from this equation that X^{NC} will approach zero in the thermodynamic limit. For a small system, X^{NC} is unequal to zero.

By introducing Eqs. (3.73) and (3.74) into Eq. (3.70), we obtain Gibbs equation for the small open system

$$dU^{\mathrm{NC}} = TdS^{\mathrm{NC}} - pdV^{\mathrm{NC}} + \boldsymbol{\mu} \cdot d\mathbf{N}^{\mathrm{NC}} \tag{3.76}$$

The corresponding Hill–Gibbs–Duhem equation is

$$dX^{\mathrm{NC}} = -S^{\mathrm{NC}}dT + V^{\mathrm{NC}}dp - \mathbf{N}^{\mathrm{NC}} \cdot d\boldsymbol{\mu} \tag{3.77}$$

This equation contains the changes due to changes in the environmental variables $T, V, \boldsymbol{\mu}$ and is more useful than Eq. (3.76). We obtain the following Maxwell relations:

$$\begin{aligned}
\left(\frac{\partial S^{\mathrm{NC}}}{\partial p}\right)_{T,\boldsymbol{\mu}} &= -\left(\frac{\partial V^{\mathrm{NC}}}{\partial T}\right)_{p,\boldsymbol{\mu}} \\
\left(\frac{\partial S^{\mathrm{NC}}}{\partial \mu_j}\right)_{T,p,\boldsymbol{\mu}_j} &= \left(\frac{\partial N_j^{\mathrm{NC}}}{\partial T}\right)_{p,\boldsymbol{\mu}} \\
\left(\frac{\partial V^{\mathrm{NC}}}{\partial \mu_j}\right)_{T,V,\boldsymbol{\mu}_j} &= -\left(\frac{\partial N_j^{\mathrm{NC}}}{\partial p}\right)_{T,\boldsymbol{\mu}}
\end{aligned} \tag{3.78}$$

From Eq. (3.77), we derive S^{NC}, V^{NC}, $\mathbf{N}^{\mathrm{NC}}$ from

$$\begin{aligned}
S^{\mathrm{NC}}(T,p,\boldsymbol{\mu}) &= -\left(\frac{\partial X^{\mathrm{NC}}}{\partial T}\right)_{p,\boldsymbol{\mu}} \\
V^{\mathrm{NC}}(T,p,\boldsymbol{\mu}) &= \left(\frac{\partial X^{\mathrm{NC}}}{\partial p}\right)_{T,\boldsymbol{\mu}} \\
N_j^{\mathrm{NC}}(T,p,\boldsymbol{\mu}) &= -\left(\frac{\partial X^{\mathrm{NC}}}{\partial \mu_j}\right)_{T,p,\boldsymbol{\mu}_j}
\end{aligned} \tag{3.79}$$

For a small system, the replica internal energy X^{NC} is unequal to zero. These equations make it possible to calculate properties that are proportional to the size of the system V^{NC} in the thermodynamic limit. In the thermodynamic limit, X^{NC} is equal to zero and the equations no longer make sense.

The change in the total entropy of the replicas follows from the Hill–Gibbs equation (3.70) and is given by

$$dS_t^{\mathrm{NC}} = \frac{1}{T}dU_t^{\mathrm{NC}} + \frac{p}{T}dV_t^{\mathrm{NC}} - \frac{\boldsymbol{\mu}}{T} \cdot d\mathbf{N}_t^{\mathrm{NC}} - \frac{X^{\mathrm{NC}}}{T}d\mathcal{N} \tag{3.80}$$

The change in the entropy per replica is given by the Gibbs equation (Eq. (3.76))

$$dS^{\mathrm{NC}} = \frac{1}{T}dU^{\mathrm{NC}} + \frac{p}{T}dV^{\mathrm{NC}} - \frac{\boldsymbol{\mu}}{T}\cdot d\mathbf{N}^{\mathrm{NC}} \tag{3.81}$$

Remark 3.11. Again, we have two equations which can serve as starting points for a non-equilibrium thermodynamic analysis of a small system with T, p, $\boldsymbol{\mu}$, $\mathcal{N}$ and the shape as environmental variables.

3.7 The Subdivision Potential

In the thermodynamic description of small systems, the subdivision potential can be used as a measure of the smallness. For five of the six ensembles treated here, we could define two thermodynamic quantities, which would be equal in a large system, but differ by the subdivision potential in a small system. The relevant expressions contain the hat-variables, and are Eqs. (3.9), (3.19), (3.35), (3.48) and (3.63), see also Table 3.2. We repeat:

$$\begin{aligned}
T^{\mathrm{MC}}\left(S^{\mathrm{MC}} - \widehat{S}^{\mathrm{MC}}\right) &= -\varepsilon^{\mathrm{MC}} \\
F^{\mathrm{C}} - \widehat{F}^{\mathrm{C}} &= \varepsilon^{\mathrm{C}} \\
G^{\mathrm{P}} - \widehat{G}^{\mathrm{P}} &= \varepsilon^{\mathrm{P}} \\
\left(p^{\mathrm{GC}} - \widehat{p}\right)V &= \varepsilon^{\mathrm{GC}} \\
(\mu_1^{\mathrm{O}} - \widehat{\mu}_1^{\mathrm{O}})N_1 &= -\varepsilon^{\mathrm{O}}
\end{aligned} \tag{3.82}$$

Each subdivision potential is here expressed in terms of a deviation between the hat-variable and its corresponding thermodynamic property. The subdivision potential can be associated with work, the reversible work needed to add one more replica to the ensemble of replicas. The summary above and Table 3.2 show that the value of ε depends much on the environmental variables in question. The subdivision potential has a positive sign, when the differential values for F, G or pV are larger than the corresponding hat-variables. The reverse sign applies to the corresponding entropy- and chemical

Table 3.2. The subdivision potential and its explicit expression for a set of ensembles.

Ensemble name	Subdivision potential	Expression	Sec.
MC	$-\varepsilon^{\mathrm{MC}}$	$T^{\mathrm{MC}}(S^{\mathrm{MC}} - \widehat{S}^{\mathrm{MC}})$	3.1
C	ε^{C}	$F^{\mathrm{C}} - \widehat{F}^{\mathrm{C}}$	3.2
P	ε^{P}	$G^{\mathrm{P}} - \widehat{G}^{\mathrm{P}}$	3.3
GC	$\varepsilon^{\mathrm{GC}}$	$(p^{\mathrm{GC}} - \widehat{p}^{\mathrm{GC}})V$	3.4
O	$-\varepsilon^{\mathrm{O}}$	$(\mu_1^{\mathrm{O}} - \widehat{\mu}_1^{\mathrm{O}})N_1$	3.5

Notes: The shape of the small system is controlled in all ensembles. The section number in the table heading refers to the section where the properties are defined.

potential differences. The difference between a hat-variable and a normal variable gives immediately the characteristic property of the small system, the subdivision potential for the ensemble in question. This quantity is basically unknown in literature, and calls for a systematic effort. The subdivision potential ε is the origin of the scaling laws in terms of the ratio surface area and volume, Ω/V, discussed in Chapters 5 and 6, and used in subsequent chapters.

As we see from Eq. (3.82), the subdivision potential depends on the choice of the set of control variables. If we include an additional variable, e.g., the surface area, in the set of control variables, the value of ε will change. The surface area normally describes some of the system's smallness. To introduce the area as a control variable will therefore eliminate a part of ε. If the surface area is the only small system variable, its introduction may even reduce ε to zero [51].

3.8 Equilibrium Between Small Systems

Consider two small systems in contact with one another. Before we can speak of equilibrium between the systems, we need to consider the systems' environments. The systems must be controlled by the same environmental variables in order for us to be able to compare them. This is so, because a small system with control of, e.g., the pressure, cannot be in equilibrium with a system with control of the volume. The same holds true for other pairs of environmental variables, (T, S) or (μ_j, N_j). The issue was discussed by Strøm *et al.* [38]. The environments of the systems therefore need to be the same.

Consider next that the number of replicas, $\mathcal{N}$, are controlled. The equilibrium conditions for a given set of environmental variables and a given shape of the small systems are of interest. The systems of interest are controlled by a microcanonical ensemble. In general, we have from Eq. (3.1)

$$\begin{aligned} dS_t^{\mathrm{MC}} &= \frac{1}{T^{\mathrm{MC}}}\mathcal{N}dU + \frac{p^{\mathrm{MC}}}{T^{\mathrm{MC}}}\mathcal{N}dV - \frac{\boldsymbol{\mu}^{\mathrm{MC}}}{T^{\mathrm{MC}}}\mathcal{N}\cdot d\mathbf{N} \\ &\quad - \frac{1}{T^{\mathrm{MC}}}\left(\varepsilon^{\mathrm{MC}} - U - p^{\mathrm{MC}}V + \boldsymbol{\mu}^{\mathrm{MC}}\cdot\mathbf{N}\right)d\mathcal{N} \\ &= \frac{1}{T^{\mathrm{MC}}}\mathcal{N}dU + \frac{p^{\mathrm{MC}}}{T^{\mathrm{MC}}}\mathcal{N}dV - \frac{\boldsymbol{\mu}^{\mathrm{MC}}}{T^{\mathrm{MC}}}\mathcal{N}\cdot d\mathbf{N} - \frac{X^{\mathrm{MC}}}{T^{\mathrm{MC}}}d\mathcal{N} \end{aligned} \tag{3.83}$$

For each of the small systems, this gives

$$\begin{aligned} dS_{t,1}^{\mathrm{MC}} &= \frac{1}{T_1^{\mathrm{MC}}}\mathcal{N}_1dU_1 + \frac{p_1^{\mathrm{MC}}}{T_1^{\mathrm{MC}}}\mathcal{N}_1dV_1 - \frac{\boldsymbol{\mu}_1^{\mathrm{MC}}}{T_1^{\mathrm{MC}}}\mathcal{N}_1\cdot d\mathbf{N}_1 - \frac{X_1^{\mathrm{MC}}}{T_1^{\mathrm{MC}}}d\mathcal{N}_1 \\ dS_{t,2}^{\mathrm{MC}} &= \frac{1}{T_2^{\mathrm{MC}}}\mathcal{N}_2dU_2 + \frac{p_2^{\mathrm{MC}}}{T_2^{\mathrm{MC}}}\mathcal{N}_2dV_2 - \frac{\boldsymbol{\mu}_2^{\mathrm{MC}}}{T_2^{\mathrm{MC}}}\mathcal{N}_2\cdot d\mathbf{N}_2 - \frac{X_2^{\mathrm{MC}}}{T_2^{\mathrm{MC}}}d\mathcal{N}_2 \end{aligned} \tag{3.84}$$

At equilibrium, we can transfer, for instance, internal energy from one system to the other, $\mathcal{N}_1dU_1 = -\mathcal{N}_2dU_2$, without changing the total entropy; $dS_{t,1}^{\mathrm{MC}} + dS_{t,2}^{\mathrm{MC}}$. It follows that one equilibrium condition is $T_1^{\mathrm{MC}} = T_2^{\mathrm{MC}}$. Similarly, transfers in volumes or component numbers imply that $p_1^{\mathrm{MC}}/T_1^{\mathrm{MC}} = p_2^{\mathrm{MC}}/T_2^{\mathrm{MC}}$ or $\boldsymbol{\mu}_1^{\mathrm{MC}}/T_1^{\mathrm{MC}} = \boldsymbol{\mu}_2^{\mathrm{MC}}/T_2^{\mathrm{MC}}$, respectively.

In addition to the equilibrium conditions derived above, small systems in equilibrium must obey a condition which refers to their replica numbers. This consideration is special for small systems. When we change $\mathcal{N}_1$ and $\mathcal{N}_2$, this must be done in such a way that the replica energy change in one system is equal, but opposite in sign to the change in replica energy of the other system. In the notation we use, this gives $X_1^{\text{MC}} d\mathcal{N}_1 = -X_2^{\text{MC}} d\mathcal{N}_2$. This again gives as equilibrium condition $T_1^{\text{MC}} = T_2^{\text{MC}}$ so that the corresponding change of $S_1^{\text{MC}} + S_2^{\text{MC}}$ is zero.

The equilibrium conditions are the same for small and large systems provided that their shape is also the same.

One aspect of small systems needs special consideration. When we know T, p, $\boldsymbol{\mu}$ and the shape, we can calculate U, V and $\mathbf{N}$. When two small systems differ in their shape, they also differ in their set of basis variables, U_1, V_1, $\mathbf{N}_1$ and U_2, V_2, $\mathbf{N}_2$. To completely equilibrate two small systems of different shapes is, though possible, difficult in practice.

The contact between two small systems is usually restricted. Consider as an example that it is only possible to transfer heat. In that case, we can only reach the condition $T_1^{\text{MC}} = T_2^{\text{MC}}$. The second system can then be considered to be a heat bath for the first. The second system can also perform as a barostat or a particle bath. Only the variable conjugate to the variable exchanged can equilibrate to the common value.

3.9 Concluding Remarks

Hill's hypothesis was used to give precise expressions for the replica energy X of various ensembles of small systems, see Table 3.1. Doing that, we found expressions for the corresponding subdivision potentials, see Table 3.2. The replica energy and the subdivision potential have their meaning on the ensemble level. They can both be connected to various types of reversible work. The ensemble of replicas is an Euler homogeneous function of first order in the number of replicas. As a consequence, the Hill–Gibbs equation applies to the ensemble of replicas. This equation contains an extra term, which is not contained in the Gibbs equation for small systems. No systematic studies of the subdivision potential are yet available. This represents a challenge.

When the subdivision potential is finite, the replica energy is connected to standard thermodynamic functions, like, for instance, the Helmholtz or Gibbs energies for the canonical and the isobaric ensembles, respectively. The ensembles of small systems in equilibrium obey equilibrium conditions which are the same as for large systems. We have studied six common ensembles. The method is of course general and can be applied to other choices. In the next chapter, we show how the replica energy X can be computed from the partition function of the corresponding statistical mechanical distribution.

Chapter 4

Statistical Mechanics

We present the basic statistical formulas and relations for six ensembles of small systems. The statistical machinery is the same for small and large systems. The partition function relates the small system replica energy to its statistical distribution in phase space. Probability distributions in phase space are given for six ensembles. Legendre transforms apply for the ensembles.

At the core of any distribution in phase space is the system energy, as expressed by its Hamiltonian. Energy levels can be discrete (as in quantum mechanics) or continuous (as in classical mechanics). The partition function is in both cases computed from the Hamiltonian. For classical systems, the main topic of this book, we present the probability distributions in phase space, and compute the replica energy in each of the six ensembles from the previous chapter, in terms of the appropriate partition function. The expressions highlight the particulars of each ensemble, as well as their differences. All describe small system behavior.

The subdivision potential of small systems, ε, and the corresponding replica energy, X, for six different ensembles were derived in Chapter 3, see Tables 3.1 and 3.2 for a summary. Here, we give the statistical foundation for each replica energy of the six ensembles. On this basis, we shall later find the equations of state in Chapter 6 for three of the ensembles; the canonical, grand canonical and nanocanonical ensembles. An ideal gas with surface adsorption will be used as example.

We adopt Hill's assumption [3, 4] that the replicas in an ensemble are independent. As we shall see, it follows that the classical description from statistical thermodynamics remains valid, also when we account for the impact of size and shape of the small systems. The Legendre transforms for partitions functions and the Liouville and Liouville–von Neumann equations also remain the same. While Legendre transforms apply on the ensemble level only, Legendre–Fenchel transforms can be used for small system properties, cf. Section 4.9.

4.1 The Hamiltonian of Small Systems

We explained in Chapters 2 and 3 that small systems have significant surface energy compared to bulk energy. The statistical behavior of the particles in a small system is affected by the presence of the surface that separates the small system from the surroundings. Close to a surface, particle interactions differ from interactions in the bulk phase. We describe the interactions, by quantum mechanics or by classical mechanics, using the appropriate Hamiltonian.

The Hamiltonian is typically a sum of kinetic and potential energy:

$$\mathcal{H}(\mathbf{p}_1, \ldots, \mathbf{p}_N, \mathbf{r}_1, \ldots, \mathbf{r}_N) = \sum_{l=1}^{N} \frac{|\mathbf{p}_l|^2}{2m_l} + \mathcal{V}(\mathbf{r}_1, \ldots, \mathbf{r}_N) \tag{4.1}$$

where $\mathbf{p}_l$ and $\mathbf{r}_l$ are momenta and positions of particle l, m_l is their mass and N is the total number of particles. The potential energy is symmetric for interchange of particles of the same component. We consider systems with n components. The number of particles of the jth component is N_j. The total number of particles is $N = \sum_{j=1}^{n} N_j$. We use the shorthand notation $\mathbf{N} = (N_1, \ldots, N_n)$.

In the quantum mechanical description, $\mathbf{p}_l$ and $\mathbf{r}_l$ are operators. We use the energy levels or the eigenvalues, $\epsilon_k(V, \mathbf{N})$, of the Hamiltonian. The eigenvalues depend also on the shape of the volume.

In the classical mechanical description, $\mathbf{p}_l$ and $\mathbf{r}_l$ are real-valued. We mostly use a sum of pair-potentials plus an external potential. The typical potential in Eq. (4.1) is

$$\mathcal{V}(\mathbf{r}_1, \ldots, \mathbf{r}_N) = \sum_{\text{pairs}} \phi_2(\mathbf{r}_i - \mathbf{r}_j) + \sum_{k=1}^{N} \phi_{\text{ext}}(\mathbf{r}_k) \tag{4.2}$$

Three-particle interactions are rarely taken into account, but are needed in a classical mechanical description of chemical reactions [46, 47, 56].

In principle, a description must keep all momenta and coordinates. According to a theorem of Berendsen, see Ciccotti and Ryckaert [57], a description using all degrees of freedom and one using a set of restricted degrees of freedom are equivalent. The theorem enables us to use the probability distributions of the restricted phase space for classical systems. The description of a molecule by its momenta p and positions r, rather than by the state of its nuclei and the electrons, is an example of such a simplification. The theorem makes it possible to use a simplified description of reactants and products in a chemical reaction [46, 47, 56]. We shall use Berendsen's theorem in Chapter 6. When particles are bound, for instance, to a surface, the number of free momenta and coordinates reduces. A particle bound to a surface lacks one degree of freedom; namely the momentum normal to the surface. A particle bound to an edge keeps only the momentum along the edge. A particle in a corner has no momentum degree of freedom at all. Following Berendsen, the ideal gas with surface adsorption can be described with a reduced number of free momenta and coordinates. This will be done in Chapter 6.

4.2 Microcanonical Ensemble: U, V, $\mathbf{N}$

The microcanonical partition function of a quantum mechanical system is

$$\Omega(U, V, \mathbf{N}) = \sum_{U < \epsilon_k(V, \mathbf{N}) < U + \Delta U} 1 \tag{4.3}$$

This is the number of states in the energy shell between U and $U + \Delta U$. The classical microcanonical partition function is

$$\Omega(U, V, \mathbf{N}) = \frac{1}{h^{3N}} \left(\prod_{k=1}^{n} \frac{1}{N_k!} \right) \int_{U<\mathcal{H}<U+\Delta U} 1 d\mathbf{p}_1 \ldots d\mathbf{p}_N d\mathbf{r}_1 \ldots d\mathbf{r}_N \tag{4.4}$$

where h is the Planck constant. For $\mathcal{N}$ independent replicas, the microcanonical partition function is given by

$$\Omega_t(U, V, \mathbf{N}, \mathcal{N}) = \Omega^{\mathcal{N}}(U, V, \mathbf{N}) \tag{4.5}$$

From the microcanonical partition function, we obtain the entropy of the ensemble of replicas

$$S_t^{\mathrm{MC}}(U, V, \mathbf{N}, \mathcal{N}) = k_{\mathrm{B}} \ln \Omega_t(U, V, \mathbf{N}, \mathcal{N}) = \mathcal{N} k_{\mathrm{B}} \ln \Omega(U, V, \mathbf{N}) \tag{4.6}$$

The replica entropy becomes

$$S^{\mathrm{MC}}(U, V, \mathbf{N}) = k_{\mathrm{B}} \ln \Omega(U, V, \mathbf{N}) \tag{4.7}$$

so that

$$S_t^{\mathrm{MC}}(U, V, \mathbf{N}, \mathcal{N}) = \mathcal{N} S^{\mathrm{MC}}(U, V, \mathbf{N}) \tag{4.8}$$

This gives a statistical mechanical basis for the results found in Section 3.1. The replica energy is given in terms of the entropy, see Eq. (3.6), and of the microcanonical partition function

$$\begin{aligned} X^{\mathrm{MC}}(U, V, \mathbf{N}) &= -T^{\mathrm{MC}}(U, V, \mathbf{N}) S^{\mathrm{MC}}(U, V, \mathbf{N}) \\ &= -k_{\mathrm{B}} T^{\mathrm{MC}}(U, V, \mathbf{N}) \ln \Omega(U, V, \mathbf{N}) \end{aligned} \tag{4.9}$$

Microcanonical phase-space distribution

The probability distribution for the classical system in the $6N$-dimensional phase space with V, U, $\mathbf{N}$ as environmental variables is

$$\begin{aligned} P_{U,V,\mathbf{N}}(\mathbf{p}_1, \ldots, \mathbf{p}_N, \mathbf{r}_1, \ldots, \mathbf{r}_N) &= \frac{1}{\Omega(U, V, \mathbf{N}) h^{3N}} \left(\prod_{k=1}^{n} \frac{1}{N_k!} \right) \\ &\times [\theta(\mathcal{H}(\mathbf{p}_1, \ldots, \mathbf{p}_N, \mathbf{r}_1, \ldots, \mathbf{r}_N) - U) \\ &\times \theta(U + \Delta U - \mathcal{H}(\mathbf{p}_1, \ldots, \mathbf{p}_N, \mathbf{r}_1, \ldots, \mathbf{r}_N))] \end{aligned} \tag{4.10}$$

Here $\Omega(U, V, \mathbf{N})$ is the classical microcanonical partition function of Eq. (4.4). Furthermore, θ is the Heaviside function, which is 0 for a negative argument and 1 for a positive argument. It follows from Eq. (4.4) that this probability distribution has been normalized to unity.

4.3 Canonical Ensemble: T, V, N

The canonical partition function is the Legendre transform of the microcanonical partition function

$$Q(T, V, \mathbf{N}) = \int dU \frac{\exp(-\beta U)\Omega(U, V, \mathbf{N})}{\Delta U} \tag{4.11}$$

With the quantum mechanical expression for Ω of Eq. (4.3), we have

$$Q(T, V, \mathbf{N}) = \sum_k \exp(-\beta\epsilon_k(V, \mathbf{N})) \tag{4.12}$$

The classical expression of Eq. (4.4) yields

$$Q(T, V, \mathbf{N}) = \frac{1}{h^{3N}} \left(\prod_{k=1}^{n} \frac{1}{N_k!} \right) \int d\mathbf{p}_1 \ldots d\mathbf{p}_N \\ \times \int_V d\mathbf{r}_1 \ldots d\mathbf{r}_N \exp(-\beta\mathcal{H}(\mathbf{p}_1, \ldots, \mathbf{p}_N, \mathbf{r}_1, \ldots, \mathbf{r}_N)) \tag{4.13}$$

where the total number of particles in the system is $\sum_{k=1}^{n} N_k = N$. Integrals over $\mathbf{p}$ are from now on always from ∞ to $-\infty$. We do not indicate this. After introducing Eq. (4.1), we integrate over the momenta and obtain

$$Q(T, V, \mathbf{N}) = \left(\prod_{k=1}^{n} \frac{1}{N_k! \Lambda_k^{3N_k}} \right) \int_V d\mathbf{r}_1 \ldots d\mathbf{r}_N \exp(-\beta\mathcal{V}(\mathbf{r}_1, \ldots, \mathbf{r}_N)) \tag{4.14}$$

where $\Lambda_k \equiv \sqrt{h^2/(2\pi m_k k_\mathrm{B} T)}$ is the mean de Broglie wavelength for the kth component.

Equation (4.14) can be written as

$$Q(T, V, \mathbf{N}) = Q_{\text{kin}}(T, V, \mathbf{N}) Q_{\text{pot}}(T, V, \mathbf{N}) \tag{4.15}$$

where the kinetic and the potential contribution to the classical canonical partition function are defined by

$$Q_{\text{kin}}(T, V, \mathbf{N}) = \prod_{k=1}^{n} \frac{V^{N_k}}{N_k! \Lambda_k^{3N_k}} \tag{4.16}$$

$$Q_{\text{pot}}(T, V, \mathbf{N}) = \frac{1}{V^N} \int_V d\mathbf{r}_1 \dots d\mathbf{r}_N \exp(-\beta \mathcal{V}(\mathbf{r}_1, \dots, \mathbf{r}_N)) \tag{4.17}$$

The powers of the volume have been chosen so as to make both Q_{kin} and Q_{pot} dimensionless without violating Eq. (4.15).

The canonical partition function for the $\mathcal{N}$ independent replicas is

$$Q_t(T, V, \mathbf{N}, \mathcal{N}) = Q^{\mathcal{N}}(T, V, \mathbf{N}) \tag{4.18}$$

The canonical partition function gives the Helmholtz energy of the ensemble of replicas from

$$F_t^{\text{C}}(T, V, \mathbf{N}, \mathcal{N}) = -k_{\text{B}} T \ln Q_t(T, V, \mathbf{N}, \mathcal{N}) = -\mathcal{N} k_{\text{B}} T \ln Q(T, V, \mathbf{N}) \tag{4.19}$$

The replica energy, which is equal to the Helmholtz energy of a single replica, becomes

$$X^{\text{C}}(T, V, \mathbf{N}) = F^{\text{C}}(T, V, \mathbf{N}) = -k_{\text{B}} T \ln Q(T, V, \mathbf{N}) \tag{4.20}$$

We also have

$$F_t^{\text{C}}(T, V, \mathbf{N}, \mathcal{N}) = \mathcal{N} F^{\text{C}}(T, V, \mathbf{N}) \tag{4.21}$$

By using Eq. (4.15) for the classical canonical partition function, we obtain the Helmholtz energy of a single replica

$$\begin{aligned} F^{\text{C}}(T, V, \mathbf{N}) &= -k_{\text{B}} T \ln Q(T, V, \mathbf{N}) \\ &= -k_{\text{B}} T \ln Q_{\text{kin}}(T, V, \mathbf{N}) - k_{\text{B}} T \ln Q_{\text{pot}}(T, V, \mathbf{N}) \\ &= F_{\text{kin}}^{\text{C}}(T, V, \mathbf{N}) + F_{\text{pot}}^{\text{C}}(T, V, \mathbf{N}) \end{aligned} \tag{4.22}$$

so that

$$F_t^{\mathrm{C}}(T, V, \mathbf{N}, \mathcal{N}) = \mathcal{N} F^{\mathrm{C}}(T, V, \mathbf{N})$$
$$= \mathcal{N} F_{\mathrm{kin}}^{\mathrm{C}}(T, V, \mathbf{N}) + \mathcal{N} F_{\mathrm{pot}}^{\mathrm{C}}(T, V, \mathbf{N}) \qquad (4.23)$$

With Stirling's approximation $N_k! \simeq (N_k/e)^{N_k}\sqrt{2\pi N_k}$, the kinetic contribution to the Helmholtz energy in Eq. (4.16) becomes

$$F_{\mathrm{kin}}^{\mathrm{C}}(T, V, \mathbf{N}) = k_{\mathrm{B}} T \sum_{k=1}^{n} \left[N_k \ln\left(\frac{N_k \Lambda_k^3}{eV}\right) + \frac{1}{2}\ln(2\pi N_k) \right]$$
$$= k_{\mathrm{B}} T \sum_{k=1}^{n} \left[V c_k \ln\left(\frac{c_k \Lambda_k^3}{e}\right) + \frac{1}{2}\ln(2\pi N_k) \right] \qquad (4.24)$$

where $c_k \equiv N_k/V$ is the particle density of component k.

Remark 4.1. The first term in the paranthesis of Eq. (4.24) gives the ideal gas contribution to the Helmholtz energy. In the thermodynamic limit, the contribution proportional to $\ln(2\pi N_k)$ is negligible compared to the first term. We discuss the relevance of this in Chapter 6, see Eq. (6.6).

Canonical phase-space distribution

The classical probability distribution in the $6N$-dimensional phase space with T, V, $\mathbf{N}$ as environmental variables is

$$P_{T,V,\mathbf{N}}(\mathbf{p}_1, \ldots, \mathbf{p}_N, \mathbf{r}_1, \ldots, \mathbf{r}_N)$$
$$= \frac{1}{Q(T, V, \mathbf{N}) h^{3N}} \left(\prod_{k=1}^{n} \frac{1}{N_k!} \right) \exp\left(-\beta \mathcal{H}\left(\mathbf{p}_1, \ldots, \mathbf{p}_N, \mathbf{r}_1, \ldots, \mathbf{r}_N\right)\right) \qquad (4.25)$$

where $Q(T, V, \mathbf{N})$ is the classical canonical partition function given in Eq. (4.13). It follows from Eq. (4.13) that this probability distribution has been normalized to unity. After introduction of Eq. (4.1) in the Hamiltonian and application of Eq. (4.15), this probability distribution can be written as the product of a kinetic and a potential energy distribution

$$
\begin{aligned}
&P_{T,V,\mathbf{N}}(\mathbf{p}_1,\ldots,\mathbf{p}_N,\mathbf{r}_1,\ldots,\mathbf{r}_N)\\
&\quad = P^{\text{kin}}_{T,V,\mathbf{N}}(\mathbf{p}_1,\ldots,\mathbf{p}_N)P^{\text{pot}}_{T,V,\mathbf{N}}(\mathbf{r}_1,\ldots,\mathbf{r}_N)
\end{aligned}
\tag{4.26}
$$

where

$$
\begin{aligned}
P^{\text{kin}}_{T,V,\mathbf{N}}(\mathbf{p}_1,\ldots,\mathbf{p}_N) &= \frac{V^N}{Q_{\text{kin}}(T,V,\mathbf{N})h^{3N}}\left(\prod_{k=1}^{n}\frac{1}{N_k!}\right)\\
&\quad\times \exp\left(-\beta\sum_{l=1}^{N}\frac{|\mathbf{p}_l|^2}{2m_l}\right)
\end{aligned}
\tag{4.27}
$$

$$
P^{\text{pot}}_{T,V,\mathbf{N}}(\mathbf{r}_1,\ldots,\mathbf{r}_N) = \frac{1}{Q_{\text{pot}}(T,V,\mathbf{N})V^N}\exp\left(-\beta\mathcal{V}\left(\mathbf{r}_1,\ldots,\mathbf{r}_N\right)\right)
\tag{4.28}
$$

From Eqs. (4.16) and (4.17) it follows that both the kinetic distribution $P^{\text{kin}}_{T,V,\mathbf{N}}$ and the potential distribution $P^{\text{pot}}_{T,V,\mathbf{N}}$ have been normalized to unity. Both distributions are symmetric for interchange of particles of the same component.

4.4 Isobaric Ensemble: T, p, $\mathbf{N}$

The isobaric partition function is the Legendre transform of the canonical partition function

$$
\Delta(T,p,\mathbf{N}) = \beta p\int_0^\infty dV\exp\left(-\beta pV\right)Q(T,V,\mathbf{N})
\tag{4.29}
$$

The factor βp has been added before the integrand in order to make Δ dimensionless, cf. Hill [3, 4]. The quantum mechanical expression for Q of Eq. (4.12) gives

$$
\Delta(T,p,\mathbf{N}) = \beta p\int_0^\infty dV\exp\left(-\beta pV\right)\sum_k\exp\left(-\beta\epsilon_k(V,\mathbf{N})\right)
\tag{4.30}
$$

The classical mechanical expression in Eq. (4.13) gives

$$\Delta(T,p,\mathbf{N}) = \frac{\beta p}{h^{3N}}\left(\prod_{k=1}^{n}\frac{1}{N_k!}\right)\int_0^\infty dV\exp\left(-\beta pV\right)\int d\mathbf{p}_1\ldots d\mathbf{p}_N$$
$$\times\int_V d\mathbf{r}_1\ldots d\mathbf{r}_N\exp\left(-\beta\mathcal{H}\left(\mathbf{p}_1,\ldots,\mathbf{p}_N,\mathbf{r}_1,\ldots,\mathbf{r}_N\right)\right) \tag{4.31}$$

The integral over V is, in both cases, over the magnitude of the volume, keeping the shape constant. This replaces the volume, V, by the pressure, p, as an environmental variable. The dependence of Δ, Q, ϵ_k and $\mathcal{H}$ on the shape of the volume is not explicitly indicated. A shape-dependence is present for all ensembles, and is always important. In the above definitions, the integral over the volume is carried out while the shape of the volume remains the same. Systems with controlled volumes, that differ from one another, can only be compared when the volumes are of the same shape. In the thermodynamic limit, the dependence on the shape of the volume disappears.

After introducing Eq. (4.1) into Eq. (4.31), we integrate over the momenta and obtain for the classical partition function

$$\Delta(T,p,\mathbf{N}) = \left(\prod_{k=1}^{n}\frac{1}{N_k!\Lambda_k^{3N_k}}\right)\beta p\int_0^\infty dV\exp\left(-\beta pV\right)$$
$$\times\int_V d\mathbf{r}_1\ldots d\mathbf{r}_N\exp\left(-\beta\mathcal{V}\left(\mathbf{r}_1,\ldots,\mathbf{r}_N\right)\right) \tag{4.32}$$

Equation (4.32) can be written as

$$\Delta(T,p,\mathbf{N}) = \Delta_{\text{kin}}(T,p,\mathbf{N})\Delta_{\text{pot}}(T,p,\mathbf{N}) \tag{4.33}$$

where the kinetic and potential contributions to the classical isobaric partition function are defined by

$$\Delta_{\text{kin}}(T,p,\mathbf{N}) = \prod_{k=1}^{n}\frac{1}{N_k!\Lambda_k^{3N_k}\left(\beta p\right)^{N_k}} \tag{4.34}$$

$$\Delta_{\text{pot}}(T,p,\mathbf{N}) = \left(\beta p\right)^{N+1}\int_0^\infty dV\exp(-\beta pV)\int_V d\mathbf{r}_1\ldots d\mathbf{r}_N$$
$$\times\exp\left(-\beta\mathcal{V}\left(\mathbf{r}_1,\ldots,\mathbf{r}_N\right)\right) \tag{4.35}$$

The powers N_k and $N+1$ of βp have been chosen so that both Δ_{kin} and Δ_{pot} are dimensionless like Δ. For small systems, the shape has an impact on Δ_{pot}, and not on Δ_{kin}.

The isobaric partition function of $\mathcal{N}$ independent replicas is

$$\Delta_t(T,p,\mathbf{N},\mathcal{N}) = \Delta^{\mathcal{N}}(T,p,\mathbf{N}) \tag{4.36}$$

From the isobaric partition function, we obtain Gibbs energy G_t^{P} of the ensemble of replicas as follows:

$$G_t^{\text{P}}(T,p,\mathbf{N},\mathcal{N}) = -k_{\text{B}}T\ln\Delta_t(T,p,\mathbf{N},\mathcal{N}) = -k_{\text{B}}T\mathcal{N}\ln\Delta(T,p,\mathbf{N}) \tag{4.37}$$

The Gibbs energy of a single replica G^{P} becomes

$$G^{\text{P}}(T,p,\mathbf{N}) = -k_{\text{B}}T\ln\Delta(T,p,\mathbf{N}) \tag{4.38}$$

so that

$$G_t^{\text{P}}(T,p,\mathbf{N},\mathcal{N}) = \mathcal{N}G^{\text{P}}(T,p,\mathbf{N}) \tag{4.39}$$

The Gibbs energy is obtained by using Eq. (4.33) as the classical isobaric partition function:

$$\begin{aligned} G_t^{\text{P}}(T,p,\mathbf{N},\mathcal{N}) &= -k_{\text{B}}T\ln\Delta_t(T,p,\mathbf{N},\mathcal{N}) \\ &= -\mathcal{N}k_{\text{B}}T\ln\Delta_{\text{kin}}(T,p,\mathbf{N}) - \mathcal{N}k_{\text{B}}T\ln\Delta_{\text{pot}}(T,p,\mathbf{N}) \end{aligned} \tag{4.40}$$

For the replica energy, which is equal to the Gibbs energy of a single replica, G^{P}, we obtain

$$\begin{aligned} X^{\text{P}}(T,p,\mathbf{N}) = G^{\text{P}}(T,p,\mathbf{N}) &= -k_{\text{B}}T\ln\Delta(T,p,\mathbf{N}) \\ &= -k_{\text{B}}T\ln\Delta_{\text{kin}}(T,p,\mathbf{N}) - k_{\text{B}}T\ln\Delta_{\text{pot}}(T,p,\mathbf{N}) \\ &= G_{\text{kin}}^{\text{P}}(T,p,\mathbf{N}) + G_{\text{pot}}^{\text{P}}(T,p,\mathbf{N}) \end{aligned} \tag{4.41}$$

It follows that

$$G_t^{\text{P}}(T,p,\mathbf{N},\mathcal{N}) = \mathcal{N}G^{\text{P}}(T,p,\mathbf{N}) = \mathcal{N}G_{\text{kin}}^{\text{P}}(T,p,\mathbf{N}) + \mathcal{N}G_{\text{pot}}^{\text{P}}(T,p,\mathbf{N}) \tag{4.42}$$

With Stirling's approximation, we obtain from Eq. (4.34) for the kinetic contribution

$$G^{\mathrm{P}}_{\mathrm{kin}}(T,p,\mathbf{N}) = k_{\mathrm{B}}T\sum_{k=1}^{n}\left[N_k \ln\frac{p\Lambda_k^3 N_k}{k_{\mathrm{B}}Te} + \frac{1}{2}\ln(2\pi N_k)\right] \tag{4.43}$$

This is the ideal gas contribution to the Gibbs energy. In the thermodynamic limit the contribution proportional to $\ln(2\pi N_k)$ is negligible compared to the other one. For a small system, this is not so. Then $G^{\mathrm{P}}_{\mathrm{pot}}$ depends on shape, while $G^{\mathrm{P}}_{\mathrm{kin}}$ does not.

Remark 4.2. This result for the Gibbs energy is particular for the isobaric ensemble: $G^{\mathrm{P}}_{\mathrm{pot}}(T,p,\mathbf{N})$ depends on the shape, $G^{\mathrm{P}}_{\mathrm{kin}}$ does not.

Isobaric phase space distribution

The classical probability distribution in the $6N$-dimensional phase space with p, T, $\mathbf{N}$ as environmental variables is

$$\begin{aligned} &P_{T,p,\mathbf{N}}(\mathbf{p}_1,\ldots,\mathbf{p}_N,\mathbf{r}_1,\ldots,\mathbf{r}_N;V) \\ &\quad= \frac{\beta p}{\Delta(T,p,\mathbf{N})h^{3N}}\prod_{k=1}^{n}\frac{1}{N_k!}\exp(-\beta pV) \\ &\qquad\times \exp\left(-\beta\mathcal{H}\left(\mathbf{p}_1,\ldots,\mathbf{p}_N,\mathbf{r}_1,\ldots,\mathbf{r}_N\right)\right) \end{aligned} \tag{4.44}$$

where $\Delta(T,p,\mathbf{N})$ is the classical isobaric partition function given in Eq. (4.31). The expression in Eq. (4.36) times $dVd\mathbf{p}_1\ldots d\mathbf{p}_N d\mathbf{r}_1\ldots d\mathbf{r}_N$ gives the probability distribution for the volume, momenta and positions of the particles. It follows from Eq. (4.31) that this probability distribution has been normalized to unity, when one integrates over these variables. After introducing Eq. (4.1), using Eqs. (4.33)–(4.35), this probability distribution can be written as the product of a kinetic and a potential distribution

$$\begin{aligned} &P_{T,p,\mathbf{N}}(\mathbf{p}_1,\ldots,\mathbf{p}_N,\mathbf{r}_1,\ldots,\mathbf{r}_N;V) \\ &\quad= P^{\mathrm{kin}}_{T,p,\mathbf{N}}(\mathbf{p}_1,\ldots,\mathbf{p}_N)P^{\mathrm{pot}}_{T,p,\mathbf{N}}(\mathbf{r}_1,\ldots,\mathbf{r}_N;V) \end{aligned} \tag{4.45}$$

where

$$P^{\text{kin}}_{T,p,\mathbf{N}}(\mathbf{p}_1,\ldots,\mathbf{p}_N) = \frac{1}{\Delta_{\text{kin}}(T,p,\mathbf{N})h^{3N}(\beta p)^N} \times \left(\prod_{k=1}^{n}\frac{1}{N_k!}\right)\exp\left(-\beta\sum_{l=1}^{N}\frac{|\mathbf{p}_l|^2}{2m_l}\right) \tag{4.46}$$

$$P^{\text{pot}}_{T,p,\mathbf{N}}(\mathbf{r}_1,\ldots,\mathbf{r}_N;V) = \frac{(\beta p)^{N+1}}{\Delta_{\text{pot}}(T,p,\mathbf{N})}\exp(-\beta pV) \times \exp\left(-\beta\mathcal{V}(\mathbf{r}_1,\ldots,\mathbf{r}_N)\right) \tag{4.47}$$

From Eqs. (4.34) and (4.35), we see that the kinetic distribution $P^{\text{kin}}_{T,p,\mathbf{N}}$ has been normalized to unity when integrating over the momenta, while the potential distribution $P^{\text{pot}}_{T,p,\mathbf{N}}$ has been normalized to unity when integrating over the volume and the positions. By integrating Eq. (4.45) over the volume and the positions, we obtain the distribution over the momenta

$$P^{\text{kin}}_{T,p,\mathbf{N}}(\mathbf{p}_1,\ldots,\mathbf{p}_N) \equiv \int_0^\infty dV \int_V d\mathbf{r}_1\ldots d\mathbf{r}_N P_{T,p,\mathbf{N}}(\mathbf{p}_1,\ldots,\mathbf{p}_N,\mathbf{r}_1,\ldots,\mathbf{r}_N;V) \tag{4.48}$$

and by integrating over the momenta we obtain the distribution over the volume and the positions

$$P^{\text{pot}}_{T,p,\mathbf{N}}(\mathbf{r}_1,\ldots,\mathbf{r}_N;V) \equiv \int d\mathbf{p}_1\ldots d\mathbf{p}_N P_{T,p,\mathbf{N}}(\mathbf{p}_1,\ldots,\mathbf{p}_N,\mathbf{r}_1,\ldots,\mathbf{r}_N;V) \tag{4.49}$$

In all ensembles, the probability distributions depend on the shape of the volume. This shape must be kept constant when V is varied. We note that both distributions are symmetric for interchange of particles of the same component.

4.5 Grand Canonical Ensemble: T, V, μ

The grand canonical partition function, is the Legendre transform of the canonical partition function

$$\Xi(T, V, \boldsymbol{\mu}) = \sum_{N_1=0}^{\infty} \cdots \sum_{N_n=0}^{\infty} \exp(\beta \boldsymbol{\mu} \cdot \mathbf{N}) Q(T, V, \mathbf{N}) \tag{4.50}$$

By using the quantum mechanical expression for Q in Eq. (4.3), we obtain

$$\Xi(T, V, \boldsymbol{\mu}) = \sum_{N_1=0}^{\infty} \cdots \sum_{N_n=0}^{\infty} \exp(\beta \boldsymbol{\mu} \cdot \mathbf{N}) \sum_{k} \exp(-\beta \epsilon_k(V, \mathbf{N})) \tag{4.51}$$

The classical mechanical expression in Eq. (4.4) gives

$$\Xi(T, V, \boldsymbol{\mu}) = \sum_{N_1=0}^{\infty} \cdots \sum_{N_n=0}^{\infty} \left[\prod_{k=1}^{n} \frac{\exp(\beta \mu_k N_k)}{h^{3N_k} N_k!} \int d\mathbf{p}_1 \ldots d\mathbf{p}_N \right.$$
$$\left. \times \int_V d\mathbf{r}_1 \ldots d\mathbf{r}_N \; \exp\left(-\beta \mathcal{H}\left(\mathbf{p}_1, \ldots, \mathbf{p}_N, \mathbf{r}_1, \ldots, \mathbf{r}_N\right)\right) \right] \tag{4.52}$$

After introducing Eq. (4.1), we integrate over the momenta and obtain for a single replica

$$\Xi(T, V, \boldsymbol{\mu}) = \sum_{N_1=0}^{\infty} \cdots \sum_{N_n=0}^{\infty} \left[\prod_{k=1}^{n} \frac{\exp\left(\beta \mu_k N_k\right)}{\Lambda_k^{3N_k} N_k!} \right.$$
$$\left. \times \int_V d\mathbf{r}_1 \ldots d\mathbf{r}_N \exp(-\beta \mathcal{V}\left(\mathbf{r}_1, \ldots, \mathbf{r}_N\right)) \right] \tag{4.53}$$

For $\mathcal{N}$ independent replicas, the grand canonical partition function is

$$\Xi(T, V, \boldsymbol{\mu}, \mathcal{N}) = \Xi^{\mathcal{N}}(T, V, \boldsymbol{\mu}) \tag{4.54}$$

From the grand canonical partition function, we obtain $\widehat{p}V\mathcal{N}$ for the ensemble of replicas

$$\widehat{p}(T, V, \boldsymbol{\mu}) V \mathcal{N} = k_{\mathrm{B}} T \ln \Xi(T, V, \boldsymbol{\mu}, \mathcal{N}) = \mathcal{N} k_{\mathrm{B}} T \ln \Xi(T, V, \boldsymbol{\mu}) \tag{4.55}$$

For the replica energy, we have

$$X^{\mathrm{GC}}(T,V,\boldsymbol{\mu}) = -\widehat{p}(T,V,\boldsymbol{\mu})V = -k_{\mathrm{B}}T\ln\Xi(T,V,\boldsymbol{\mu}) \tag{4.56}$$

The logarithm of the grand canonical partition function of a single replica gives thus the integral pressure $\widehat{p}$, and *not* the differential pressure $p^{\mathrm{GC}} \equiv \partial\left(\widehat{p}V\right)/\partial V$.

In the ideal gas limit, the interaction potential is zero. For the classical grand canonical partition function of a single replica, we then have

$$\Xi(T,V,\boldsymbol{\mu}) = \prod_{k=1}^{n}\sum_{N_k=0}^{\infty}\frac{\exp(\beta\mu_k N_k)V^{N_k}}{\Lambda_k^{3N_k}N_k!} = \prod_{k=1}^{n}\exp\left[\frac{\exp(\beta\mu_k)V}{\Lambda_k^3}\right] \tag{4.57}$$

In the ideal gas limit, we obtain

$$\widehat{p}(T,V,\boldsymbol{\mu})V = k_{\mathrm{B}}T\sum_{k=1}^{n}\frac{\exp(\beta\mu_k)V}{\Lambda_k^3} \tag{4.58}$$

By using Eq. (3.52), this gives

$$\widehat{p}(T,V,\boldsymbol{\mu})V = k_{\mathrm{B}}T\sum_{k=1}^{n}N_j^{\mathrm{GC}}(T,V,\boldsymbol{\mu}) = k_{\mathrm{B}}TN^{\mathrm{GC}}(T,V,\boldsymbol{\mu}) \tag{4.59}$$

Remark 4.3. We have seen that the integral pressure, $\widehat{p}$, satisfies the ideal gas law in the ideal gas limit, while the differential pressure, p^{GC}, does *not* satisfy this law in the ideal gas limit when the system is small. These results are particular for the grand canonical ensemble.

Grand canonical phase-space distribution

The classical probability distribution in the $6N$-dimensional phase space with T, V, $\boldsymbol{\mu}$ as environmental variables is

$$P_{T,V,\boldsymbol{\mu}}(\mathbf{p}_1,\ldots,\mathbf{p}_N,\mathbf{r}_1,\ldots,\mathbf{r}_N;\mathbf{N})$$
$$= \frac{1}{\Xi(T,V,\boldsymbol{\mu})}\prod_{k=1}^{n}\frac{\exp\left(\beta\mu_k N_k\right)}{h^{3N_k}N_k!}\exp\left(-\beta\mathcal{H}\left(\mathbf{p}_1,\ldots,\mathbf{p}_N,\mathbf{r}_1,\ldots,\mathbf{r}_N\right)\right) \tag{4.60}$$

By introducing the definition of the grand canonical partition function in Eq. (4.52), we find that this probability distribution is normalized when we integrate over momenta and positions of particles, summing over possible numbers of particles of all the components.

The distribution over the positions and $\mathbf{N}$ is obtained from Eq. (4.60) as follows:

$$\begin{aligned} P^{\text{pot}}_{T,V,\boldsymbol{\mu}}&(\mathbf{r}_1,\ldots,\mathbf{r}_N;\mathbf{N}) \\ &= \int d\mathbf{p}_1\ldots d\mathbf{p}_N P_{T,V,\boldsymbol{\mu}}(\mathbf{p}_1,\ldots,\mathbf{p}_N,\mathbf{r}_1,\ldots,\mathbf{r}_N;\mathbf{N}) \\ &= \frac{1}{\Xi(T,V,\boldsymbol{\mu})}\prod_{k=1}^{n}\frac{\exp(\beta\mu_k N_k)}{\Lambda_k^{3N_k}N_k!}\exp(-\beta\mathcal{V}(\mathbf{r}_1,\ldots,\mathbf{r}_N)) \end{aligned} \tag{4.61}$$

This probability distribution is normalized when we integrate over positions of particles, summing over possible numbers of particles of all components.

4.6 Osmotic Ensemble: T, p, N_1, μ_1

The osmotic partition function is the Legendre transform of the isobaric partition function

$$\Upsilon(T,p,N_1,\boldsymbol{\mu}_1) = \sum_{N_2=0}^{\infty}\cdots\sum_{N_n=0}^{\infty}\exp\left(\beta(\boldsymbol{\mu}_1\cdot\mathbf{N}_1)\right)\Delta(T,p,\mathbf{N}) \tag{4.62}$$

By introducing the quantum mechanical expression for Δ in Eq. (4.30), we obtain

$$\begin{aligned} \Upsilon(T,p,N_1,\boldsymbol{\mu}_1) = \beta p\sum_{N_2=0}^{\infty}&\cdots\sum_{N_n=0}^{\infty}\exp(\beta(\mu_2N_2+\cdots+\mu_nN_n)) \\ &\times\int_0^{\infty}dV\exp\left(-\beta pV\right)\sum_k\exp\left(-\beta\epsilon_k(V,N_j)\right) \end{aligned} \tag{4.63}$$

The classical mechanical expression of Eq. (4.31) gives

$$\Upsilon(T,p,N_1,\boldsymbol{\mu}_1) = \frac{\beta p}{N_1!h^{3N_1}} \sum_{N_2=0}^{\infty} \cdots \sum_{N_n=0}^{\infty} \left[\left(\prod_{k=2}^{n} \frac{\exp(\beta\mu_k N_k)}{N_k!h^{3N_k}} \right) \right.$$
$$\times \int_0^{\infty} dV \exp(-\beta pV) \int d\mathbf{p}_1 \ldots d\mathbf{p}_N \int_V d\mathbf{r}_1 \ldots d\mathbf{r}_N$$
$$\left. \times \exp\left(-\beta\mathcal{H}\left(\mathbf{p}_1,\ldots,\mathbf{p}_N,\mathbf{r}_1,\ldots,\mathbf{r}_N\right)\right) \right] \tag{4.64}$$

After introducing Eq. (4.1), we integrate over the momenta and obtain

$$\Upsilon(T,p,N_1,\boldsymbol{\mu}_1) = \frac{\beta p}{N_1!\Lambda_1^{3N_1}} \sum_{N_2=0}^{\infty} \cdots \sum_{N_n=0}^{\infty} \left[\left(\prod_{k=2}^{n} \frac{\exp(\beta\mu_k N_k)}{N_k!\Lambda_k^{3N_k}} \right) \right.$$
$$\times \int_0^{\infty} dV \exp(-\beta pV) \int_V d\mathbf{r}_1 \ldots d\mathbf{r}_N$$
$$\left. \times \exp(-\beta\mathcal{V}\left(\mathbf{r}_1,\ldots,\mathbf{r}_N\right)) \right] \tag{4.65}$$

The integral over V is, in both cases, over the magnitude of the volume, keeping the shape constant. This replaces the volume V by the pressure p as control variable. The dependence of Υ, Δ, ϵ_k and $\mathcal{H}$ on the volume shape is not explicitly indicated. This dependence is always important for small systems. Therefore, a shape-dependence is present for all ensembles. When we compare results obtained at different volumes, we must do this for the same shape. In the above definitions, the integral over the volume is carried out keeping the shape constant. In the thermodynamic limit, the dependence on the volume shape disappears.

For $\mathcal{N}$ independent replicas, the osmotic partition function is

$$\Upsilon(T,p,N_1,\boldsymbol{\mu}_1,\mathcal{N}) = \Upsilon^{\mathcal{N}}(T,p,N_1,\boldsymbol{\mu}_1) \tag{4.66}$$

From the osmotic partition function, we obtain the total replica energy X_t^{O} of the ensemble of replicas by

$$X_t^{\mathrm{O}}(T,p,N_1,\boldsymbol{\mu}_1,\mathcal{N}) = \mathcal{N}X^{\mathrm{O}}(T,p,N_1,\boldsymbol{\mu}_1)$$
$$= -k_{\mathrm{B}}T\ln\Upsilon(T,p,N_1,\boldsymbol{\mu}_1,\mathcal{N}) = -\mathcal{N}k_{\mathrm{B}}T\ln\Upsilon(T,p,N_1,\boldsymbol{\mu}_1) \tag{4.67}$$

The replica energy of a single replica and the integral chemical potential for the first component becomes

$$X^{\mathrm{O}}\left(T,p,N_1,\boldsymbol{\mu}_1\right)=\widehat{\mu}_1^{\mathrm{O}}\left(T,p,N_1,\boldsymbol{\mu}_1\right)N_1=-k_{\mathrm{B}}T\ln\Upsilon(T,p,N_1,\boldsymbol{\mu}_1) \tag{4.68}$$

In the ideal gas limit the interaction potential is zero. The classical osmotic partition function of a single replica becomes

$$\begin{aligned}\Upsilon(T,p,N_1,\boldsymbol{\mu}_1)&=\frac{\beta p}{N_1!\Lambda_1^{3N_1}}\times\sum_{N_2=0}^{\infty}\cdots\sum_{N_n=0}^{\infty}\left[\left(\prod_{k=2}^{n}\frac{\exp(\beta\mu_k N_k)}{N_k!\Lambda_k^{3N_k}}\right)\right.\\&\quad\left.\times\int_0^{\infty}dV\exp(-\beta pV)V^N\right]=\frac{1}{N_1!\Lambda_1^{3N_1}(\beta p)^{N_1}}\\&\quad\times\sum_{N_2=0}^{\infty}\cdots\sum_{N_n=0}^{\infty}N!\prod_{k=2}^{n}\frac{\exp(\beta\mu_k N_k)}{N_k!\Lambda_k^{3N_k}(\beta p)^{N_k}}\end{aligned} \tag{4.69}$$

For the integral chemical potential of the first component, we obtain for the ideal gas, using Stirlings approximation $\left[N!\simeq(N/e)^N\sqrt{2\pi N}\right]$:

$$\begin{aligned}\widehat{\mu}_1^{\mathrm{O}}\left(T,p,N_1,\boldsymbol{\mu}_1\right)N_1&=k_{\mathrm{B}}T\left[N_1\ln\left(\frac{N_1\Lambda_1^3\beta p}{e}\right)+\frac{1}{2}\ln\left(2\pi N_1\right)\right]\\&\quad-k_{\mathrm{B}}T\ln\left[\sum_{N_2=0}^{\infty}\cdots\sum_{N_n=0}^{\infty}N!\prod_{k=2}^{n}\frac{\exp(\beta\mu_k N_k)}{N_k!\Lambda_k^{3N_k}(\beta p)^{N_k}}\right]\end{aligned} \tag{4.70}$$

The differential chemical potential for the first component in the ideal gas limit becomes

$$\begin{aligned}\mu_1^{\mathrm{O}}\left(T,p,N_1,\mu_1\right)&=\frac{\partial(\widehat{\mu}_1^{\mathrm{O}}N_1)}{\partial N_1}=k_{\mathrm{B}}T\left[\ln\left(N_1\Lambda_1^3\beta p\right)+\frac{1}{2N_1}\right]\\&\quad-k_{\mathrm{B}}T\frac{\partial}{\partial N_1}\ln\left[\sum_{N_2=0}^{\infty}\cdots\sum_{N_n=0}^{\infty}N!\prod_{k=2}^{n}\frac{\exp(\beta\mu_k N_k)}{N_k!\Lambda_k^{3N_k}(\beta p)^{N_k}}\right]\end{aligned} \tag{4.71}$$

Remark 4.4. The integral chemical potential, $\widehat{\mu}_1^{\mathrm{O}}$, of component 1 in a mixture of small systems, applies to an ideal gas. The differential chemical potential, μ_1^{O}, applies to the same. The two expressions for the same component differ for a small system.

Osmotic phase-space distribution

The classical probability distribution in the $6N$-dimensional phase space with T, p, N_1, $\boldsymbol{\mu}_1$ as environmental variables, is

$$P_{T,p,N_1,\boldsymbol{\mu}_1}(\mathbf{p}_1,\ldots,\mathbf{p}_N,\mathbf{r}_1,\ldots,\mathbf{r}_N;V,\mathbf{N}_1)=\frac{\beta p\exp(-\beta pV)}{\Upsilon(T,p,N_1,\boldsymbol{\mu}_1)N_1!h^{3N_1}}$$

$$\times\prod_{k=2}^{n}\frac{\exp\left(\beta\mu_k N_k\right)}{h^{3N_k}N_k!}\exp\left(-\beta\mathcal{H}\left(\mathbf{p}_1,\ldots,\mathbf{p}_N,\mathbf{r}_1,\ldots,\mathbf{r}_N\right)\right) \qquad (4.72)$$

The probability distribution is normalized when we integrate over momenta, and positions of particles, summing over possible numbers of particles of components $2,\ldots,n$, and integrating over the volume. This can be seen when we introduce the definition of the osmotic partition function into Eq. (4.64). The distribution over the positions, V, and $\mathbf{N}_1$ is found from Eq. (4.72) as follows:

$$\begin{aligned}
&P^{\mathrm{pot}}_{T,p,N_1,\boldsymbol{\mu}_1}(\mathbf{r}_1,\ldots,\mathbf{r}_N;V,\mathbf{N}_1)\\
&\quad=\int d\mathbf{p}_1\ldots d\mathbf{p}_N P_{T,p,N_1,\boldsymbol{\mu}_1}(\mathbf{p}_1,\ldots,\mathbf{p}_N,\mathbf{r}_1,\ldots,\mathbf{r}_N;V,\mathbf{N}_1)\\
&\quad=\frac{\beta p\exp(-\beta pV)}{\Upsilon(T,p,N_1,\boldsymbol{\mu}_1)N_1!\Lambda_1^{3N_1}}\prod_{k=2}^{n}\frac{\exp(\beta\mu_k N_k)}{\Lambda_k^{3N_k}N_k!}\exp(-\beta\mathcal{V}\left(\mathbf{r}_1,\ldots,\mathbf{r}_N\right))
\end{aligned} \qquad (4.73)$$

This probability distribution is normalized for integration over positions of particles, summing over possible numbers of particles of components 2 to n, and integration over the volume.

4.7 Nanocanonical Ensemble: T, p, $\boldsymbol{\mu}$

Consider next a small system that can exchange volume, energy and particles of all components with the environment. This small system is completely open. It is called nanocanonical following Hill, who gives a thorough discussion of this case in his books [3, 4]. The ensemble is indicated by superscript NC.

The nanocanonical partition function is the Legendre transform of the isobaric partition function

$$\Psi(T,p,\boldsymbol{\mu}) = \sum_{N_1=0}^{\infty} \cdots \sum_{N_n=0}^{\infty} \exp(\beta\boldsymbol{\mu}\cdot\mathbf{N})\Delta(T,p,\mathbf{N}) \tag{4.74}$$

The same result can be obtained from the Legendre transform of the grand canonical partition function

$$\Psi(T,p,\boldsymbol{\mu}) = \beta p \int_0^{\infty} dV \exp(-\beta pV)\,\Xi(T,V,\boldsymbol{\mu}) \tag{4.75}$$

By using the quantum mechanical expression for Δ in Eq. (4.30), we obtain

$$\begin{aligned}\Psi(T,p,\boldsymbol{\mu}) = \beta p \sum_{N_1=0}^{\infty} \cdots \sum_{N_n=0}^{\infty} \exp(\beta\boldsymbol{\mu}\cdot\mathbf{N}) \\ \times \int_0^{\infty} dV \exp(-\beta pV) \sum_k \exp(-\beta\epsilon_k(V,\mathbf{N}))\end{aligned} \tag{4.76}$$

The classical mechanical expression of Eq. (4.31) gives

$$\begin{aligned}\Psi(T,p,\boldsymbol{\mu}) = \beta p \sum_{N_1=0}^{\infty} \cdots \sum_{N_n=0}^{\infty} \Bigg[\left(\prod_{k=1}^{n} \frac{\exp(\beta\mu_k N_k)}{N_k!h^{3N_k}}\right) \int_0^{\infty} dV \exp(-\beta pV) \\ \times \int d\mathbf{p}_1 \ldots d\mathbf{p}_N \int_V d\mathbf{r}_1 \ldots d\mathbf{r}_N \exp(-\beta\mathcal{H}(\mathbf{p}_1,\ldots,\mathbf{p}_N,\mathbf{r}_1,\ldots,\mathbf{r}_N))\Bigg]\end{aligned} \tag{4.77}$$

After introducing Eq. (4.1), we integrate over the momenta and obtain

$$\Psi(T,p,\boldsymbol{\mu}) = \beta p \sum_{N_1=0}^{\infty} \cdots \sum_{N_n=0}^{\infty} \left[\left(\prod_{k=1}^{n} \frac{\exp(\beta\mu_k N_k)}{N_k!\Lambda_k^{3N_k}} \right) \right.$$
$$\left. \times \int_0^{\infty} dV \exp(-\beta pV) \int_V d\mathbf{r}_1 \ldots d\mathbf{r}_N \exp(-\beta\mathcal{V}(\mathbf{r}_1,\ldots,\mathbf{r}_N)) \right] \tag{4.78}$$

The integral over V is over the magnitude of the volume, keeping the shape constant. This replaces the volume as the control variable by the pressure p. The dependence of Ψ, Δ, ε_k and $\mathcal{H}$ on the shape of the volume is not explicitly indicated. This dependence is important for small systems, as the shape-dependence is present in all ensembles. In the above definitions, the integral over the volume is carried out, keeping the shape unaltered. In the thermodynamic limit, the dependence on the shape of the volume disappears.

For $\mathcal{N}$ independent replicas, the nanocanonical partition function is given by

$$\Psi(T,p,\boldsymbol{\mu},\mathcal{N}) = \Psi^{\mathcal{N}}(T,p,\boldsymbol{\mu}) \tag{4.79}$$

From the nanocanonical partition function, we obtain the total replica energy X_t^{NC} of the ensemble of replicas by

$$\begin{aligned} X_t^{\mathrm{NC}}(T,p,\boldsymbol{\mu},\mathcal{N}) &= \mathcal{N} X^{\mathrm{NC}}(T,p,\boldsymbol{\mu}) = -k_{\mathrm{B}}T \ln \Psi(T,p,\boldsymbol{\mu},\mathcal{N}) \\ &= -\mathcal{N} k_{\mathrm{B}}T \ln \Psi(T,p,\boldsymbol{\mu}) \end{aligned} \tag{4.80}$$

The replica energy of a single replica is

$$X^{\mathrm{NC}}(T,p,\boldsymbol{\mu}) = -k_{\mathrm{B}}T \ln \Psi(T,p,\boldsymbol{\mu}) \tag{4.81}$$

In the ideal gas, the interaction potential is zero, and the classical nanocanonical partition function of a single replica becomes

$$\Psi(T,p,\boldsymbol{\mu})$$
$$= \beta p \sum_{N_1=0}^{\infty} \cdots \sum_{N_n=0}^{\infty} \left[\left(\prod_{k=1}^{n} \frac{\exp(\beta\mu_k N_k)}{N_k!\Lambda_k^{3N_k}} \right) \int_0^{\infty} dV \exp(-\beta pV) V^N \right]$$

$$= \sum_{N_1=0}^{\infty} \cdots \sum_{N_n=0}^{\infty} \frac{N!}{(\beta p)^N} \prod_{k=1}^{n} \frac{\exp(\beta\mu_k N_k)}{N_k!\Lambda_k^{3N_k}} = \sum_{N=0}^{\infty} \left[\sum_{k=1}^{n} \frac{\exp(\beta\mu_k)}{\beta p \Lambda_k^3}\right]^N$$

$$= \left(1 - \sum_{k=1}^{n} \frac{\exp(\beta\mu_k)}{\beta p \Lambda_k^3}\right)^{-1} \tag{4.82}$$

The single replica energy is

$$X^{\mathrm{NC}}(T, p, \boldsymbol{\mu}) = k_{\mathrm{B}} T \ln \left(1 - \sum_{k=1}^{n} \frac{\exp(\beta\mu_k)}{\beta p \Lambda_k^3}\right) \tag{4.83}$$

Nanocanonical phase-space distribution

The classical probability distribution in the $6N$-dimensional phase space with p, T, $\boldsymbol{\mu}$ as environmental variables is

$$P_{T,p,\boldsymbol{\mu}}(\mathbf{p}_1, \ldots, \mathbf{p}_N, \mathbf{r}_1, \ldots, \mathbf{r}_N; V, \mathbf{N})$$
$$= \frac{\beta p \exp(-\beta p V)}{\Psi(T, p, \boldsymbol{\mu})} \prod_{k=1}^{n} \frac{\exp(\beta\mu_k N_k)}{h^{3N_k} N_k!} \exp(-\beta \mathcal{H}(\mathbf{p}_1, \ldots, \mathbf{p}_N, \mathbf{r}_1, \ldots, \mathbf{r}_N)) \tag{4.84}$$

By introducing the definition of the nanocanonical partition function in Eq. (4.77), we find a probability distribution. The distribution is normalized when we integrate over momenta and positions of particles, summing over possible numbers of particles of the components, and integrating over the volume.

For the distribution over the positions, V and $\mathbf{N}$, Eq. (4.84) gives

$$P_{T,p,\boldsymbol{\mu}}^{\mathrm{pot}}(\mathbf{r}_1, \ldots, \mathbf{r}_N; V, \mathbf{N})$$
$$= \int d\mathbf{p}_1 \ldots d\mathbf{p}_N P_{T,p,\boldsymbol{\mu}}(\mathbf{p}_1, \ldots, \mathbf{p}_N, \mathbf{r}_1, \ldots, \mathbf{r}_N; V, \mathbf{N})$$
$$= \frac{\beta p \exp(-\beta p V)}{\Psi(T, p, \boldsymbol{\mu})} \prod_{k=1}^{n} \frac{\exp(\beta\mu_k N_k)}{\Lambda_k^{3N_k} N_k!} \exp(-\beta \mathcal{V}(\mathbf{r}_1, \ldots, \mathbf{r}_N)) \tag{4.85}$$

This probability distribution is normalized when we integrate over positions of particles, summing over possible numbers of particles of all the components, and integrating over the volume.

4.8 Legendre Transforms for Partition Functions

We have seen so far in this chapter that the partition functions of the different ensembles can be related by Legendre transforms. The procedure is general and can be illustrated in more detail by an example. For instance, Eq. (4.50) gives the grand canonical partition function as the Legendre transform of the canonical partition function

$$\Xi(T,V,\boldsymbol{\mu}) = \sum_{N_1=0}^{\infty} \cdots \sum_{N_n=0}^{\infty} \exp(\beta\boldsymbol{\mu}\cdot\mathbf{N})Q(T,V,\mathbf{N}) \tag{4.86}$$

We know that the logarithm of Q gives the Helmholtz energy F^{C}, cf. Eq. (4.20). It follows that the canonical partition function can be found from the equation of state for the Helmholtz energy

$$Q(T,V,\mathbf{N}) = \exp\left[-\beta F^{\mathrm{C}}(T,V,\mathbf{N})\right] \tag{4.87}$$

By introducing this into Eq. (4.50), we obtain

$$\Xi(T,V,\boldsymbol{\mu}) = \sum_{N_1=0}^{\infty} \cdots \sum_{N_n=0}^{\infty} \exp\left[\beta\boldsymbol{\mu}\cdot\mathbf{N} - \beta F^{\mathrm{C}}(T,V,\mathbf{N})\right] \tag{4.88}$$

The logarithm of Ξ gives the compressional energy (or the negative grand potential), cf. Eq. (4.56). This results in

$$\begin{aligned}\widehat{p}(T,V,\boldsymbol{\mu})V &= k_{\mathrm{B}}T\ln\Xi(T,V,\boldsymbol{\mu})\\ &= k_{\mathrm{B}}T\ln\left\{\sum_{N_1=0}^{\infty} \cdots \sum_{N_n=0}^{\infty} \exp\left[\beta\boldsymbol{\mu}\cdot\mathbf{N} - \beta F^{\mathrm{C}}(T,V,\mathbf{N})\right]\right\}\end{aligned} \tag{4.89}$$

In this manner, we can obtain the equation of state for the grand canonical integral pressure $\widehat{p}(T,V,\boldsymbol{\mu})$ if we know the equation of state for the canonical Helmholtz energy $F^{\mathrm{C}}(T,V,\mathbf{N})$.

All partition functions are related by Legendre transforms, so it is possible to present many such relations. We have just shown the procedure for one case, but we can use the same procedure to transform results from one set to another set of control variables. By using the relations of Sections 4.2 and 4.3, we obtain the following relation between the equations of state for the microcanonical entropy and the canonical Helmholtz energy of a small system:

$$F^{\mathrm{C}}(T,V,\mathbf{N}) = -k_{\mathrm{B}}T\ln\left[\frac{1}{\Delta U}\int dU\exp\left(-\beta\left[U - TS^{\mathrm{MC}}(U,V,\mathbf{N})\right]\right)\right] \tag{4.90}$$

By using the relations of Sections 4.3 and 4.4, we obtain a relation between the canonical Helmholtz energy and the isobaric Gibbs energy of a small system as follows:

$$G^{\mathrm{P}}(T,p,\mathbf{N}) = -k_{\mathrm{B}}T\ln\left[\beta p\int dV\exp\left(-\beta\left[pV + F^{\mathrm{C}}(T,V,\mathbf{N})\right]\right)\right] \tag{4.91}$$

By using the relations of Sections 4.4 and 4.6, we obtain a relation between the isobaric Gibbs energy and the integral chemical potential of the solute of a small system as follows:

$$\widehat{\mu}_1^{\mathrm{O}}(T,p,N_1,\boldsymbol{\mu}_1)N_1 = -k_{\mathrm{B}}T\ln\left\{\sum_{N_2=0}^{\infty}\cdots\sum_{N_n=0}^{\infty}\exp\left\{\beta\left[\boldsymbol{\mu}_1\cdot\mathbf{N}_1 - G^{\mathrm{P}}(T,p,\mathbf{N})\right]\right\}\right\} \tag{4.92}$$

And finally, by using the relations of Sections 4.4–4.6, we obtain a relation between the nanocanonical subdivision potential, the isobaric Gibbs energy and the grand canonical integral pressure of a small system as follows:

$$\varepsilon^{\mathrm{NC}}(T,p,\boldsymbol{\mu}) = X^{\mathrm{NC}}(T,p,\boldsymbol{\mu}) = -k_{\mathrm{B}}T\ln\left[\sum_{N_1=0}^{\infty}\cdots\sum_{N_n=0}^{\infty}\exp\left(\beta\left[\boldsymbol{\mu}\cdot\mathbf{N} - G^{\mathrm{P}}(T,p,\mathbf{N})\right]\right)\right] \tag{4.93}$$

$$= -k_{\mathrm{B}}T\ln\left[\beta p\int dV\exp\left(-\beta V\left[p - \widehat{p}(T,V,\boldsymbol{\mu})\right]\right)\right] \tag{4.94}$$

The relations given above provide systematic links between variables of small systems that are controlled in different ways. The relations follow from a statistical mechanical description which holds true for small as well as for large systems. Once we know the equation of state for one set of control variables, we can use the above identities to obtain the equation of state for another set of control variables. We shall use such transformations in Chapter 6 on the ideal gas with surface energies.

4.9 Legendre–Fenchel Transforms

As a system shrinks in size, and the energy ceases to be extensive in the volume, the thermodynamic theory as we know it from Gibbs ceases to apply [4, 20]. We have, for instance, shown [49] that the concept of pressure changes to include integral and differential pressures. This leads to an extension of Gibbs thermodynamics.

The statistical mechanical machinery remains the same, however. We have seen in the previous section that Legendre transforms apply to partition functions and produce relations between equations of state that have different sets of control variables. From these relations, we can obtain transforms for thermodynamic energies that are of wider applicability, provided that the probability distribution in question is sharply peaked. These are the so-called *Legendre–Fenchel transforms* [27, 54, 58].

Consider as an example Eq. (4.89). When the distribution over $\mathbf{N}$ is sharply peaked, we can approximate the contribution of the sums by the contribution of a peak. This results in

$$\widehat{p}(T, V, \boldsymbol{\mu})V = \max_{\mathbf{N}} \left[\boldsymbol{\mu} \cdot \mathbf{N} - F^{\mathrm{C}}(T, V, \mathbf{N})\right] \tag{4.95}$$

which is an example of a Legendre–Fenchel transform [58]. The corresponding Legendre transform is

$$\widehat{p}(T, V, \boldsymbol{\mu})V = \boldsymbol{\mu} \cdot \mathbf{N}(T, V, \boldsymbol{\mu}) - F^{\mathrm{C}}(T, V, \mathbf{N}(T, V, \boldsymbol{\mu})) \tag{4.96}$$

In a similar manner, we have from Eqs. (4.90)–(4.94) the following Legendre–Fenchel transforms:

$$F^{\mathrm{C}}(T, V, \mathbf{N}) = \max_{U} \left[U - TS^{\mathrm{MC}}(U, V, \mathbf{N})\right] \tag{4.97}$$

$$G^{\mathrm{P}}(T,p,\mathbf{N}) = \max_{V} \left[pV + F^{\mathrm{C}}(T,V,\mathbf{N})\right] \tag{4.98}$$

$$\widehat{\mu}_1^{\mathrm{O}}(T,p,N_1,\boldsymbol{\mu}_1)N_1 = \max_{\mathbf{N}_1} \left[\boldsymbol{\mu}_1 \cdot \mathbf{N}_1 - G^{\mathrm{P}}(T,p,\mathbf{N})\right] \tag{4.99}$$

$$\begin{aligned}\varepsilon^{\mathrm{NC}}(T,p,\boldsymbol{\mu}) &= X^{\mathrm{NC}}(T,p,\boldsymbol{\mu}) \\ &= \max_{\mathbf{N}} \left[\boldsymbol{\mu} \cdot \mathbf{N} - G^{\mathrm{P}}(T,p,\mathbf{N})\right] &(4.100)\\ &= \max_{V} \{V\left[p - \widehat{p}(T,V,\boldsymbol{\mu})\right]\} &(4.101)\end{aligned}$$

The corresponding Legendre transforms are

$$F^{\mathrm{C}}(T,V,\mathbf{N}) = U(T,V,\mathbf{N}) - TS^{\mathrm{MC}}(U(T,V,\mathbf{N}),V,\mathbf{N}) \tag{4.102}$$

$$G^{\mathrm{P}}(T,p,\mathbf{N}) = pV(T,p,\mathbf{N}) + F^{\mathrm{C}}(T,V(T,p,\mathbf{N}),\mathbf{N}) \tag{4.103}$$

$$\begin{aligned}&\widehat{\mu}_1^{\mathrm{O}}(T,p,N_1,\boldsymbol{\mu}_1)N_1 \\ &\quad = \boldsymbol{\mu}_1 \cdot \mathbf{N}_1(T,p,N_1,\boldsymbol{\mu}_1) - G^{\mathrm{P}}(T,p,N_1,\mathbf{N}_1(T,p,N_1,\boldsymbol{\mu}_1))\end{aligned} \tag{4.104}$$

$$\begin{aligned}\varepsilon^{\mathrm{NC}}(T,p,\boldsymbol{\mu}) &= X^{\mathrm{NC}}(T,p,\boldsymbol{\mu}) \\ &= \boldsymbol{\mu} \cdot \mathbf{N}(T,p,\boldsymbol{\mu}) - G^{\mathrm{P}}(T,p,\mathbf{N}(T,p,\boldsymbol{\mu})) &(4.105)\\ &= V(T,p,\boldsymbol{\mu})\left[p - \widehat{p}(T,V(T,p,\boldsymbol{\mu}),\boldsymbol{\mu})\right] &(4.106)\end{aligned}$$

Equation (4.95) is a Legendre–Fenchel transformation, which transforms from control variables $T, V, \mathbf{N}$ to $T, V, \boldsymbol{\mu}$. The Legendre–Fenchel transform works in both directions, so we also have

$$F^{\mathrm{C}}(T,V,\mathbf{N}) = \max_{\boldsymbol{\mu}} \left[\boldsymbol{\mu} \cdot \mathbf{N} - \widehat{p}(T,V,\boldsymbol{\mu})V\right] \tag{4.107}$$

This can be done provided that the probability distribution over $\boldsymbol{\mu}$ is sharply peaked. Similarly, we can use the following reverse transforms of Eqs. (4.98)–(4.101) to give

$$-F^{\mathrm{C}}(T,V,\mathbf{N}) = \max_{p} \left[pV - G^{\mathrm{P}}(T,p,\mathbf{N})\right] \tag{4.108}$$

$$G^{\mathrm{P}}(T,p,\mathbf{N}) = \max_{\boldsymbol{\mu}_1} \left[\boldsymbol{\mu}_1 \cdot \mathbf{N}_1 - \widehat{\mu}_1^{\mathrm{O}}(T,p,N_1,\boldsymbol{\mu}_1)N_1\right] \tag{4.109}$$

$$G^{\mathrm{P}}(T,p,\mathbf{N}) = \max_{\boldsymbol{\mu}} \left[\boldsymbol{\mu}\cdot\mathbf{N} - \varepsilon^{\mathrm{NC}}(T,p,\boldsymbol{\mu})\right] \tag{4.110}$$

$$V\widehat{p}(T,V,\boldsymbol{\mu}) = \max_{p} \left[Vp - \varepsilon^{\mathrm{NC}}(T,p,\boldsymbol{\mu})\right] \tag{4.111}$$

What is characteristic is that the Legendre–Fenchel transform is always a convex function. If $F^{\mathrm{C}}(T,V,\mathbf{N})$ is not a convex function of V, the transformed function $G^{\mathrm{P}}(T,p,\mathbf{N})$ is a convex function of p, see Eq. (4.98). The reverse transform of $G^{\mathrm{P}}(T,p,\mathbf{N})$ given in Eq. (4.108) does not give $F^{\mathrm{C}}(T,V,\mathbf{N})$, it gives the convex envelope of $F^{\mathrm{C}}(T,V,\mathbf{N})$, and this is typical.

In Chapter 21 on isometric and isotensional stretching of polymers, we apply the Legendre–Fenchel transform to stretching of small polymers, and transform the Helmholtz stretching energy for a controlled length into the Gibbs stretching energy for a controlled force. The reverse transform gives a convex envelope of the Helmholtz stretching energy [27, 54]. Galteland *et al.* [59] found similar transformation properties for Helmholtz and Gibbs energies in a single fluid which was confined to a slit pore. The Legendre–Fenchel transform captured the layering transitions in the slit. For an in-depth discussion of the thermodynamic Legendre–Fenchel transformations, and their application in both directions, we refer to the literature [58].

4.10 The Liouville–von Neumann and the Liouville equation

The Liouville–von Neumann equation for quantum mechanical systems and the Liouville equation for classical systems give the time evolution of the statistical descriptions. This part of the statistical mechanical theory applies also unaltered to small systems. We give the equations on the quantum mechanical and classical mechanical form, respectively.

In quantum mechanics, we use the density operator ρ which is Hermitian and has a trace equal to one:

$$\mathrm{Tr}(\rho) = 1 \tag{4.112}$$

The expectation value of an observable B is

$$\langle B \rangle = \mathrm{Tr}(B\rho) \tag{4.113}$$

and the time evolution of the density operator is

$$\frac{d}{dt}\rho = -\frac{i}{\hbar}\left[\mathcal{H}, \rho\right] \equiv -\frac{i}{\hbar}\left[\mathcal{H}\rho - \rho\mathcal{H}\right] \tag{4.114}$$

This is the Liouville–von Neumann equation. The equilibrium density matrix for systems where the temperature is controlled is given by

$$\rho_{\mathrm{eq}} = \frac{\exp\left(-\beta\mathcal{H}\right)}{\mathrm{Tr}\left[\exp\left(-\beta\mathcal{H}\right)\right]} \tag{4.115}$$

where $\beta = 1/k_{\mathrm{B}}T$. The entropy of the system is given by

$$S = S_{\mathrm{eq}} - k_{\mathrm{B}}\,\mathrm{Tr}\left[\rho \ln\left(\rho_{\mathrm{eq}}^{-1}\rho\right)\right] \tag{4.116}$$

The Hamiltonian depends on the environmental variables, like for instance the volume and the shape of the volume, and so does the density operator. Bedeaux and Mazur [60] added a dissipative contribution to the Liouville–von Neumann equation in their construction of the entropy production of quantum mechanical systems.

For classical cases, we consider $P(\mathbf{p}_1, \ldots, \mathbf{p}_N, \mathbf{r}_1, \ldots, \mathbf{r}_N, t; Y)$, as the time-dependent distribution in phase space. The Liouville equation is

$$\frac{\partial P}{\partial t} = \sum_{j=1}^{N}\left(\frac{\partial \mathcal{H}}{\partial \mathbf{r}_j}\cdot\frac{\partial P}{\partial \mathbf{p}_j} - \frac{\partial \mathcal{H}}{\partial \mathbf{p}_j}\cdot\frac{\partial P}{\partial \mathbf{r}_j}\right) \equiv \left[\mathcal{H}, P\right] \tag{4.117}$$

Given the Hamiltonian in Eq. (4.1), this can be written as

$$\frac{\partial P}{\partial t} = -\sum_{j=1}^{N}\left(\mathbf{F}_{j,Y}\cdot\frac{\partial P}{\partial \mathbf{p}_j} + \frac{\mathbf{p}_j}{m_j}\cdot\frac{\partial P}{\partial \mathbf{r}_j}\right) \tag{4.118}$$

where

$$\mathbf{F}_{j,Y}\left(\mathbf{r}_1, \ldots, \mathbf{r}_N\right) = -\frac{\partial}{\partial \mathbf{r}_j}\mathcal{V}\left(\mathbf{r}_1, \ldots, \mathbf{r}_N\right) \tag{4.119}$$

is the force on particle j. This force will depend on Y, when the chemical potentials, rather than the number of particles, are environmental variables. The equilibrium distribution for systems, where T, V, N are controlled, is given by

$$P_{\text{eq}} = \frac{\exp(-\beta\mathcal{H})}{Q(T,V,N)} \tag{4.120}$$

4.11 Concluding Remarks

We have shown how to apply the well-known statistical mechanical toolbox to small systems. Results were presented for the six sets of environmental control variables of Chapter 3. A statistical basis of the system's thermodynamic variables has thus been provided. Among the small system's thermodynamic variables that become available through the procedure, we have found the hat-variables $\widehat{p}$, $\widehat{H}$, $\widehat{\mu}$ and the subdivision potential ε. Their values are characteristic for the ensemble they belong to. The integral pressure in the grand canonical ensemble satisfies the ideal gas law in the ideal gas limit. The differential pressure does not satisfy this limit value. We have shown that small system contributions are inherent in the potential energy contributions.

Like in classical thermodynamics, statistical mechanics provides the path from molecular models to thermodynamic properties, here of small systems. Probability distributions in phase space were given for six ensembles. Legendre transforms apply for the ensemble, while Legendre–Fenchel transforms may have a more general applicability.

Unlike in classical thermodynamics, shape is a variable in small systems thermodynamics. Therefore, it must be kept constant when the volume is varied. We shall apply these formulas to an ideal gas in a cubic box in Chapter 6 where systems have surface-, edge- or corner-energies, and find the equation of state under three conditions. The resulting pair-correlation functions are discussed in Chapter 8.

Chapter 5

The Small System Method

We present a method derived from a hypothesis of a size-dependent scaling law and describe how we can use it to obtain properties of small systems as a function of system size. The method is called the small system method. The scaling law gives properties of large systems by extrapolation. Using molecular simulations, we demonstrate how the method can be used in practice.

We present a method for the determination of thermodynamic data of small systems, the so-called *small system method* (SSM). The method derives from a hypothesis of a size-dependent scaling law and can therefore also be used to obtain properties of large systems. The scaling law, presented as a hypothesis, is verified for an ideal gas with surface adsorption in the following chapter and is supported by many numerical results [1, 2, 31, 32, 48].

The SSM has so far been applied to results from molecular simulations, and we shall also demonstrate the method using such applications. In principle, SSM can also be applied to experiments. Whenever one can identify individual atoms or molecules, for instance, by atomic force microscopy, transmission and scanning electron microscopy, the method may be relevant. This remains to be investigated, however.

The purpose of this chapter is to give practical insight into how the formulas of the previous and up-coming chapters can be applied to provide precise and useful data for small systems. We show how size and shape effects appear when the properties are sampled from molecular simulations. The SSM is particular in the sense that it provides thermodynamic properties of several systems from one single simulation. We give a prescription on how to use it. The size-dependent scaling law is then central.

5.1 Size-Dependent Scaling: A Hypothesis

Scaling laws are attractive because they facilitate the determination of thermodynamic properties across different length scales. We verify such relations for an ideal gas with surface energies in Chapter 6. Before we derive the statistical mechanical expressions in Chapter 6, we approach the scaling problem in a more qualitative manner.

We present the scaling law as a hypothesis. The hypothesis is that *the density of any additive thermodynamic property can be written as an expansion in the system's inverse characteristic size.* The characteristic size of a three-dimensional system can be defined in two ways, either by $L \equiv \sqrt[3]{V}$ or by $L_\Omega \equiv V/\Omega$, where Ω is the surface area of the system. From now on, we refer to the hypothesis as the *small system method*, or SSM, for short. Results obtained so far, see e.g., [1, 2, 31, 32, 48], make us believe that the hypothesis has a rather general validity.

Consider a variable, A, which is additive, in the sense that we can calculate A as the sum of bulk and surface contributions, and possibly also line and corner contributions [48]. The hypothesis says: In a small three-dimensional system, we can expand the density of A, $a \equiv A/V$, in the inverse of L as

$$a = \frac{A}{V} = \alpha^\infty + \alpha^s L^{-1} + \alpha^{se} L^{-2} + \alpha^{sc} L^{-3} + \cdots \qquad (5.1)$$

The first term, α^∞, is the value of the density a in the large system limit, $L \to \infty$. The second term, proportional to α^s, gives the correction to A proportional to the surface area of the small system. For a flat surface element, it can be interpreted in terms of the surface excess density of A. All coefficients in Eq. (5.1), except α^∞, depend on the shape of the volume. The terms in the expansion will be explicitly determined for an ideal gas with surface energy in Chapter 6. The expansion has empirical support from the work of Qiao *et al.* [61].

Alternatively, one may expand in the inverse of L_Ω, which gives

$$a = \frac{A}{V} = \alpha^\infty + \alpha_\Omega^s L_\Omega^{-1} + \alpha_\Omega^{se} L_\Omega^{-2} + \alpha_\Omega^{sc} L_\Omega^{-3} + \cdots \tag{5.2}$$

where

$$\alpha_\Omega^s = \alpha^s \frac{L_\Omega}{L} = \alpha^s \frac{V^{2/3}}{\Omega}, \quad \alpha_\Omega^{se} = \alpha^{se}\left(\frac{L_\Omega}{L}\right)^2 = \alpha^{se}\left(\frac{V^{2/3}}{\Omega}\right)^2$$
$$\alpha_\Omega^{sc} = \alpha^{sc}\left(\frac{L_\Omega}{L}\right)^3 = \alpha^{sc}\left(\frac{V^{2/3}}{\Omega}\right)^3 \tag{5.3}$$

There is an advantage of using L_Ω rather than L: The coefficient α_Ω^s gives the contribution from a flat surface and is *independent of the shape of the small volume* [48].

A surface may contribute to higher order terms in the expansion if the surface is curved. The third term, proportional to α^{se}, is the first term that has a contribution due to edges (contact lines). When the surface is curved, the lowest order correction contributes also to α^{se}. The surface tension, for instance, obtains a correction proportional to the Tolman length [52]. The fourth term, proportional to α^{sc}, is the first term in the expansion that is due to corners. For the surface tension, the corresponding correction is described by the surface rigidities.

We expect that an excess at edges contributes to α^{se} and that an excess at corners contributes to α^{sc}. We discuss Eqs. (5.1) and (5.2), and how we can obtain the large system value, α^∞, later in this chapter. When we use Eq. (5.1), we sample sub-volumes of size V in a reservoir, in practice a box of volume V_r, and plot the results as a function of $1/L$. By extrapolating the $1/L$-dependence to large L's, we obtain results that apply in the thermodynamic limit. Alternatively, we can plot results as a function of $1/L_\Omega$. All coefficients in Eq. (5.2), except α^∞ and α_Ω^s, depend on the shape of the volume.

The SSM can be used to investigate properties of surfaces. The surface is chosen to be planar. It can have any simply connected, non-fractal shape. The surface area is Ω_c and the circumference is L_c. The lengths L and L_Ω are defined by $L \equiv \sqrt{\Omega_c}$ and $L_\Omega \equiv \Omega_c/L_c$. In the

SSM, we compare the densities $a \equiv A/\Omega_c$. The different surface areas are isomorphic. The density can again be expanded in $1/L$ or $1/L_\Omega$ just like Eqs. (5.1) and (5.2). The relations between the coefficients are now

$$\alpha_\Omega^s = \alpha^s \frac{L_\Omega}{L} = \alpha^s \frac{\Omega_c^{1/2}}{L_c}, \quad \alpha_\Omega^{se} = \alpha^{se}\left(\frac{L_\Omega}{L}\right)^2 = \alpha^{se}\frac{\Omega_c}{L_c^2}$$
$$\alpha_\Omega^{sc} = \alpha^{sc}\left(\frac{L_\Omega}{L}\right)^3 = \alpha^{sc}\frac{\Omega_c^{3/2}}{L_c^3} \tag{5.4}$$

Again, α_Ω^s is independent of the shape of the surface. In Chapters 14 and 15, we use this to find, respectively, the system's enthalpies and the surface thermodynamic factors for CO_2 on graphite. An alternative way to obtain the thermodynamic properties of a small system is the overlapping distribution method discussed by Bråten *et al.* [62].

A discussion of **Hadwiger's theorem** [63] is in place. This theorem states that any geometric functional of a system that is translationally invariant, additive, and continuous can be written as a linear combination of the four Minkowski functionals: volume, surface area, and the integrated total and Gaussian curvatures. After dividing such a property by its volume, Hadwiger's theorem describes the density $a = A/V$ by a constant plus contributions which vary with L^{-1}, L^{-2}, and L^{-3}, as in Eq. (5.1). It does not give terms of any higher power in $1/L$ and is thus at variance with Eq. (5.1). We find higher order terms to be important in Chapter 6. Blokhuis [64] supported this view for molecular systems. He showed that Hadwiger's theorem does not properly account for changes in positional correlations arising from curvature changes close to the surface. He found it therefore necessary to add a term proportional to the integral of the square of the total curvature of the surface. Such a contribution is expected from the Helfrich energy of the surface [64]. It adds a contribution to the L^{-3} term, and it may also contribute to terms that are of higher order in $1/L$. We shall see when Hadwiger's theorem applies and does not apply in Chapter 6.

5.2 Fluctuation Formulas

In SSM, we calculate properties using sub-system sampling. We compute fluctuations in the variables of interest in a sub-volume V. The fluctuations give the thermodynamic properties of the sub-system. There are many such properties, and we discuss them extensively in Chapters 9 and 10. Here, we restrict the discussion to two types of correlations.

The particle number correlation density, $B_{jk}(T, V, \boldsymbol{\mu})$, is used in the computation of Kirkwood–Buff integrals (see Chapter 10). This correlation density is defined as

$$B_{jk}(T, V, \boldsymbol{\mu}) = \frac{k_B T}{V}\left(\frac{\partial \langle N_j \rangle}{\partial \mu_k}\right)_{T,V,\boldsymbol{\mu}_k} = \frac{\langle N_j N_k \rangle - \langle N_j \rangle \langle N_k \rangle}{V} \tag{5.5}$$

We compute B_{jk} from the average of the product of the particle numbers j and k minus the product of the average particle numbers divided by the volume. The average number of a single particle at constant chemical potentials $\boldsymbol{\mu}_k \equiv (\mu_1, \ldots, \mu_{k-1}, \mu_{k+1}, \ldots, \mu_n)$ and temperature is

$$\langle N_j \rangle = \left(\frac{\partial \widehat{p} V}{\partial \mu_j}\right)_{T,V,\boldsymbol{\mu}_j} \tag{5.6}$$

where n is the number of components. It follows from Eq. (5.5) that the matrix of B_{jk} is symmetric. For a fluid of identical particles, Eq. (5.5) simplifies to

$$B(T, V) = \frac{\langle N^2 \rangle - \langle N \rangle^2}{V} \tag{5.7}$$

The matrix B_{jk} is related to the matrix of thermodynamic factors, see Chapter 10.

The second correlation gives the heat capacity of the small system at constant volume and chemical potentials. It derives from fluctuations in the energy, here denoted by the Hamiltonian, $\mathcal{H}$, of the sub-system.

$$C_V(T, V, \boldsymbol{\mu}) \equiv \left(\frac{\partial U}{\partial T}\right)_{V,\boldsymbol{\mu}}$$
$$= \frac{1}{k_{\mathrm{B}}T^2}\left\{\left[\langle \mathcal{H}^2\rangle - \langle \mathcal{H}\rangle^2\right] + \boldsymbol{\mu}\cdot\left[\langle \mathbf{N}\mathcal{H}\rangle - \langle \mathbf{N}\rangle\langle \mathcal{H}\rangle\right]\right\}$$
$$= C_{V,\mathcal{H}}(T, V, \boldsymbol{\mu}) + C_{V,N}(T, V, \boldsymbol{\mu}) \tag{5.8}$$

In Section 8.2, we write B_{jk} as B_{jk}^{GC} and C_V as C_V^{GC}. The procedure to find B_{jk} and $C_{V,\mathcal{H}}$ will be illustrated in the following section. From these properties, several other thermodynamic properties can be derived.

5.3 A System of Repulsive Particles

In order to demonstrate the SSM, we consider for simplicity a fluid of repulsive particles. The fluid contains one or two components. The particles interact with the so-called Weeks–Chandler–Anderson potential (WCA) [65]. This potential contains the repulsive part of the Lennard-Jones potential only. We refer to Appendix D for its definition. It is computationally time-effective to use this interaction potential, which has a finite range.

In order to obtain the properties of the small system as functions of density, the reduced densities were varied between 0.1 and 1.0 in increments of 0.1. The number densities were $c_{i,r} = N_{i,r}/V_r$, where $N_{i,r}$ is the number of particles of component i and V_r is the volume of the reservoir. The reduced particle masses in the pure fluid and in the binary mixture of WCA-particles were always $m^* = 1$. Furthermore, we used as particle diameters $\sigma_{11} = \sigma_{12} = \sigma_{22} = 1$, and energies $\varepsilon_{11} = \varepsilon_{22} = 1$, while $\varepsilon_{12} = 0.1$, all in reduced units, cf. Appendix D.

The simulations in this chapter were done with the WCA-potential, using the LAMMPS open source molecular dynamics simulations code [66]. The reservoir at start contained from 1000 to 64000 particles of a single component. The binary fluid contained a total of 81000 particles, where 54000 particles were of type 1 and 27000 particles were of type 2. When pair potentials are used, the energy is divided equally between pairs. A particle inside the sub-volume obtains half the energy of the interaction energy with particles outside.

A system was generated by placing particles on a grid. In the next step, thermostats were applied. In the case of the mixture, a barostat was also used. The simulations were run for one million integration steps with a reduced time step of 0.001. Data were written to file every 100th integration step, so the analysis was done on 10000 "snapshots" of the system state. This was done for a system with a reduced density of 0.6.

5.4 The Small System in a Large Reservoir

A small system can exist and be sampled within a reservoir of a few thousand particles. Within the reservoir, we can always choose a sub-volume which does not satisfy periodic boundary conditions and as a consequence has particle number density correlations that differ from those of the reservoir.

In order to observe such behavior, we start simulations by choosing a large, usually cubic, box to represent the environment. We apply periodic boundary conditions to the box. It is therefore a system in the thermodynamic limit. The equilibrium simulation of the large box can be progressed, using either a molecular dynamics or a Monte Carlo algorithm. According to the ergodic hypothesis, these methods give the same results.

A small system is created as a sub-volume within such a large simulation box. We discuss first the conditions of the large reservoir (the environment) before we proceed to specify how we obtain the properties of the small system.

Choosing the reservoir control variables

A typical reservoir has a few thousand particles in a cubic box subject to periodic boundary conditions. Possible control variables for the large simulation box (the reservoir) are canonical $T_r, V_r, \mathbf{N}_r$, micro-canonical $U_r, V_r, \mathbf{N}_r$, isobaric $T_r, p_r, \mathbf{N}_r$, or grand canonical $T_r, V_r, \boldsymbol{\mu}_r$, see Chapters 3 and 4. Subscript r indicates a property of the reservoir. The reservoir possesses the chosen control variables throughout the whole simulation. Averages that apply to the small volume are computed from these controlled variables. As long as the reservoir is in the thermodynamic limit, we are free to choose whichever ensemble that is practical for the reservoir. Simulations done with constant $T_r, V_r, \boldsymbol{\mu}_r$ are much more time-consuming than others.[a] Except in a few cases, see, e.g., Ref. [1], it is unfeasible to run simulations with these variables. The case is used mainly to verify equivalence of approaches. For instance, we will find the same large volume value by extrapolation with data from different volume shapes.

The SSM is next applied to a series of sub-volumes that are all generated within the large box. Thanks to the reservoir, the sub-volume will always be in the grand canonical ensemble $T = T_r, V, \boldsymbol{\mu} = \boldsymbol{\mu}_r$. The sub-volume must not only have the variables $T, V, \boldsymbol{\mu}$ controlled. Also, the shape must be controlled, cf. Chapters 2 and 3. The variables must be constant over the duration of the simulation. Variables, like the internal energy, U, pressure, p, and number of particles, $\mathbf{N}$, of the sub-volume will, however, fluctuate and the formulas above will apply.

Choosing the small system size

The system to be investigated has a volume V sufficiently much smaller than V_r. This sub-volume can exchange particles and energy freely with the reservoir and has therefore $T = T_r, V, \boldsymbol{\mu} = \boldsymbol{\mu}_r$ as control variables. The reservoir determines T and $\boldsymbol{\mu}$. The $\mathcal{N}$ replicas of the small system are thus in the grand canonical ensemble. This situation is illustrated in Fig. 5.1.

The figure shows the reservoir which consists of numerous boxes repeated in a periodic manner, as indicated by horizontal and vertical lines. The small system occupies normally a subspace in one box, here indicated by a square with dashed lines. Since the position of the small system is chosen at random (black diamond), the system will

[a]Much of the time issue with the grand canonical ensemble stems from the fact that it must be done using Monte Carlo simulations, which are notoriously difficult to run on several cores on the computer at the same time.

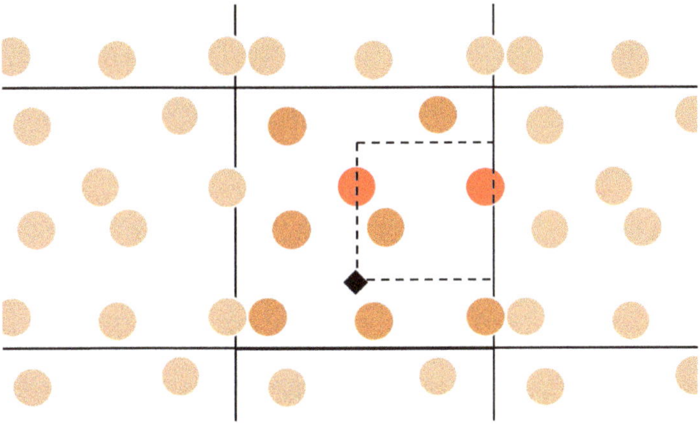

Figure 5.1. A small cubic system embedded in a large reservoir. The reservoir has periodic boundary conditions, and is a system in the thermodynamic limit. The reservoir controls the variables of the small system. The small system has excess surface energy because it can violate periodic boundary conditions, as illustrated by the square with dashed lines.

not be subject to periodic boundary conditions. This is illustrated by two particles of a darker shade. The configurations of these particles in the small box do not obey periodic boundary conditions. But that configuration is allowed during small system sampling.

The interactions between particles in the small box determine the probability for a particle to be close to the box surface, edge or corner. This probability differs from the probability to be far from the surface. The thermodynamic properties of the small system therefore differ from those in the thermodynamic limit. We formulate the difference in probability in terms of surface, $U_k^s(T, \boldsymbol{\mu})$, edge, $U_k^e(T, \boldsymbol{\mu})$ and corner, $U_k^c(T, \boldsymbol{\mu})$, energies of component k for the ideal gas in a cube in Chapter 6.

Correcting for the reservoir size

The small system volume must be sufficiently smaller than the reservoir volume [1, 67, 68] for SSM to apply. This is illustrated in Fig. 5.2, for a single WCA-liquid with density $c^* = 0.6$ and temperature $T^* = 2.0$. Figure 5.2 shows how the particle number

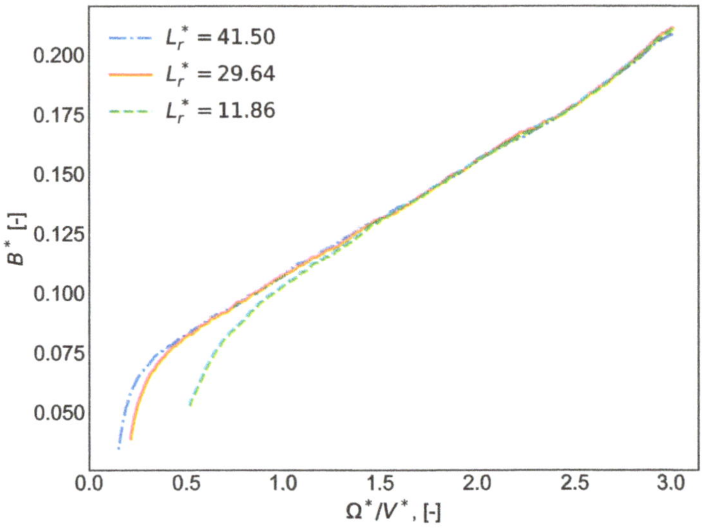

Figure 5.2. The dimensionless particle correlation density B^* of a fluid of reduced density 0.6, sampled by a small spherical volume embedded in a cubic reservoir with sides $L_r^* = 11.86$, 29.64, and 41.50.

correlation density B^* changes as the size of the reservoir changes from 11.86 to 41.50 in reduced units. The property is plotted vs the inverse characteristic size $L_\Omega^{-1} = \Omega/V$ for a wide range of values.

Each reservoir provides a curve with a linear regime for small values of $L_\Omega^{-1} = \Omega/V$. The linear regimes overlap, showing that B^* is independent of the reservoir size in this regime. The linear regimes can be extrapolated to small values of Ω/V. The extrapolated value is the thermodynamic limit value or the bulk value of the fluid. In the figure, these extrapolated values coincide, as they should. At the same time, when Ω/V becomes smaller, and the sub-volume volume approaches the size of the reservoir, the curves start to deviate from the straight line. The smaller the reservoir becomes, the earlier the deviation starts.

If the system volume becomes too close to the reservoir volume, the reservoir is no longer able to control the variables of the small system. The small system is then no longer grand canonical, and the method fails. Strøm *et al.* [48] found that the following formula was able to correct for the V/V_r–dependence in a single fluid:

$$a = \frac{A}{V} = \alpha^{\infty} + \alpha_{\Omega}^{s} L_{\Omega}^{-1} + \alpha_{\Omega}^{se} L_{\Omega}^{-2} + \alpha_{\Omega}^{sc} L_{\Omega}^{-3} + \alpha^{r} \frac{V}{V_r} \qquad (5.9)$$

This formula made it possible to fit results from differently sized reservoirs, also in the non-linear region. We refer to the original work for further details. Deviations from the linear regime may also occur for very small volumes, i.e. when Ω/V becomes large. Such deviations are typically due to curvature of spheres, and edge and corner effects of cubes, see Section 5.1.

The shape of the small system

The shape of a small system can be controlled, e.g., by confinement. A film, for instance, can be controlled by positioning adjacent walls. When we want to replace V by the pressure p as an environmental variable for a small volume, we must keep the shape of the small system unaltered in order to have a well-defined procedure. Small systems with different volumes can only be directly compared if the shape is the same. We need to specify the small system shape: spherical, cubic, tetrahedral, etc.

The sub-volume to be sampled can have any shape (sphere, cube, etc.). Properties of interest are first sampled for a given shape and size. The size of the small volume is then slightly increased, without changing the shape, and the sampling is repeated. The increase in volume can be done over and over again until the size L of the small embedded volume approaches the size of the reservoir. The front cover of this book illustrates this.

Choosing the small system position and sampling frequency

In principle, any point inside the reservoir can be selected as a position of a sub-volume. The sub-volume position in the reservoir is selected at random, see Fig. 5.3. The natural choice for the position of a sphere is the center of the sphere. Random positions of the center need be sampled. For a cube, a corner can be used to position the sub-volume, as was done in Figs. 5.1 and 5.3. Sub-volumes can be

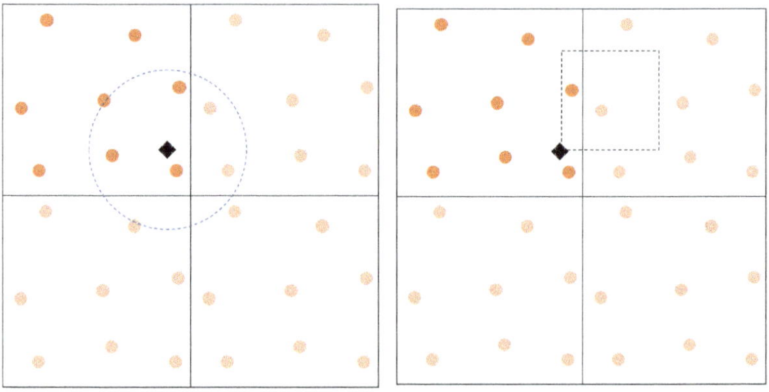

(a) (b)

Figure 5.3. The small system can be a sphere (a) or a cube (b). The position of the sphere is defined by the center, and the number of atoms inside the sphere is counted, while gradually increasing the volume. The position of the cube is here defined by one corner. To sample different cubic volumes, we vary the side of the cube.

inserted repeatedly, while sampling a new section of the simulation box. Small volumes can well stick out of the collection of periodic replicas of reservoirs, see Fig. 5.3. In principle, we must "exhaust" the available information in the system (sample all microstates), while we avoid to do time integration. The typical largest number of sub-volumes, that exhausts the information, is given by V_r/V.

Figure 5.4 shows the convergence of the particle number correlation density B^* in Eq. (5.7) toward a steady state, as the number of snapshots increases in a molecular dynamics simulation of a WCA fluid. The simulations were done for three densities: 0.1, 0.5, and 1.0 (reduced units), and the ratio between surface area and volume, Ω/V in reduced units, was approximately 1 to enable a comparison of results. The small system was a sphere and the sampling was done in a random way: 1, 10 or 100 times for each snapshot.

The value of B^* is shown as a function of the number of snapshots. The figure shows that we hardly improve the convergence by increasing the number of sampled embedded volumes beyond 500 snapshots. In most cases, the property B^* converged already after

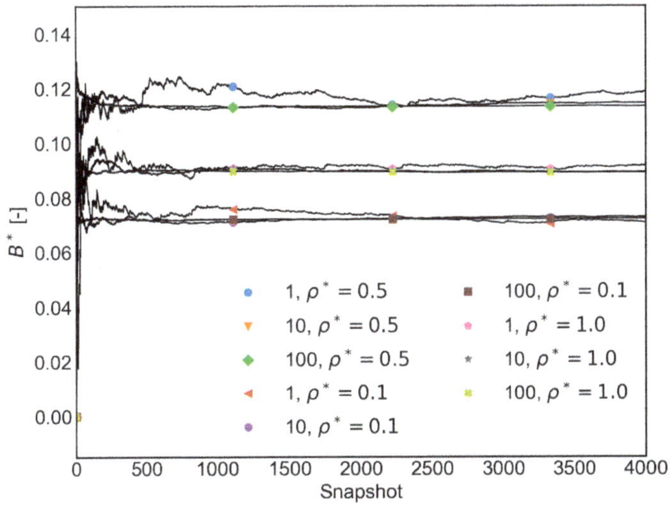

Figure 5.4. The value of the reduced particle number correlation density, B^*, in a series of snapshots from the sampling period. The small system is a sphere and the sampling is done randomly 1, 10 or 100 times for each snapshot. The ratio between surface area and volume, Ω^*/V^* in reduced units, was approximately 1 in all cases. Results are shown for three reduced fluid densities, $\rho^* = 0.1$, 0.5 and 1.0.

2000 snapshots. It is also apparent from this figure that we are able to exhaust the information available by embedding 10 small systems in each snapshot. The number of embedded small volumes depend on the system under consideration. If the simulation is time-consuming, it is worth spending more time on extracting information from each of the snapshots. However, if the computational effort to integrate the system in time is small, it is preferable to sample a smaller number of embedded volumes in each snapshot.

5.5 The Sampling Routine: A Prescription

On the basis of the previous section, we summarize the SSM sampling procedure as follows:

1. Generate a configuration of particles using a molecular dynamics simulation of the reservoir. The particles should have positions

corresponding to a state in an ensemble. The choice of ensemble for the reservoir is not important for the outcome of the method.

2. At a random position in the simulation box, insert a small open volume, i.e. a small system. The position of the small system can be defined in different ways. For a spherical volume, the system position is naturally the center of the sphere, while for a cubic shape, it is easier to choose one of the corners as the system's random position.
3. Around the random position chosen in step 2, generate subsequently larger and larger sub-volumes by either changing the radius (of a sphere) or the side length (of a cube).
4. In each of the generated sub-volumes, count the number of particles of each type that are confined to the volume and tally these. Also, calculate the internal energy associated with the particles in this particular sub-volume. When pair potentials are used, the energy is divided equally between pairs. A particle inside the sub-volume obtains half the energy of the interaction energy with particles outside.
5. Repeat procedure steps 1–4 several times, choosing new positions for the sub-volume each time. Keep the same configuration of the reservoir. Make sure not to sample several times the same configuration, as the sample value then may converge to a value which is not necessarily the small system value.

The sampling routine becomes efficient when the number of snapshots is tuned with the number of randomly embedded volumes. In the example of Fig. 5.4, 10 snapshots were sufficient to obtain values for Ω^*/V^* near unity. For Monte Carlo as well as molecular dynamics simulations, it is necessary to generate new configurations of the reservoir and repeat steps 1–5 a few times before the results converge.

5.6 Examples

Single component

Results for the reduced particle number correlation density, B^*, and the heat capacity at constant volume that derives from fluctuations

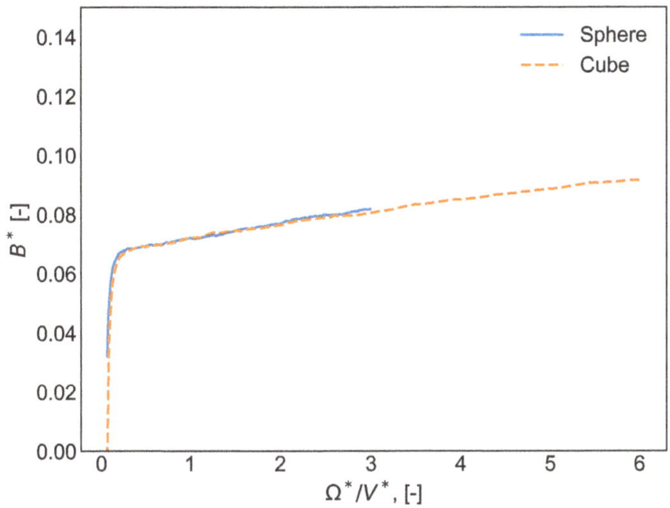

Figure 5.5. Reduced particle number correlation density, B^*, from small system sampling with cubic (red) and spherical (blue) sub-volumes. Values are plotted vs the inverse of the characteristic system length, Ω^*/V^*. The curves overlap for intermediate values of Ω^*/V^*.

in the Hamiltonian, $C^*_{V,\mathcal{H}}$, are given for the single WCA fluid and fluid mixture in Figs. 5.5 and 5.6. Figure 5.5 shows the dimensionless particle correlation for a sphere and a cube. As in Fig. 5.2, we see that the lines coincide for $1 < \Omega^*/V^* < 3$. This implies that the sphere and the cube give the same result within the accuracy. For even larger values of Ω^*/V^*, we may find deviations from the straight line due to corner and edges of a cube and due to the sphere curvature. At smaller values of Ω^*/V^*, the deviation due to V/V_r becomes sizable. Also, here, we can extrapolate the lines for the cube and the sphere to the same limiting value, which is the value for the bulk WCA-fluid. This confirms that the SSM procedure is sound.

The positive slope of the line in Fig. 5.5 is particular for the WCA-fluid. The line provides information on how B changes, as the system becomes smaller. With the repulsive particles, we find that B increases. The value of B is related to the finite-volume Kirkwood–Buff integrals, see Chapter 10. In Section 6.3, we explain that c/B gives the thermodynamic factor.

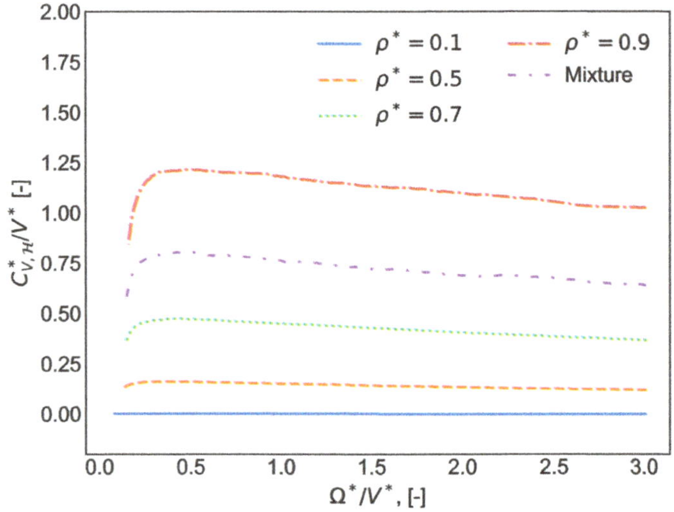

Figure 5.6. The contribution to the reduced heat capacity at constant volume from the Hamiltonian, $C_{V,\mathcal{H}}(T,V,\boldsymbol{\mu})^*/V^*$, for WCA-particles in a small sphere plotted as a function of the reduced inverse characteristic length. Results are shown for different densities and for a mixture.

The contributions to the heat capacity of a small system of volume V presents a similar picture, see Fig. 5.6. The results for $C_{V,\mathcal{H}}$ were computed for a sphere. Reduced densities varied between 0.1 and 0.9. The higher the density is, the larger is the size-dependent contribution to the heat capacity. The effect of system smallness is thus best noticed at high densities. All curves have a linear regime which allows extrapolation to the thermodynamic limit value for the fluid density in question. We conclude that the small system of the repelling particles used here has a smaller heat capacity than the bulk fluid has. More systematic work is needed to reveal the physical meaning of these data.

Binary do not highlight

Thermodynamic data for mixtures are also available by SSM. Results are shown in Fig. 5.7 for the WCA-mixture defined in Section 5.3. The elements of the reduced particle number correlation density matrix, $B_{11}, B_{12} = B_{21}$ and B_{22}, are plotted. The matrix element

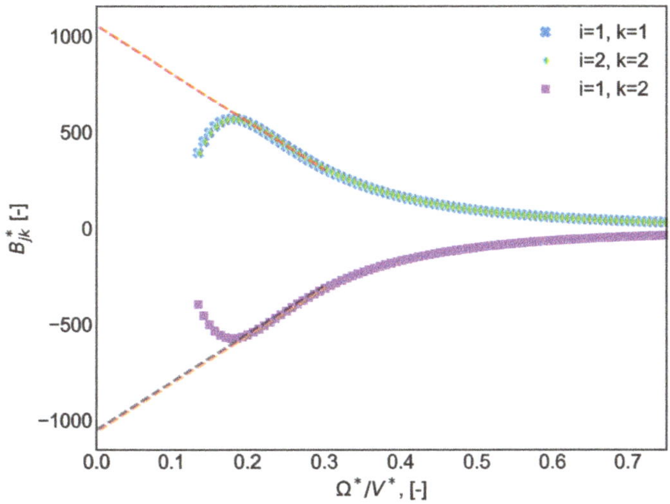

Figure 5.7. Elements of the particle correlation density matrix, B_{jk}, in a binary mixture of WCA-particles as a function of characteristic length (reduced). The mixture is kept at a temperature of $T^* = 2.0$ and a composition of 70% particles of type 1 and 30% particles of type 1. Data were sampled with a sphere, and small systems were embedded 10 times for each snapshot, cf. Fig. 5.4.

B_{ij} was calculated directly from the particle number fluctuations in the small volume. Values of B^*_{ij} are plotted as function of the reduced inverse characteristic size Ω^*/V^*. The figure shows that $B^*_{11} = B^*_{22}$ (top curve) as expected for particles with the same mass, reduced diameter, $\sigma_{11} = \sigma_{12} = \sigma_{22}$, and reduced potential depth, $\varepsilon_{11} = \varepsilon_{22}$. We note that $\varepsilon_{12} = 0.1\varepsilon_{11} = 0.1\varepsilon_{22}$. The interaction coefficient (see the lower curve) is symmetric, $B^*_{12} = B^*_{21}$. For the system parameters considered, $B^*_{12} = B^*_{21} = -B^*_{11} = -B^*_{22}$. Both curves have a linear region which can be used to find the thermodynamic limit value by extrapolation. All curves approach zero for large values of Ω^*/V^*. In Section 6.3, we explain how the inverse of the matrix B_{ij} times one of the concentrations gives the matrix of thermodynamic factors.

One contribution to the heat capacity of a small system of volume V for the binary mixture is given in Fig. 5.6. It behaves in the same way as the corresponding contribution of a single fluid.

5.7 Concluding Remarks

We have seen above, through a few examples, how the SSM makes thermodynamic properties available for small systems. As examples, we computed B and $C_{V,\mathcal{H}}$ for a single WCA fluid, and B_{ij} and $C_{V,\mathcal{H}}$ for a mixture. From these properties, we can compute the remaining thermodynamic variables of the fluids.

We have presented the rationale and procedure of the SSM method. The method can be applied to small systems embedded in larger reservoirs. The number of particles of any type inside the small volume is counted along with their energy. From these numbers, we compute the particle and energy density correlation functions.

When the properties are plotted as a function of the inverse small characteristic length, L_Ω^{-1}, we can determine the thermodynamic limit value α^∞. The line itself and its slope α_Ω^s give information on the small system. Both α^∞ and α_Ω^s are independent of the shape of the volume. All higher order coefficients depend on the shape of the volume.

The SSM has so far been used with considerable success, not only for three-dimensional systems [1, 2, 31, 32, 48, 67, 68] but also for surfaces, see Chapters 14 and 15. The method offers a new possibility for systematic investigations of small system thermodynamic properties. Much work to elucidate the thermodynamics of small systems remains to be done.

Chapter 6

Ideal Gas Mixture with Surface Energy

We compute the equations of state for an ideal gas mixture with surface adsorption energy. The system is controlled, respectively, by a canonical, grand canonical or nanocanonical ensemble. Statistical mechanical tools apply to an ensemble of small volumes. The equation of state is in general a function of the system size, shape and the environmental variables; creating a scaling law. Hadwiger's theorem applies only in special cases.

The ideal gas model is an important and useful model in classical thermodynamics. This remains so for small systems. In large systems, interactions between gas and wall particles are negligible. In small systems, however, they are not. The difference between the small and the large systems is the energy contained in the interactions with the walls.

Remark 6.1. The small volume can be a parallelepiped, which has surfaces, edges and corners. It can also be a spheroid, which has a curved surface. Two or one-dimensional systems are both relevant for the analysis.

We shall see in this chapter how small system-effects can dominate, when the volume contains a few molecules at room temperature. This leads to equations of state that are particular for the system in question. We want to derive such equations for an ideal gas with surface adsorption, controlled by the canonical, grand canonical, and nanocanonical ensembles. Doing this, we can prove Eq. (5.1) for these special cases.

Consider therefore an ideal gas mixture in a small volume, V. The ensembles will be used to show how surfaces, edges and corners contribute to the thermodynamic properties of the small systems. The dimensionality of the system is set by d. We shall examine cubes ($d = 3$), squares ($d = 2$) and straight lines ($d = 1$), because the partition functions can be calculated in these cases.

For the canonical and grand canonical ensembles, the characteristic length of the volume is defined by $L \equiv \sqrt[d]{V}$, which is equal to the side of the cubes and the squares, and the length of the straight lines. For the nanocanonical ensemble, the characteristic length is defined by $L \equiv \sqrt[d]{k_{\mathrm{B}}T/p}$.

Remark 6.2. The volume of cubes, squares and straight lines is equal to L^d. The surface area of the cube, the length of the boundary of the square and the number of end points of the straight line are proportional to L^{d-1}. There is a proportionality factor c_s and we write $\Omega = c_s L^{d-1}$. We have $c_s = 6$ for a cube, 4 for a square and 2 for a straight line. The length of the edges for a cube and the number of corners of a square is proportional to L^{d-2}. By using the proportionality factor c_e, we can write this as $L_e = c_e L^{d-2}$. We have $c_e = 12$ for a cube, 4 for a square and 0 for a straight line. For the cube, the number of corners is $c_c = 8$. For squares and straight lines, $c_c = 0$. The alternative characteristic length for cubes, squares and straight lines, is given by $L_\Omega \equiv V/\Omega = L/c_s$.

Remark 6.3. When the surface tension depends on the curvature, the linear contribution in the curvature is given for $d = 3$ by the Tolman length and the quadratic contribution, by two rigidities. The contributions due to these terms are similar to the edge and corner effects of the cube.

In particular, we want to show how the properties depend on the characteristic length of the system. We recapitulate Eq. (5.1) from Chapter 5

$$a = \frac{A}{V} = \alpha^{\infty} + \alpha^{s} L^{-1} + \alpha^{se} L^{-2} + \alpha^{sc} L^{-3} + \cdots \tag{6.1}$$

Equivalently, we could use $L_\Omega \equiv V/\Omega$ as characteristic length and expand a in $1/L_\Omega$ by using Eq. (5.2). The relation between L and L_Ω for cubes, squares and straight lines is $L_\Omega = L/c_s$.

We focus on the contributions from the volume and the surface of the cube, the surface area and the edge length of squares, and the length and the number of corners or straight lines. Other contributions will be neglected in this chapter. This corresponds to taking c_s finite and $c_e = c_c = 0$. The contributions from finite values of c_e and c_c, which complicate the expressions considerably, are given in Appendices A and B at the end of this book. Qiao *et al.* [61] give an equation of state for a confined medium which can be expanded in $1/L_\Omega$. Bråten *et al.* [69] present a model for these systems that also contains curvature contributions. For simplicity, we do not consider quantum mechanical effects.

6.1 A Model for Surface Energy

A position probability-difference

A small system has significant surface-, edge- and/or corner energies. In order to bring out the impact of the surface, we distinguish between particles close to and far away from the surface. The probability for a particle to be close to a surface (edge or corner) of a small parallelepiped differs from the probability to be away from these positions. We account for the *probability-difference* by saying that a particle k close to the surface obtains a surface energy, U_k^s (an edge energy U_k^e, or a corner energy U_k^c).

We assume now that the energy associated with surface adsorption does not depend on the volume of the small system. It may depend on the temperature T and the concentrations $\mathbf{n} = \mathbf{N}/V$ when T, V and $\mathbf{N}$ are controlled; alternatively, the temperature T and chemical potentials $\boldsymbol{\mu}$ when T, V and $\boldsymbol{\mu}$ are controlled; or the temperature T, the pressure p and chemical potentials $\boldsymbol{\mu}$ when T, p and $\boldsymbol{\mu}$ are controlled.

The surface-, edge-, and corner energies will contribute significantly to the thermodynamic properties of the small system. They become negligible in the large system limit (Fig. 6.1).

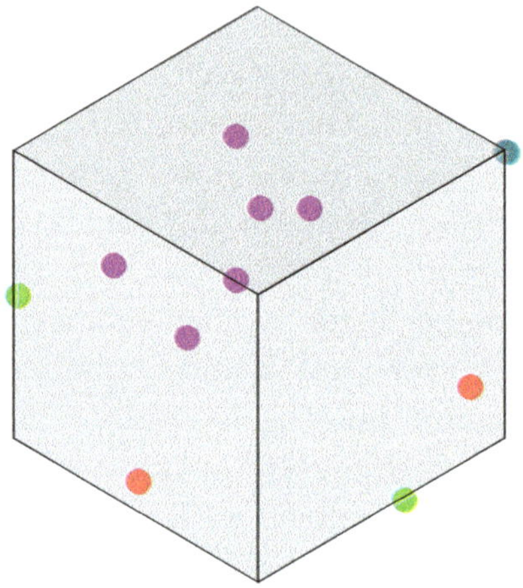

Figure 6.1. Ideal gas in a cube with surface (red), edge (green) and corner (blue) energies.

Relevance of small volume corrections

When does the surface contribution to the energy give a contribution comparable to the energy of the volume? We can use the above model for surface adsorption to find an answer with water as example. Small system corrections become important when L approaches the maximum of $c_s\Lambda_k \exp(\beta U_k^s)$, see Eq. (6.2). The corrections depend on the temperature through the factor $\beta = 1/(k_B T)$.

In order to obtain numerical insight, we multiply the surface tension of water with the diameter of a water molecule squared $(3.3\ \text{Å})^2$. This gives a measure of U_k^s. At room temperature, βU_k^s is nearly 2. For a cube, $c_s = 6$, and $c_s \exp(\beta U_k^s)$ becomes 50. The mean de Broglie wavelength at room temperature is about 0.2 Å. The value of the second term in the parenthesis of Eq. (6.2) is 10% when $L \approx 10\,\text{nm}$. It becomes equal in magnitude to the thermodynamic limit value, when $L \approx 1\,\text{nm}$. Given the size of the molecule, this corresponds to a volume of roughly 27×10^3 or 27 particles at room temperature, respectively. For lower temperatures, the small system effect will be noticeable already for larger volumes. The integral pressure may similarly start to deviate from the differential pressure at these conditions.

6.2 Canonical Ensemble: T, V, N

The derivation of the equation of state starts with the appropriate partition function. For the canonical ensemble, the one-particle partition function for component k in a small system with a surface is obtained from statistical mechanics [13] as follows:

$$Q_{1,k}(T,V) = \frac{V}{\Lambda_k^d} + \frac{\Omega}{\Lambda_k^{d-1}} \exp\left(\beta U_k^s\right) = \frac{V}{\Lambda_k^d}\left[1 + c_s \frac{\Lambda_k}{L}\exp(\beta U_k^s)\right] \tag{6.2}$$

where we note that $c_s/L = 1/L_\Omega$. The $\mathbf{N}$-particle canonical partition function becomes, cf. Chapter 4:

$$Q(T,V,\mathbf{N}) = \prod_{k=1}^{n} Q_k(T,V,N_k) = \prod_{k=1}^{n} \frac{1}{N_k!} Q_{1,k}^{N_k}(T,V) \tag{6.3}$$

The Helmholtz energy of a single replica is

$$\begin{aligned} F^{\mathrm{C}}(T,V,\mathbf{N}) &= -k_{\mathrm{B}}T \ln Q(T,V,\mathbf{N}) \\ &= -k_{\mathrm{B}}T \sum_{k=1}^{n} \ln Q_k(T,V,N_k) = \sum_{k=1}^{n} F_k^{\mathrm{C}}(T,V,N_k) \end{aligned} \tag{6.4}$$

With Stirling's formula for $N_k!$, we find

$$\begin{aligned} F_k^{\mathrm{C}}(T,V,N_k) &= k_{\mathrm{B}}T \ln N_k! - N_k k_{\mathrm{B}}T \ln Q_{1,k}(T,V) \\ &= N_k k_{\mathrm{B}}T \ln\left(\frac{N_k}{e}\right) + \frac{1}{2}k_{\mathrm{B}}T \ln(2\pi N_k) \\ &\quad - N_k k_{\mathrm{B}}T \ln Q_{1,k}(T,V) \end{aligned} \tag{6.5}$$

By introducing Eq. (6.2) into this, we obtain

$$\begin{aligned} F_k^{\mathrm{C}}(T,V,N_k) = k_{\mathrm{B}}T \Bigg\{ & N_k \ln\left(\frac{N_k \Lambda_k^d}{Ve}\right) + \frac{1}{2}\ln(2\pi N_k) \\ & - N_k \ln\left[1 + c_s \frac{\Lambda_k}{L}\exp(\beta U_k^s)\right] \Bigg\} \end{aligned}$$

$$= k_{\mathrm{B}}TV\left\{n_k \ln\left(\frac{n_k\Lambda_k^d}{e}\right) + \frac{1}{2V}\ln(2\pi n_k\Lambda_k^d)\right.$$
$$\left. + \frac{d}{2V}\ln\left(\frac{L}{\Lambda_k}\right) - n_k \ln\left[1 + c_s\frac{\Lambda_k}{L}\exp(\beta U_k^s)\right]\right\}$$
$$\approx k_{\mathrm{B}}TN_k\left\{\ln\left(\frac{n_k\Lambda_k^d}{e}\right) - c_s\frac{\Lambda_k}{L}\exp(\beta U_k^s)\right\} \qquad (6.6)$$

where $n_k \equiv N_k/V$. In the last approximation, we kept only one contribution which is linear in the inverse length of the system. This is the dominant term when L is large compared to $\Lambda_k \exp(\beta U_k^s)$. The term proportional to $V^{-1}\ln(L/\Lambda_k)$ in the third line of the equation is negligible, when compared to the $1/L$ contribution.

The exact expression for the Helmholtz energy of a canonical ensemble is not a simple function of $1/L$. It agrees only approximately with Hadwiger's theorem [63], see Chapter 5. This applies also to other thermodynamic variables in the canonical ensemble. But the result in Eq. (6.6) shows that we can use the $1/L$-expansion without invoking the Hadwiger theorem.

When U_k^s depends on T and n_k only, and not on the volume, we find for p^{C}, $s^{\mathrm{C}} \equiv S^{\mathrm{C}}/V$, μ_j^{C} from Eqs. (3.18)–(3.24):

$$p^{\mathrm{C}}(T, V, \mathbf{N}) = -\left(\frac{\partial F^{\mathrm{C}}}{\partial V}\right)_{T,\mathbf{N}}$$
$$= \frac{k_{\mathrm{B}}T}{V}\sum_{k=1}^{n} N_k\left[1 - \frac{c_s\frac{\Lambda_k}{L}\exp\left(\beta U_k^s\right)}{1 + c_s\frac{\Lambda_k}{L}\exp(\beta U_k^s)}\left(\frac{1}{d} + \beta\sum_{j=1}^{n} n_j\frac{\partial U_k^s}{\partial n_j}\right)\right]$$
$$\approx k_{\mathrm{B}}T\sum_{k=1}^{n} n_k\left[1 - c_s\frac{\Lambda_k}{L}\exp(\beta U_k^s)\left(\frac{1}{d} + \beta\sum_{j=1}^{n} n_j\frac{\partial U_k^s}{\partial n_j}\right)\right] \qquad (6.7)$$

where $n_k \equiv N_k/V$.

$$s^{\mathrm{C}}(T, V, \mathbf{N}) = -\frac{1}{V}\left(\frac{\partial F^{\mathrm{C}}}{\partial T}\right)_{V,\mathbf{N}}$$
$$= -k_{\mathrm{B}}\sum_{k=1}^{n} n_k\left[\ln(n_k\Lambda_k^d/e) - \frac{d}{2} + \frac{1}{2N_k}\ln(2\pi N_k)\right]$$

$$
\begin{aligned}
&+ k_{\mathrm{B}} \sum_{k=1}^{n} n_k \ln \left[1 + c_s \frac{\Lambda_k}{L} \exp\left(\beta U_k^s\right)\right] \\
&- \frac{c_s k_{\mathrm{B}}}{2} \sum_{k=1}^{n} n_k \frac{\Lambda_k}{L} \left[1 + c_s \frac{\Lambda_k}{L} \exp(\beta U_k^s)\right]^{-1} \\
&\times \left[1 + \frac{2}{k_{\mathrm{B}}} \left(\frac{U_k^s}{T} - \frac{\partial U_k^s}{\partial T}\right)\right] \exp(\beta U_k^s) \\
\approx\; & -k_{\mathrm{B}} \sum_{k=1}^{n} n_k \left[\ln(n_k \Lambda_k^d / e) - \frac{d}{2}\right] \\
&+ \frac{c_s k_{\mathrm{B}}}{2} \sum_{k=1}^{n} n_k \frac{\Lambda_k}{L} \left[1 - \frac{2}{k_{\mathrm{B}}} \left(\frac{U_k^s}{T} - \frac{\partial U_k^s}{\partial T}\right)\right] \exp(\beta U_k^s)
\end{aligned}
\tag{6.8}
$$

In the last equality, we neglected all contributions smaller than those proportional to $1/L$. The term proportional to $N_k^{-1} \ln N_k$ in the first square bracket is proportional to $L^{-3} \ln(L/\Lambda_k)$ and was therefore neglected.

$$
\begin{aligned}
\mu_j^{\mathrm{C}}(T, V, \mathbf{N}) = \left(\frac{\partial F^{\mathrm{C}}}{\partial N_j}\right)_{T,V,\mathbf{N}_j} &= k_{\mathrm{B}} T \ln(n_j \Lambda_j^d) + \frac{k_{\mathrm{B}} T}{2 N_j} \\
&- k_{\mathrm{B}} T \ln \left[1 + c_s \frac{\Lambda_j}{L} \exp(\beta U_j^s)\right] \\
&- k_{\mathrm{B}} T \sum_{k=1}^{n} n_k \frac{\partial}{\partial n_j} \ln \left[1 + c_s \frac{\Lambda_k}{L} \exp\left(\beta U_k^s\right)\right] \\
&\approx k_{\mathrm{B}} T \ln(n_j \Lambda_j^d) - c_s k_{\mathrm{B}} T \frac{\Lambda_j}{L} \exp(\beta U_j^s) \\
&- c_s \sum_{k=1}^{n} n_k \frac{\Lambda_k}{L} \frac{\partial U_k^s}{\partial n_j} \exp(\beta U_k^s)
\end{aligned}
\tag{6.9}
$$

In the last approximation, we kept only the contributions linear in the inverse linear size of the system. These are then dominant terms when L becomes large. The term proportional to N_j^{-1} is proportional to L^{-3} and was therefore neglected. From the unexpanded expressions for p^{C}, s^{C} and μ_j^{C}, it follows that the surface contribution also contributes to the amplitude of the terms proportional to $(\Lambda_k/L)^2$ and $(\Lambda_k/L)^3$ and higher powers. Of course, the edges (Tolman length) and the corners (rigidities) also contribute to these higher-order amplitudes, see Appendices A and B.

By using Eqs. (3.18) and (3.19), we find for the subdivision potential ε^{C}:

$$\begin{aligned}
\varepsilon^{\mathrm{C}}/V &= (F^{\mathrm{C}} - \widehat{F}^{\mathrm{C}})/V = (F^{\mathrm{C}} + p^{\mathrm{C}}V - \boldsymbol{\mu}^{\mathrm{C}} \cdot \mathbf{N})/V \\
&= k_{\mathrm{B}}T \sum_{k=1}^{n} \left\{ \frac{1}{2V} \ln\left(\frac{2\pi N_k}{e}\right) \right. \\
&\quad - n_k \frac{c_s \frac{\Lambda_k}{L} \exp(\beta U_k^s)}{1 + c_s \frac{\Lambda_k}{L} \exp(\beta U_k^s)} \left(\frac{1}{d} + \beta \sum_{j=1}^{n} n_j \frac{\partial U_k^s}{\partial n_j} \right) \\
&\quad \left. + n_k \sum_{j=1}^{n} n_j \frac{\partial}{\partial n_j} \ln\left[1 + c_s \frac{\Lambda_k}{L} \exp(\beta U_k^s)\right] \right\}
\end{aligned} \tag{6.10}$$

We see that $\varepsilon^{\mathrm{C}}/V$ goes to zero in the thermodynamic limit, and observe that $c_s/L = 1/L_\Omega$.

The matrix of thermodynamic factors, Γ_{mj}^{C}, is defined by

$$\begin{aligned}
\Gamma_{mj}^{\mathrm{C}}(T, V, \mathbf{N}) &\equiv n_m A_{mj}^{\mathrm{C}}(T, V, \mathbf{N}) \\
&\equiv \frac{N_m}{k_{\mathrm{B}}T} \left(\frac{\partial \mu_j^{\mathrm{C}}}{\partial N_m} \right)_{T,V,\mathbf{N}_m} = \frac{n_m}{k_{\mathrm{B}}T} \left(\frac{\partial \mu_j^{\mathrm{C}}}{\partial n_m} \right)_{T,V,\mathbf{N}_m}
\end{aligned} \tag{6.11}$$

This results in

$$\begin{aligned}
&\Gamma_{mj}^{\mathrm{C}}(T, V, \mathbf{N}) \\
&= \delta_{mj} \left(1 - \frac{1}{2N_m} \right) - n_m \frac{\partial}{\partial n_m} \left\{ \ln\left[1 + c_s \frac{\Lambda_j}{L} \exp(\beta U_j^s)\right] \right. \\
&\quad \left. + \sum_{k=1}^{n} n_k \frac{\partial}{\partial n_j} \ln\left[1 + c_s \frac{\Lambda_k}{L} \exp(\beta U_k^s)\right] \right\}
\end{aligned}$$

$$\approx \delta_{mj}\left(1-\frac{1}{2N_m}\right)-c_s\frac{n_m}{k_\mathrm{B}T}\left[\frac{\Lambda_j}{L}\frac{\partial U_j^s}{\partial n_m}\exp(\beta U_j^s)\right.$$
$$\left.+\frac{\Lambda_m}{L}\frac{\partial U_m^s}{\partial n_j}\exp(\beta U_m^s)+\sum_{k=1}^{n} n_k\frac{\Lambda_k}{L}\frac{\partial^2}{\partial n_m \partial n_j}\exp(\beta U_k^s)\right] \quad (6.12)$$

We kept the last contribution as a differential. The contribution due to this differential contains terms proportional to the square of the first-order derivatives and terms proportional to the second-order derivatives. The $1/L$-dependence demonstrated for all thermodynamic functions treated so far in this section motivate their plots vs. $1/L$ in the scaling laws we discussed in Chapter 5. From the Maxwell relations from Eq. (3.20) or from Eq. (6.12), we obtain the symmetry relation

$$\Gamma_{jm}^{\mathrm{C}}(T,V,\mathbf{N})N_m = \Gamma_{mj}^{\mathrm{C}}(T,V,\mathbf{N})N_j \quad (6.13)$$

The thermodynamic factor is extensively discussed in Chapter 10 for solutions and in Chapter 15 for surface adsorption.

6.3 Grand Canonical Ensemble: T, V, $\boldsymbol{\mu}$

When the system is controlled by the grand canonical ensemble, two pressures are relevant; the differential pressure, p, and the integral pressure, $\widehat{p}$. For each of the pressures, there is a particular equation of state, as we shall see in what follows.

The grand canonical partition function is obtained from statistical mechanics as follows:

$$\Xi(T,V,\boldsymbol{\mu}) = \prod_{k=1}^{n}\sum_{N_k=0}^{\infty}\exp(\beta\mu_k N_k)Q_k(T,V,N_k)$$
$$= \exp\left(\sum_{k=1}^{n}\exp(\beta\mu_k)Q_{1,k}(T,V)\right) \quad (6.14)$$

where we used Eq. (6.3). By introducing the expressions above, thermodynamic properties can be derived in terms of T, V, $\boldsymbol{\mu}$, U_k^s.

From Eq. (4.56), we find for the integral pressure

$$\widehat{p}(T,V,\boldsymbol{\mu}) = \frac{k_{\mathrm{B}}T}{V}\ln\Xi(T,V,\boldsymbol{\mu}) = \frac{k_{\mathrm{B}}T}{V}\sum_{k=1}^{n}\exp(\beta\mu_k)Q_{1,k}(T,V)$$
$$= k_{\mathrm{B}}T\sum_{k=1}^{n}\frac{\exp(\beta\mu_k)}{\Lambda_k^d}\left[1 + c_s\frac{\Lambda_k}{L}\exp(\beta U_k^s)\right] \tag{6.15}$$

This is the equation of state for the integral pressure for a system controlled by the grand canonical ensemble. The equation of state reduces to the ideal gas law in the large system limit, i.e., when $\widehat{p} \to p^{\mathrm{GC}}$.

The equation of state for the differential pressure in a system controlled by the grand canonical ensemble is given by

$$p^{\mathrm{GC}}(T,V,\boldsymbol{\mu}) = \left(\frac{\partial \widehat{p}V}{\partial V}\right)_{T,\boldsymbol{\mu}}$$
$$= k_{\mathrm{B}}T\sum_{k=1}^{n}\frac{\exp(\beta\mu_k)}{\Lambda_k^d}\left[1 + c_s\frac{d-1}{d}\frac{\Lambda_k}{L}\exp(\beta U_k^s)\right] \tag{6.16}$$

where we used that U_k^s does not depend on the volume.

By using Eq. (3.46), we determine the subdivision potential, $\varepsilon^{\mathrm{GC}}$:

$$\varepsilon^{\mathrm{GC}}/V = p^{\mathrm{GC}} - \widehat{p}$$
$$= -k_{\mathrm{B}}T\frac{c_s}{Ld}\sum_{k=1}^{n}\frac{\exp[\beta(\mu_k + U_k^s)]}{\Lambda_k^{d-1}} \tag{6.17}$$

In the thermodynamic limit $\varepsilon^{\mathrm{GC}}/V$ approaches zero. The alternative inverse characteristic length is given by $c_s/L = 1/L_\Omega$.

The entropy density $s^{\mathrm{GC}} = S^{\mathrm{GC}}/V$ and the particle densities $n_j^{\mathrm{GC}} = N_j^{\mathrm{GC}}/V$ become

$$s^{\mathrm{GC}}(T,V,\boldsymbol{\mu}) = \left(\frac{\partial \widehat{p}}{\partial T}\right)_{V,\boldsymbol{\mu}} = k_{\mathrm{B}}\sum_{k=1}^{n}\frac{\exp(\beta\mu_k)}{\Lambda_k^d}\left\{\left[1 + \frac{d}{2} - \beta\mu_k\right]\right.$$
$$\left. + c_s\frac{\Lambda_k}{L}\left[\frac{d+1}{2} - \beta(\mu_k + U_k^s) + \frac{\partial U_k^s}{k_{\mathrm{B}}\partial T}\right]\exp(\beta U_k^s)\right. \tag{6.18}$$

and

$$n_j^{\text{GC}}(T,V,\boldsymbol{\mu}) = \left(\frac{\partial \widehat{p}}{\partial \mu_j}\right)_{T,V,\boldsymbol{\mu}_j} = \frac{\exp(\beta\mu_j)}{\Lambda_j^d}\left[1 + c_s\frac{\Lambda_j}{L}\exp(\beta U_j^s)\right]$$
$$+ c_s \sum_{k=1}^{n} \frac{\exp(\beta\mu_k)}{\Lambda_k^d}\frac{\Lambda_k}{L}\frac{\partial U_k^s}{\partial \mu_j}\exp(\beta U_k^s) \tag{6.19}$$

In the thermodynamic limit, Eq. (6.19) gives the ideal gas law, as it should; $n_j = \exp(\beta\mu_j)\Lambda_j^{-d}$.

For the inverse matrix of thermodynamic factors, we find

$$\left(\boldsymbol{\Gamma}^{-1}\right)_{jm}^{\text{GC}}(T,V,\boldsymbol{\mu})n_j^{\text{GC}}(T,V,\boldsymbol{\mu}) \equiv B_{jm}^{\text{GC}}(T,V,\boldsymbol{\mu})$$
$$\equiv \frac{k_{\text{B}}T}{V}\left(\frac{\partial N_j^{\text{GC}}}{\partial \mu_m}\right)_{T,V,\boldsymbol{\mu}_m} = k_{\text{B}}T\left(\frac{\partial n_j^{\text{GC}}}{\partial \mu_m}\right)_{T,V,\boldsymbol{\mu}_m}$$
$$= \delta_{mj}n_j^{\text{GC}}\left[1 + c_s\frac{\Lambda_j}{L}\exp(\beta U_j^s)\right] + c_s\frac{\exp(\beta\mu_j)}{\Lambda_j^d}\frac{\Lambda_j}{L}\frac{\partial U_j^s}{\partial \mu_m}\exp(\beta U_j^s)$$
$$+ c_s\frac{\exp(\beta\mu_m)}{\Lambda_m^d}\frac{\Lambda_m}{L}\frac{\partial U_m^s}{\partial \mu_j}\exp(\beta U_m^s)$$
$$+ c_s\sum_{k=1}^{n}\frac{\exp(\beta\mu_k)}{\Lambda_k^d}\frac{\Lambda_k}{L}\left(\frac{\partial U_k^s}{\partial \mu_j}\frac{\partial U_k^s}{\partial \mu_m} + \frac{\partial^2 U_k^s}{\beta\partial\mu_j\partial\mu_m}\right)\exp(\beta U_k^s) \tag{6.20}$$

In Chapter 5, we explained how to calculate the particle number fluctuation correlation density matrix B_{jm}^{GC} using the small system method. We invert the whole matrix $\boldsymbol{\Gamma}^{-1}$ in order to obtain $\boldsymbol{\Gamma}$. The matrix elements of $\boldsymbol{\Gamma}$ are the thermodynamic factors. From the Maxwell relations for Eq. (3.49), or from Eq. (6.20), we obtain the symmetry relation

$$B_{mj}^{\text{GC}}(T,V,\boldsymbol{\mu}) = (\boldsymbol{\Gamma}^{-1})_{mj}^{\text{GC}}(T,V,\boldsymbol{\mu})n_m^{\text{GC}}(T,V,\boldsymbol{\mu})$$
$$= (\boldsymbol{\Gamma}^{-1})_{jm}^{\text{GC}}(T,V,\boldsymbol{\mu})n_j^{\text{GC}}(T,V,\boldsymbol{\mu}) = B_{jm}^{\text{GC}}(T,V,\boldsymbol{\mu}) \tag{6.21}$$

The system is ideal when $(\boldsymbol{\Gamma}^{-1})_{jm}^{\text{GC}}n_j^{\text{GC}} = \delta_{jm}n_j^{\text{GC}}$. For the ideal gas with surface energies, this occurs in the thermodynamic limit, when

$L \to \infty$. For the enthalpy density, $h^{\mathrm{GC}} \equiv H^{\mathrm{GC}}/V$, Eq. (3.53) gives

$$h^{\mathrm{GC}}(T,V,\boldsymbol{\mu}) = Ts^{\mathrm{GC}}(T,V,\boldsymbol{\mu}) + \sum_{j=1}^{n} \mu_j n_j^{\mathrm{GC}}(T,V,\boldsymbol{\mu}) = \sum_{k=1}^{n} \frac{\exp(\beta\mu_k)}{\Lambda_k^d}$$
$$\times \left\{ 1 + \frac{d}{2} + c_s \frac{\Lambda_k}{L} \left[\frac{d+1}{2} - \beta U_k^s + \frac{\partial U_k^s}{k_{\mathrm{B}} \partial T} + \beta \sum_{j=1}^{n} \mu_j \frac{\partial U_k^s}{\partial \mu_j} \right] \exp(\beta U_k^s) \right\} \tag{6.22}$$

We conclude that the quantities $\widehat{p}$, p^{GC}, s^{GC}, n_j^{GC}, $(\boldsymbol{\Gamma}^{-1})_{jm}^{\mathrm{GC}} n_j^{\mathrm{GC}}$ and h^{GC} are all linear in $1/L$, and there are no contributions to the higher-order terms. We show in Appendices A and B that edges only contribute to terms proportional to $(1/L)^2$ and that corners only contribute to terms proportional to $(1/L)^3$. These variations are special for systems with control of $(T, V, \boldsymbol{\mu})$. The same can e.g. not be observed for control variables $(T, V, \mathbf{N})$.

Remark 6.4. Hadwigers theorem, see Chapter 5, applies to all variables, $\widehat{p}$, p^{GC}, s^{GC}, n_j^{GC}, $(\boldsymbol{\Gamma}^{-1})_{jm}^{\mathrm{GC}} n_j^{\mathrm{GC}}$ and h^{GC} when they are controlled in the grand canonical ensemble.

Ratios of densities are not linear in $1/L$. Therefore, we provide the expression for $(\boldsymbol{\Gamma}^{-1})_{jm}^{\mathrm{GC}} n_j^{\mathrm{GC}} = B_{jm}^{\mathrm{GC}}$ rather than for $(\boldsymbol{\Gamma}^{-1})_{jm}^{\mathrm{GC}}$ for use in the finite-volume Kirkwood–Buff integrals [70] of Chapter 10.

6.4 Nanocanonical Ensemble: T, p, $\boldsymbol{\mu}$

The nanocanonical ensemble controls the variables $T, p, \boldsymbol{\mu}$. The average volume and number of particles of each component are then finite and can be calculated. We shall see how here.

We start with the nanocanonical partition function as given by Eq. (4.75)

$$\Psi(T,p,\boldsymbol{\mu}) = \beta p \int_0^{\infty} dV \exp(-\beta p V) \Xi(T,V,\boldsymbol{\mu}) \tag{6.23}$$

By introducing Eqs. (6.14) and (6.2), we obtain

$$\Psi(T,p,\boldsymbol{\mu}) = \beta p \int_0^\infty dV \exp(-\beta pV) \exp\left(\sum_{k=1}^{n} \exp(\beta\mu_k) Q_{1,k}(T,V)\right)$$

$$= \beta p \int_0^\infty dV \exp\left\{-\beta pV \left(1 - \sum_{k=1}^{n} \frac{\exp(\beta\mu_k)}{\beta p \Lambda_k^d}\right) + c_s(\beta pV)^{1-1/d} \sum_{k=1}^{n} \frac{\exp\left[\beta(\mu_k + U_k^s)\right]}{(\beta p \Lambda_k^d)^{1-1/d}}\right\} \tag{6.24}$$

By expanding the exponent with the surface contribution, we obtain

$$\Psi(T,p,\boldsymbol{\mu}) \approx \beta p \int_0^\infty dV \exp\left\{-\beta pV \left(1 - \sum_{k=1}^{n} \frac{\exp(\beta\mu_k)}{\beta p \Lambda_k^d}\right)\right\} \times \left\{1 + c_s(\beta pV)^{1-1/d} \sum_{k=1}^{n} \frac{\exp\left[\beta(\mu_k + U_k^s)\right]}{(\beta p \Lambda_k^d)^{1-1/d}}\right\} \tag{6.25}$$

This can be integrated and results in

$$\Psi(T,p,\boldsymbol{\mu}) = \left(1 - \sum_{k=1}^{n} \frac{\exp(\beta\mu_k)}{\beta p \Lambda_k^d}\right)^{-1} + c_s \Gamma\left(2 - \frac{1}{d}\right) \times \left(1 - \sum_{k=1}^{n} \frac{\exp(\beta\mu_k)}{\beta p \Lambda_k^d}\right)^{-2+1/d} \sum_{k=1}^{n} \frac{\exp\left[\beta\left(\mu_k + U_k^s\right)\right]}{(\beta p \Lambda_k^d)^{1-1/d}} \tag{6.26}$$

where Γ is the gamma function. The replica energy is accordingly

$$X^{\mathrm{NC}}(T,p,\boldsymbol{\mu}) = -k_{\mathrm{B}}T \ln \Psi(T,p,\boldsymbol{\mu}) = k_{\mathrm{B}}T \ln\left(1 - \sum_{k=1}^{n} \frac{\exp\left(\beta\mu_k\right)}{\beta p \Lambda_k^d}\right) - k_{\mathrm{B}}T c_s \Gamma\left(2 - \frac{1}{d}\right)\left(1 - \sum_{k=1}^{n} \frac{\exp\left(\beta\mu_k\right)}{\beta p \Lambda_k^d}\right)^{-1+1/d} \times \sum_{k=1}^{n} \frac{\exp\left[\beta(\mu_k + U_k^s)\right]}{(\beta p \Lambda_k^d)^{1-1/d}} \tag{6.27}$$

The replica energy is equal to the subdivision potential in this ensemble, $\varepsilon^{\mathrm{NC}}(T,p,\boldsymbol{\mu}) = X^{\mathrm{NC}}(T,p,\boldsymbol{\mu})$.

We explained in Section 6.1 that the surface energy may depend on $T,p,\boldsymbol{\mu}$. This will complicate a further analysis. For simplicity, we assume the surface energy to be constant. This is sufficient to bring out how the volume and the particle number depend on $T,p,\boldsymbol{\mu}$.

The average volume of a single replica becomes

$$\begin{aligned}
V^{\mathrm{NC}}(T,p,\boldsymbol{\mu}) &= \frac{\partial X^{\mathrm{NC}}(T,p,\boldsymbol{\mu})}{\partial p} \\
&= \frac{1}{\beta p}\sum_{k=1}^{n}\frac{\exp(\beta\mu_k)}{\beta p\Lambda_k^d}\left(1-\sum_{k=1}^{n}\frac{\exp(\beta\mu_k)}{\beta p\Lambda_k^d}\right)^{-1} \\
&\quad - \frac{c_s}{\beta p}\Gamma\left(2-\frac{1}{d}\right)\left(\frac{1}{d}-1\right)\left(1-\sum_{k=1}^{n}\frac{\exp(\beta\mu_k)}{\beta p\Lambda_k^d}\right)^{-2+1/d} \\
&\quad \times \sum_{k=1}^{n}\frac{\exp\left[\beta\left(\mu_k+U_k^s\right)\right]}{\left(\beta p\Lambda_k^d\right)^{1-1/d}}
\end{aligned} \tag{6.28}$$

The average number of particles in a single replica is likewise

$$\begin{aligned}
N_j^{\mathrm{NC}}(T,p,\boldsymbol{\mu}) &= -\frac{\partial X^{\mathrm{NC}}(T,p,\boldsymbol{\mu})}{\partial \mu_j} = \left(1-\sum_{k=1}^{n}\frac{\exp(\beta\mu_k)}{\beta p\Lambda_k^d}\right)^{-1}\frac{\exp(\beta\mu_j)}{\beta p\Lambda_j^d} \\
&\quad + c_s\Gamma\left(2-\frac{1}{d}\right)\left(1-\frac{1}{d}\right)\frac{\exp(\beta\mu_j)}{\beta p\Lambda_j^d} \\
&\quad \times \left(1-\sum_{k=1}^{n}\frac{\exp(\beta\mu_k)}{\beta p\Lambda_k^d}\right)^{-2+1/d}\sum_{k=1}^{n}\frac{\exp\left[\beta\left(\mu_k+U_k^s\right)\right]}{\left(\beta p\Lambda_k^d\right)^{1-1/d}} \\
&\quad + c_s\Gamma\left(2-\frac{1}{d}\right)\left(1-\sum_{k=1}^{n}\frac{\exp(\beta\mu_k)}{\beta p\Lambda_k^d}\right)^{-1+1/d} \\
&\quad \times \frac{\exp\left[\beta\left(\mu_j+U_j^s\right)\right]}{\left(\beta p\Lambda_j^d\right)^{1-1/d}}
\end{aligned} \tag{6.29}$$

These equations show how each choice of $T, p, \boldsymbol{\mu}$ determines the average volume and number of particles of each component.

The thermodynamic limit-expressions of the above formulas are interesting. In this limit, the contributions due to the surface are negligible, and $T, p, \boldsymbol{\mu}$ are no longer independent. By applying the Hill–Gibbs–Duhem equation, we can calculate p as a function of $T, \boldsymbol{\mu}$. Also, V^{NC} and N_j^{NC} are infinitely large in the thermodynamic limit. So the common factor, the inverse parenthesis in Eqs. (6.28) and (6.29), must be equal to zero. It follows that in the thermodynamic limit

$$p(T, \boldsymbol{\mu}) = \sum_{k=1}^{n} \frac{\exp(\beta\mu_k)}{\beta\Lambda_k^d} = k_{\mathrm{B}}T \sum_{k=1}^{n} n_k = k_{\mathrm{B}}Tn \tag{6.30}$$

This is simply the ideal gas law.

The subdivision potential per unit of volume, $\varepsilon^{\mathrm{NC}}/V^{\mathrm{NC}}$, is equal to the replica energy per unit of volume, $X^{\mathrm{NC}}/V^{\mathrm{NC}}$. The ratio approaches zero in the thermodynamic limit because V^{NC} goes to infinity.

Remark 6.5. Hadwiger's theorem does not apply or is not relevant for the nanocanonical ensemble.

6.5 Concluding Remarks

We have shown in this Chapter that small systems, which are *clusters* of ideal gas controlled in different ways, obtain different coefficients for the scaling law given in Eq. (5.1). There are, for instance, three expressions for the pressure with terms linear in $1/L$. For the canonical ensemble, we find from Eq. (6.7)

$$p^{\mathrm{C}}(T, V, \mathbf{N}) \approx k_{\mathrm{B}}T \sum_{k=1}^{n} n_k \left[1 - c_s \frac{\Lambda_k}{L} \exp\left(\beta U_k^s\right) \left(\frac{1}{d} + \beta \sum_{j=1}^{n} n_j \frac{\partial U_k^s}{\partial n_j}\right)\right] \tag{6.31}$$

For the grand canonical ensemble, we obtain from Eqs. (6.15) and (6.16):

$$\widehat{p}(T,V,\boldsymbol{\mu}) = k_{\rm B}T\sum_{k=1}^{n}\frac{\exp(\beta\mu_k)}{\Lambda_k^d}\left[1 + c_s\frac{\Lambda_k}{L}\exp(\beta U_k^s)\right]$$

$$p^{\rm GC}(T,V,\boldsymbol{\mu}) = k_{\rm B}T\sum_{k=1}^{n}\frac{\exp(\beta\mu_k)}{\Lambda_k^d}\left[1 + c_s\frac{d-1}{d}\frac{\Lambda_k}{L}\exp(\beta U_k^s)\right] \tag{6.32}$$

The terms proportional to $1/L$ contain different coefficients. The first equation is *the equation of state for the canonical ensemble.* The next two expressions are equations of state for the grand canonical ensemble, one for integral and one for the differential pressure. All three reduce to the ideal gas law in the thermodynamic limit.

It follows that $s^{\rm C} \neq s^{\rm GC}$ and $h^{\rm C} \neq h^{\rm GC}$. The subdivision potentials of the different ensembles are all different, $\varepsilon^{\rm C} \neq \varepsilon^{\rm GC} \neq \varepsilon^{\rm NC}$, as expected from Chapter 3. All subdivision potentials approach zero in the thermodynamic limit.

The equation of state for a cluster of an ideal gas with surface energy (a small volume) clearly depends on the environmental variables. In the large system limit, the equations above become the same, as they should. They reduce to the ideal gas law.

Hadwiger's theorem [63] applies *only* for the grand canonical ensemble. But when we only need the expansion in $1/L$, this has no practical consequence.

Chapter 7

Internal Energy

We show how to obtain the partial molar internal energy of a component in a mixture, using the Small System Method of Chapter 5. For a one-component small system, we give the integral relation that expresses the internal energy in terms of a pair correlation function.

Thermodynamic descriptions of mixtures, central to chemistry and chemical engineering [11], contain partial molar properties. In this chapter, we derive expressions for partial internal energies of mixtures which are confined to a small volume.[a] The partial internal energy is a function of temperature T, volume V and composition. Also for small systems, it can be expressed by correlations between the internal energy and particle number density. For small systems, the pair correlation function depends on the particle coordinates, not only the interparticle distance.

Laboratory experiments or process operations take place mostly at constant T and p. In a molecular dynamics simulation, the variable set T, V is more natural, in particular for small systems. We can transform between different variable sets in the thermodynamic limit. The Small System Method, discussed in Chapter 5, can be used to obtain limit values.

In this chapter, we examine the internal energy U^{GC} of a small system controlled by the variables T, V, $\boldsymbol{\mu}$ in a grand canonical ensemble. For large systems, the internal energy is written in terms

[a]The partial internal energy can be related to the partial molar internal energy, see Eq. (7.5).

of its partial molar energies. For small systems, this is not possible, as we will discuss. We show how to express the internal energy in terms of pair correlation functions.

We shall see how the total internal energy as well as the partial molar energies can be determined for small systems. We provide the basic formulas in Section 7.1. In Section 7.2, we derive an integral relation for the internal energy of a small system in terms of pair correlation functions. Formulas for the transformation from $T, V, \boldsymbol{\mu}$ to $T, p, \mathbf{N}$ in the thermodynamic limit are provided in Appendix C.

7.1 Partial Internal Energies in a Small System

The following expressions for particle number and internal energy are correct for large as well as small systems, see Section 9.4:

$$N_j^{\mathrm{GC}}(T,V,\boldsymbol{\mu}) = \left(\frac{\partial \hat{p}V}{\partial \mu_j}\right)_{T,V,\boldsymbol{\mu}_j} = -k_{\mathrm{B}}T\left(\frac{\partial \ln \Xi(T,V,\boldsymbol{\mu})}{\partial \mu_j}\right)_{T,V,\boldsymbol{\mu}_j} \tag{7.1}$$

$$\begin{aligned} U^{\mathrm{GC}}(T,V,\boldsymbol{\mu}) &= \langle \mathcal{H} \rangle^{\mathrm{GC}} \\ &= \left(\frac{\partial\left(\beta \hat{p}V\right)}{\partial \beta}\right)_{V,\beta\boldsymbol{\mu}} = -\left(\frac{\partial \ln \Xi(T,V,\boldsymbol{\mu})}{\partial \beta}\right)_{V,\beta\boldsymbol{\mu}} \end{aligned} \tag{7.2}$$

The symbol $\langle \ldots \rangle^{\mathrm{GC}}$ is used for the grand canonical ensemble average.

The second-order derivatives are

$$\left(\frac{\partial N_j^{\mathrm{GC}}}{\partial \mu_i}\right)_{T,V,\boldsymbol{\mu}_i} = \beta\left(\langle N_j N_i \rangle^{\mathrm{GC}} - N_j^{\mathrm{GC}} N_i^{\mathrm{GC}}\right) \tag{7.3}$$

$$\left(\frac{\partial U^{\mathrm{GC}}}{\partial \mu_j}\right)_{T,V,\boldsymbol{\mu}_j} = \beta\left(\langle \mathcal{H} N_j \rangle^{\mathrm{GC}} - U^{\mathrm{GC}} N_j^{\mathrm{GC}}\right) \tag{7.4}$$

where we used Eq. (9.41). The derivatives on the left-hand side are both additive quantities in the sense that they are sums of bulk and surface contributions. It is therefore possible to use Eqs. (5.1) and (5.2) in the Small System Method to study them, see Chapter 5.

Equation (7.4) can be used to find the derivatives of the internal energy with respect to N_j^{GC} instead of μ_j. We use the identity

$$\left(\frac{\partial U^{\mathrm{GC}}}{\partial N_j^{\mathrm{GC}}}\right)_{T,V,\boldsymbol{\mu}_j} = \left(\frac{\partial U^{\mathrm{GC}}}{\partial \mu_j}\right)_{T,V,\boldsymbol{\mu}_j} \left(\frac{\partial N_j^{\mathrm{GC}}}{\partial \mu_j}\right)^{-1}_{T,V,\boldsymbol{\mu}_j} \tag{7.5}$$

It follows that

$$\left(\frac{\partial U^{\mathrm{GC}}}{\partial N_j^{\mathrm{GC}}}\right)_{T,V,\boldsymbol{\mu}_j} = \frac{\langle \mathcal{H} N_j \rangle^{\mathrm{GC}} - U^{\mathrm{GC}} N_j^{\mathrm{GC}}}{\left\langle N_j^2 \right\rangle^{\mathrm{GC}} - (N_j^{\mathrm{GC}})^2} \tag{7.6}$$

This is the partial molar internal energy of component j as a function of $T, V, \boldsymbol{\mu}$. The quantity $(\partial U^{\mathrm{GC}}/\partial N_j^{\mathrm{GC}})_{T,V,\boldsymbol{\mu}_j}$ is not additive, however. But the numerator as well as the denominator on the right-hand side of Eq. (7.6) are additive. We can make use of this to find thermodynamic data for large systems, in the following procedure:

1. We use the Small System Method described in Chapter 5 for each additive term separately.
2. We find the thermodynamic limit-value of each term by extrapolation.
3. From the limit-values, we find the partial molar properties for other ensembles, e.g., when $T, p, \mathbf{N}$ are controlled, from Legendre transforms, cf. Section 4.8. This transformation is further explained in Appendix C. We will apply this procedure to the enthalpy H in Chapters 11 and 14.

7.2 An Integral Relation for the Internal Energy

In the thermodynamic limit, we find the internal energy in terms of an integral of the product of the pair interaction potential and the pair correlation function. This possibility exists also for small systems.

Consider a grand canonical ensemble where the temperature, the volume and the chemical potentials are constant. The internal energy for a one-component system is

$$U^{GC}(T,V,\mu) = \frac{1}{\Xi(T,V,\mu)} \sum_{N=0}^{\infty} \frac{\exp(\beta\mu N)}{h^{3N} N!} \int d\mathbf{p}_1 \ldots d\mathbf{p}_N \times \int_V d\mathbf{r}_1 \ldots d\mathbf{r}_N \mathcal{H} \exp(-\beta\mathcal{H}) \tag{7.7}$$

where the grand canonical partition function is given by

$$\Xi(V,T,\mu) = \sum_{N=0}^{\infty} \frac{\exp(\beta\mu N)}{h^{3N} N!} \int d\mathbf{p}_1 \ldots d\mathbf{p}_N \int_V d\mathbf{r}_1 \ldots d\mathbf{r}_N \exp(-\beta\mathcal{H}) \tag{7.8}$$

We do not denote the dependence of $\mathcal{H}$ on $\mathbf{p}_1, \ldots, \mathbf{p}_N, \mathbf{r}_1, \ldots, \mathbf{r}_N$ for ease of notation. After introducing Eq. (4.1), we integrate over the momenta and obtain for a single replica

$$\Xi(V,T,\mu) = \sum_{N=0}^{\infty} \frac{\exp(\beta\mu N)}{\Lambda^{3N} N!} \int_V d\mathbf{r}_1 \ldots d\mathbf{r}_N \exp(-\beta\mathcal{V}) \tag{7.9}$$

We also do not indicate the dependence of $\mathcal{V}$ on $\mathbf{r}_1, \ldots, \mathbf{r}_N$ for ease of notation. After introducing Eq. (4.1) into Eq. (7.7), we obtain

$$U^{GC}(T,V,\mu) = U^{GC}_{kin}(T,V,\mu) + U^{GC}_{pot}(T,V,\mu) \tag{7.10}$$

where

$$\begin{aligned} U^{GC}_{kin}(T,V,\mu) &= \frac{1}{\Xi(T,V,\mu)} \sum_{N=0}^{\infty} \frac{\exp(\beta\mu N)}{h^{3N} N!} \int d\mathbf{p}_1 \ldots d\mathbf{p}_N \sum_{j=1}^{N} \frac{|\mathbf{p}_j|^2}{2m} \\ &\quad \times \exp\left(-\beta \sum_{k=1}^{N} \frac{|\mathbf{p}_k|^2}{2m}\right) \int_V d\mathbf{r}_1 \ldots d\mathbf{r}_N \exp(-\beta\mathcal{V}) \\ &= \frac{3k_BT}{2\Xi(T,V,\mu)} \sum_{N=0}^{\infty} \frac{\exp(\beta\mu N)}{\Lambda^{3N} N!} N \int_V d\mathbf{r}_1 \ldots d\mathbf{r}_N \exp(-\beta\mathcal{V}) \\ &= \frac{3}{2} k_BT N^{GC}(T,V,\mu) \end{aligned} \tag{7.11}$$

and

$$U_{pot}^{\mathrm{GC}}(T,V,\mu) = \frac{1}{\Xi(T,V,\mu)} \sum_{N=0}^{\infty} \frac{\exp(\beta\mu N)}{h^{3N} N!} \int d\mathbf{p}_1 \dots d\mathbf{p}_N$$

$$\times \exp\left(-\beta \sum_{k=1}^{N} \frac{|\mathbf{p}_k|^2}{2m}\right) \int_V d\mathbf{r}_1 \dots d\mathbf{r}_N \mathcal{V} \exp(-\beta\mathcal{V})$$

$$= \frac{1}{\Xi(T,V,\mu)} \sum_{N=0}^{\infty} \frac{\exp(\beta\mu N)}{\Lambda^{3N} N!} \int_V d\mathbf{r}_1 \dots d\mathbf{r}_N \mathcal{V} \exp(-\beta\mathcal{V}) \tag{7.12}$$

In the absence of external potentials, the potential energy is the sum of pair potentials, see Eq. (4.2). This gives

$$U_{pot}^{\mathrm{GC}}(T,V,\mu) = \frac{1}{\Xi(T,V,\mu)} \sum_{N=0}^{\infty} \frac{\exp(\beta\mu N)}{\Lambda^{3N} N!} N(N-1)$$

$$\times \int_V d\mathbf{r}_1 \dots d\mathbf{r}_N \varphi(r_{12}) \exp(-\beta\mathcal{V})$$

$$= \frac{1}{\Xi(T,V,\mu)} \int_V d\mathbf{r}_1 d\mathbf{r}_2 \varphi(r_{12}) \sum_{N=0}^{\infty} \frac{\exp(\beta\mu N)}{\Lambda^{3N} N!} N(N-1)$$

$$\times \int_V d\mathbf{r}_3 \dots d\mathbf{r}_N \exp(-\beta\mathcal{V}) \tag{7.13}$$

For a small system, the pair correlation function for the grand canonical ensemble is given in Chapter 8, see Eq. (8.12)

$$g_2^{\mathrm{GC}}(\mathbf{r}_1,\mathbf{r}_2;T,V,\mu) = \frac{V^2}{\Xi(T,V,\mu)\langle N\rangle(\langle N\rangle - 1)} \sum_{N=0}^{\infty} \frac{\exp(\beta\mu N)}{\Lambda^{3N} N!}$$

$$\times N(N-1) \int_V d\mathbf{r}_3 \dots d\mathbf{r}_N \exp(-\beta\mathcal{V}) \tag{7.14}$$

This then gives

$$
\begin{aligned}
U_{pot}^{\mathrm{GC}}(T, V, \mu) &= \frac{\langle N \rangle (\langle N \rangle - 1)}{V^2} \int_V d\mathbf{r}_1 d\mathbf{r}_2 \varphi(r_{12}) g_2^{\mathrm{GC}}(\mathbf{r}_1, \mathbf{r}_2; T, V, \mu) \\
&= c(c - 1/V) \int_V d\mathbf{r}_1 d\mathbf{r}_2 \varphi(r_{12}) g_2^{\mathrm{GC}}(\mathbf{r}_1, \mathbf{r}_2; T, V, \mu)
\end{aligned}
$$

where $c = c(T, V, \mu) \equiv \langle N \rangle / V$. The expression shows the particular feature of a small system; the expression depends on both positions of the pair members, not only on the interparticle distance. In the thermodynamic limit, we may count on the interpartical dependence only. Similar expressions can be derived for other ensembles, and other pair correlation functions. The pair correlation functions are presented in more detail in Chapter 8.

7.3 Concluding Remarks

Following the procedure from Chapter 5, we have shown how to obtain partial molar internal energies for small systems in the grand canonical ensemble. An integral relation for the internal energy of a small system was given for one component with the environmental variables of the grand canonical ensemble. Unlike the corresponding relation for systems in the thermodynamic limit, the pair correleation function for the small system depends on the separate positions of the pair members. Other ensembles have similar relations. This will be further discussed and applied in Chapters 8–11 and 14.

Chapter 8

Small System Correlation Functions

We saw in Chapter 4 that the ensemble distribution of a small system depends on the variables that control it. Expressions for six ensembles were given. We proceed here to define the pair correlation functions for the same six ensembles of small systems. They differ from the corresponding distributions for large systems by their dependency on two positions rather than one.

The statistical mechanical apparatus presented in Chapter 4 makes it possible to define and sometimes calculate correlation functions. For a small system this function becomes system-specific, because the system is not extensive in V, like a large system is. The lack of extensivity is due to a relatively large surface-to-volume ratio. We have shown in Chapter 7 how to address this for the internal energy through a definition of the system's correlation function. We restricted ourselves to show this for classical systems, using Eq. (4.1) for the Hamiltonian. The Hamiltonian is a sum of kinetic energies and a potential energy, which contains inter-particle interactions as well as external interactions due, for instance, to the wall.

Here we present the probability distributions in phase space for the various ensembles. These were first given in Chapter 4. We shall use the notation which is summarized in Table 8.1. Here $\mathbf{p}_1, \ldots, \mathbf{p}_N$ and $\mathbf{r}_1, \ldots, \mathbf{r}_N$ are the momenta and positions of all $N = \sum_{i=1}^{n} N_i$ particles. The $2dN$-dimensional space spanned by

Table 8.1. Symbols used for the partition function and the probability distribution of six ensembles.

Ensemble	Partition function	Probability distribution
Microcanonical	Ω	$P_{U,V,\mathbf{N}}(\mathbf{p}_1,\ldots,\mathbf{p}_N,\mathbf{r}_1,\ldots,\mathbf{r}_N)$
Canonical	Q	$P_{T,V,\mathbf{N}}(\mathbf{p}_1,\ldots,\mathbf{p}_N,\mathbf{r}_1,\ldots,\mathbf{r}_N)$
Isobaric	Δ	$P_{T,p,\mathbf{N}}(\mathbf{p}_1,\ldots,\mathbf{p}_N,\mathbf{r}_1,\ldots,\mathbf{r}_N;V)$
Grand canonical	Ξ	$P_{T,V,\boldsymbol{\mu}}(\mathbf{p}_1,\ldots,\mathbf{p}_N,\mathbf{r}_1,\ldots,\mathbf{r}_N;\mathbf{N})$
Osmotic	Γ	$P_{T,p,N_1,\boldsymbol{\mu}_1}(\mathbf{p}_1,\ldots,\mathbf{p}_N,\mathbf{r}_1,\ldots,\mathbf{r}_N;V,\mathbf{N}_1)$
Nanocanonical	Ψ	$P_{T,p,\boldsymbol{\mu}}(\mathbf{p}_1,\ldots,\mathbf{p}_N,\mathbf{r}_1,\ldots,\mathbf{r}_N;V,\mathbf{N})$

$\mathbf{p}_1,\ldots,\mathbf{p}_N,\mathbf{r}_1,\ldots,\mathbf{r}_N$, is the *phase space*. All ensemble distributions are symmetric for the interchange of particles of the same component. The ensemble distributions were all chosen so that they are normalized. The bottom four ensembles in the table have an added variable to the probability distribution of momenta and positions. For the isobaric ensemble, P gives also the probability that the volume is V. For the GC ensemble, P gives also the probability that the number of particles of the components is $\mathbf{N}$. For the osmotic ensemble, P gives also the probability that the volume and the number of particles of components 2 to n is given by V and $\mathbf{N}_1 \equiv (N_2,\ldots,N_n)$. For the nanocanonical ensemble, P gives also the probability that the volume and the number of particles of the components are given by V and $\mathbf{N}$. All probability distributions depend also on the shape of the volume, as stated in Chapter 4.

From the probability distributions, we obtain a reduced distribution by integrating any P' over a subset of variables. The Maxwell velocity distribution in Section 8.1 is an example of such a reduced distribution. It does not depend on the ensemble of a small system, however. We proceed to define one- and two-particle densities (Section 8.2) and pair correlation functions for all ensembles (Section 8.3). The pair correlation functions depend in general on the environmental variables, including on the shape, of a small system. Most characteristically, they depend on two vectorial particle positions.

8.1 The Maxwell Velocity Distribution

The Maxwell distribution for particles is a distribution of velocities of particles with mass m_l [13]:

$$P_{\text{Maxwell}}(\mathbf{p}) = (2\pi m_l k_{\text{B}} T)^{-d/2} \exp\left(-\frac{|\mathbf{p}|^2}{2m_l k_{\text{B}} T}\right) \tag{8.1}$$

The Maxwell distribution depends neither on the ensemble, nor on the size and the shape of the system. When a small system is in local equilibrium, we can use the velocity distribution to determine the local temperature.

With the exception of the microcanonical ensemble, all ensemble distributions are proportional to a product of the factor $\exp(-|\mathbf{p}_l|^2/2m_l k_{\text{B}} T)$ times, a factor that only depends on the positions. The exponential factor is unaltered by integration over all other momenta and positions. It follows that the distribution over the momenta is the same for all these ensembles, and given by the Maxwell distribution

$$P_{T,l}^{\text{kin}}(\mathbf{p}) = P_{\text{Maxwell}}(\mathbf{p}) \tag{8.2}$$

The result is also correct for the microcanonical ensemble. The proof is, however, less straightforward, but not different from the proof in the thermodynamic limit.

Remark 8.1. The fact that the Maxwell distribution is also valid for small systems is very important. It implies, for instance, that we can use the velocity distribution in molecular dynamics simulations to determine the kinetic temperature. In equilibrium this temperature is equal to the temperature of the system. This is of great practical importance, as we obtain a way to determine T.

8.2 One- and Two-Particle Densities

Consider a multicomponent system. The one-particle density of component l is given by

$$\begin{aligned} c_{l,X}(\mathbf{r}) &= \int dY \int d\mathbf{p}_1 \cdots d\mathbf{p}_N \int_V d\mathbf{r}_1 \cdots d\mathbf{r}_N \\ &\quad \times \sum_{i \,\epsilon\, comp\ l} \delta(\mathbf{r} - \mathbf{r}_i) P(\mathbf{p}_1, \ldots, \mathbf{p}_N, \mathbf{r}_1, \ldots, \mathbf{r}_N; Y) \\ &= \int dY N_l \int d\mathbf{p}_1 \cdots d\mathbf{p}_N \\ &\quad \times \int_V d\mathbf{r}_2 \cdots d\mathbf{r}_N P(\mathbf{p}_1, \ldots, \mathbf{p}_N, \mathbf{r}, \mathbf{r}_2, \ldots, \mathbf{r}_N; Y) \end{aligned} \tag{8.3}$$

The particles are labeled such that particle 1 is from component l, which was used in the last identity. The symbol X is a shorthand notation for the environmental variables. The symbol Y is a shorthand notation for the variables of the ensemble distribution. The integral is a sum, if this variable is a number of particles. The N_l can either be one of the Y variables, as in the grand canonical ensemble, or one of the X variables, as in the canonical ensemble. The ensemble distribution P is normalized, so

$$\frac{1}{\langle V \rangle} \int d\mathbf{r} c_{l,X}(\mathbf{r}) = \frac{\langle N_l \rangle}{\langle V \rangle} \equiv c_{l,X} \tag{8.4}$$

where $\langle \cdots \rangle$ is the average over the ensemble.

When the pressure is controlled, there is usually a narrow probability distribution of volumes around the average volume $\langle V \rangle$. This is the reason why we integrate over all values of $\mathbf{r}$. The average volume of the small system is given by

$$\begin{aligned} \langle V \rangle &= \int dY V \int d\mathbf{p}_1 \cdots d\mathbf{p}_N \\ &\quad \times \int_V d\mathbf{r}_1 \cdots d\mathbf{r}_N P(\mathbf{p}_1, \ldots, \mathbf{p}_N, \mathbf{r}_1, \ldots, \mathbf{r}_N; Y) \end{aligned} \tag{8.5}$$

When V is one of the environmental variables, $\langle V \rangle = V$. Similarly, $\langle N_l \rangle = N_l$ if N_l is one of the environmental variables.

In large systems, the one-particle density $c_{l,X}(\mathbf{r})$ of component l is constant and equal to $c_{l,X} = \langle N_l \rangle / \langle V \rangle$. In small systems, the surface contributions are important. The one-particle density of component l close to the surface is then, in general, very different from $c_{l,X}$.

Remark 8.2. For small systems, the one-particle density dependence on $\mathbf{r}$ is important. A large percentage of positions in a small system is close to the surface, and the one-particle density is therefore here different from the average in a large part of the small system.

The two-particle density of components l and k is given by

$$\begin{aligned} c_{l,k,X}(\mathbf{r},\mathbf{r}') &= \int dY \int d\mathbf{p}_1 \cdots d\mathbf{p}_N \int_V d\mathbf{r}_1 \cdots d\mathbf{r}_N \sum_{i \in \, comp \; l} \delta(\mathbf{r}-\mathbf{r}_i) \\ &\quad \times \sum_{j \in \, comp \; k} \delta\left(\mathbf{r}'-\mathbf{r}_j\right) P(\mathbf{p}_1,\ldots,\mathbf{p}_N,\mathbf{r}_1,\ldots,\mathbf{r}_N;Y) \\ &= \int dY \; N_l(N_k-\delta_{lk}) \times \int d\mathbf{p}_1 \cdots d\mathbf{p}_N \\ &\quad \times \int_V d\mathbf{r}_3 \cdots d\mathbf{r}_N P(\mathbf{p}_1,\ldots,\mathbf{p}_N,\mathbf{r},\mathbf{r}',\mathbf{r}_3,\ldots,\mathbf{r}_N;Y) \end{aligned} \tag{8.6}$$

where the particles are labeled such that particle 1 is from component l and particle 2 is from component k in the last identity. The definition in Eq. (8.6) applies to all ensembles in Table 8.1. The expression is also valid for two particles of the same component, $l = k$.

By integrating Eq. (8.6) over $\mathbf{r}$ and $\mathbf{r}'$, we obtain

$$\int_V d\mathbf{r}d\mathbf{r}' c_{l,k,X}(\mathbf{r},\mathbf{r}') = \langle N_l \left(N_k - \delta_{lk}\right)\rangle \tag{8.7}$$

In the absence of correlations and excess densities, it follows that

$$c_{l,k,X}(\mathbf{r},\mathbf{r}') = \frac{\langle N_l \rangle \left(\langle N_k \rangle - \delta_{lk}\right)}{\langle V \rangle^2} = c_{l,X}\left(c_{k,X} - \delta_{lk}/\langle V \rangle\right) \tag{8.8}$$

Remark 8.3. For a small system, the surface contributions are central. The two-particle densities of the same or different components depend on both $\mathbf{r}_1$ and $\mathbf{r}_2$ in a large part of the small system. They depend on both positions and not only on the distance r_{12}.

8.3 Pair Correlation Functions

We define the pair correlation functions for all ensembles by

$$g_{l,k,X}(\mathbf{r},\mathbf{r}') \equiv \frac{c_{l,k,X}(\mathbf{r},\mathbf{r}')}{c_{l,X}\left(c_{k,X} - \delta_{lk}/\langle V\rangle\right)} \tag{8.9}$$

For an ideal gas in a large volume, the pair correlation function is equal to unity for all particle positions. This expresses the fact that the particles experience no correlations and no surface effects. Deviations in the pair correlation function from 1 can be due to both correlations and surface effects.

Remark 8.4. When the average volume is so small that its diameter is comparable to the correlation length, the pair correlation functions depend on both vectorial positions, $\mathbf{r}, \mathbf{r}'$. The value will be very different from 1 for a large fraction of the possible positions. This is not a reason to change the definition of the pair correlation functions given in Eq. (8.9).

When the size of the average volume is finite, the pair correlation function does not approach unity for distances large compared to the particle diameter. The pair correlation function will in fact differ from 1 by a term proportional to $1/\langle V\rangle$. This leads to serious convergence problems in the calculation of Kirkwood–Buff integrals, when one uses finite volume pair correlation functions. We show in Chapter 10 that it is possible and easier to use the fluctuations in the total number of particles in the small volume of the components to obtain these integrals.

We proceed to give explicit expressions for the pair correlation functions for all but the microcanonical ensemble. In all expressions, the particles are labeled such that particle 1 is from component l and particle 2 is from component k. The expressions apply also to two particles of the same component.

Canonical ensemble: T, V, $\mathbf{N}$

With the probability distribution in the $2dN$-dimensional phase space with T, V, $\mathbf{N}$ as environmental variables given in Section 4.3, and the above definitions, it follows that the pair correlation function is given by

$$g_{l,k;T,V,\mathbf{N}}(\mathbf{r}_1,\mathbf{r}_2) = \frac{1}{Q_{\text{pot}}(T,V,\mathbf{N})V^{N-2}} \times \int_V d\mathbf{r}_3 \cdots d\mathbf{r}_N \exp(-\beta\mathcal{V}(\mathbf{r}_1,\ldots,\mathbf{r}_N)) \quad (8.10)$$

Isobaric ensemble: T, p, $\mathbf{N}$

Considering a system with the environmental variables p, T, $\mathbf{N}$ and the probability distribution in the $2dN$-dimensional phase space from Section 4.4. with the above definitions, we obtain the pair correlation function of the isobaric ensemble from

$$g_{l,k;T,p,\mathbf{N}}(\mathbf{r}_1,\mathbf{r}_2) = \frac{(\beta p)^{N+1}}{\Delta_{\text{pot}}(T,p,\mathbf{N})\,\langle V(T,p,\mathbf{N})\rangle^2} \times \int_0^\infty dV \exp(-\beta pV) \times \int_V d\mathbf{r}_3 \cdots d\mathbf{r}_N \exp(-\beta\mathcal{V}(\mathbf{r}_1,\ldots,\mathbf{r}_N)) \quad (8.11)$$

Grand canonical ensemble T, V, $\boldsymbol{\mu}$

The pair correlation function with T, V, $\boldsymbol{\mu}$ as environmental variables is

$$g_{l,k;T,V,\boldsymbol{\mu}}(\mathbf{r}_1,\mathbf{r}_2) = \frac{V^2}{\Xi(T,V,\boldsymbol{\mu})\,\langle N_l\rangle\,(\langle N_k\rangle - \delta_{lk})} \sum_{N_1,\ldots,N_n} N_l\,(N_k - \delta_{lk}) \times \prod_{j=1}^{n} \frac{\exp(\beta\mu_j N_j)}{\Lambda_j^{3N_j} N_j!} \int_V d\mathbf{r}_3 \cdots d\mathbf{r}_N \exp(-\beta\mathcal{V}(\mathbf{r}_1,\ldots,\mathbf{r}_N)) \quad (8.12)$$

Osmotic ensemble T, p, N_1, $\boldsymbol{\mu}_1$

The pair correlation function with T, p, N_1, $\boldsymbol{\mu}_1$ as environmental variables is

$$g_{l,k;T,p,N_1,\boldsymbol{\mu}_1}(\mathbf{r}_1,\mathbf{r}_2) = \frac{\beta p \langle V\rangle^2}{\Upsilon(T,p,N_1,\boldsymbol{\mu}_1)N_1!\Lambda_1^3\langle N_l\rangle(\langle N_k\rangle-\delta_{lk})}$$
$$\times\int_0^\infty dV\exp(-\beta pV)\sum_{N_2,\ldots,N_n}N_l\left(N_k-\delta_{lk}\right)$$
$$\times\prod_{j=2}^{n}\frac{\exp(\beta\mu_j N_j)}{\Lambda_j^{3N_j}N_j!}\int_V d\mathbf{r}_3\cdots d\mathbf{r}_N\exp(-\beta\mathcal{V}(\mathbf{r}_1,\ldots,\mathbf{r}_N)) \quad (8.13)$$

Both l and k can be equal to 1.

Nanocanonical ensemble T, p, $\boldsymbol{\mu}$

The pair correlation function with T, p, $\boldsymbol{\mu}$ as environmental variables is

$$g_{l,k;T,p,\boldsymbol{\mu}}(\mathbf{r}_1,\mathbf{r}_2) = \frac{\beta p\langle V\rangle^2}{\Psi(T,p,\boldsymbol{\mu})\langle N_l\rangle(\langle N_k\rangle-\delta_{lk})}\int_0^\infty dV\exp(-\beta pV)$$
$$\times\sum_{N_1,\ldots,N_n}N_l\left(N_k-\delta_{lk}\right)\prod_{j=1}^{n}\frac{\exp(\beta\mu_j N_j)}{\Lambda_j^{3N_j}N_j!}$$
$$\times\int_V d\mathbf{r}_3\cdots d\mathbf{r}_N\exp(-\beta\mathcal{V}(\mathbf{r}_1,\ldots,\mathbf{r}_N)) \quad (8.14)$$

8.4 Concluding Remarks

The presentation of the pair correlation distributions above extends Chapter 4 on statistical mechanics. They could be used to compute thermodynamic variables from molecular properties. As clearly indicated in the explicit expressions for the pair correlation functions, Eqs. (8.10)–(8.14), all pair correlation functions depend on two vectorial positions, $\mathbf{r}_1, \mathbf{r}_2$, in a large part of the volume. This is, for instance, important in Chapter 10 where we define Kirkwood–Buff integrals for finite volumes. We show in that chapter that it is possible and easier to use the fluctuations in the total number of particles in the small volume of the components to obtain these integrals.

Chapter 9

Fluctuations

Thermodynamic functions were linked to partition functions of the various ensembles in Chapter 4. Here, we connect these functions to fluctuations in internal energy, volume and particle numbers. Thereby, we connect macroscopic properties of small systems to particle fluctuations.

Chapter 3 gave the thermodynamic expression for the replica energy, and Chapter 4 linked the replica energy to its statistical mechanical description. We considered six ensembles, cf. Table 3.1, and shall continue to do this. For each ensemble, we now show how to calculate the average of a thermodynamic variable from the fluctuations in the small volume; in other words, we provide links to the fluctuations.

This chapter builds on Chapter 4. It provides the fundamental basis for expressions used in Chapter 5, and in the chapters that follow. For each ensemble, we take the partition function from Chapter 4, and use it to relate the internal energy, heat capacity, compressibility, and thermodynamic factor to properties that fluctuate on the molecular level.

We thus provide the path between molecular energy levels, their populations and their macro-level expressions. The formulas can be used with quantum mechanics or classical mechanics. They enable us to confidently use the Small System Method of Chapter 5, to determine small- and large-scale thermodynamic properties. An overview of the main resulting formulas is provided in Table 9.1.

Table 9.1. Ensemble and corresponding partition function, fluctuating property and formula that links the property to the fluctuations.

Ensemble	Partition function	Property	Link
MC	$\Omega(U,V,\mathbf{N})$	N/A	N/A
C	$Q(T,V,\mathbf{N})$	C_V^{C}	$[\langle\mathcal{H}^2\rangle^{\mathrm{C}} - (\langle\mathcal{H}\rangle^{\mathrm{C}})^2]/k_{\mathrm{B}}T^2$
P	$\Delta(T,p,\mathbf{N})$	$1/\kappa_T^{\mathrm{P}}$	$k_{\mathrm{B}}T[\langle V^2\rangle^{\mathrm{P}} - (\langle V\rangle^{\mathrm{P}})^2]/\langle V\rangle^{\mathrm{P}}$
GC	$\Xi(T,V,\boldsymbol{\mu})$	B_{jk}^{GC}	$(\langle N_j N_k\rangle^{\mathrm{GC}} - N_j^{\mathrm{GC}} N_k^{\mathrm{GC}})/V$
		$C_{V,H}^{\mathrm{GC}}$	$[\langle\mathcal{H}^2\rangle^{\mathrm{GC}} - (\langle\mathcal{H}\rangle^{\mathrm{GC}})^2]/k_{\mathrm{B}}T^2$

Note: N/A means not applicable.

9.1 Microcanonical Ensemble: U, V, $\mathbf{N}$

The microcanonical partition function is not of interest in the context of the present chapter, as it has no fluctuation in the energy, volume or particle number. We include the partition function for the MC ensemble only for the sake of completeness.

The microcanonical partition function on quantum mechanical form is

$$\Omega(U,V,\mathbf{N}) = \sum_{U<\epsilon_k(V,\mathbf{N})<U+\Delta U} 1 \tag{9.1}$$

Alternatively, within classical mechanics, we have

$$\Omega(U,V,\mathbf{N}) = \frac{1}{h^{3N}} \prod_{k=1}^{n} \frac{1}{N_k!} \int_{U<\mathcal{H}<U+\Delta U} 1 \; d\mathbf{p}_1 \cdots d\mathbf{p}_N d\mathbf{r}_1 \cdots d\mathbf{r}_N \tag{9.2}$$

For the entropy of a single replica, we obtain as equation of state

$$S^{\mathrm{MC}}(U,V,\mathbf{N}) = k_{\mathrm{B}} \ln \Omega(U,V,\mathbf{N}) \tag{9.3}$$

By using Eq. (3.10), we obtain the thermodynamic potentials

$$\frac{1}{T^{\mathrm{MC}}} = \left(\frac{\partial S^{\mathrm{MC}}}{\partial U}\right)_{V,\mathbf{N}} = k_{\mathrm{B}} \left(\frac{\partial \ln \Omega(U,V,\mathbf{N})}{\partial U}\right)_{V,\mathbf{N}} \tag{9.4}$$

$$\frac{p^{\mathrm{MC}}}{T^{\mathrm{MC}}} = \left(\frac{\partial S^{\mathrm{MC}}}{\partial V}\right)_{U,\mathbf{N}} = k_{\mathrm{B}} \left(\frac{\partial \ln \Omega(U,V,\mathbf{N})}{\partial V}\right)_{U,\mathbf{N}} \tag{9.5}$$

$$\frac{\mu_j^{\mathrm{MC}}}{T^{\mathrm{MC}}} = -\left(\frac{\partial S^{\mathrm{MC}}}{\partial N_j}\right)_{U,V,\mathbf{N}_j} = -k_{\mathrm{B}} \left(\frac{\partial \ln \Omega(U,V,\mathbf{N})}{\partial N_j}\right)_{U,V,\mathbf{N}_j} \tag{9.6}$$

9.2 Canonical Ensemble: T, V, N

The quantum mechanical canonical partition function is given by Eq. (4.12)

$$Q(T,V,\mathbf{N}) = \sum_k \exp(-\beta \epsilon_k(V,\mathbf{N})) \tag{9.7}$$

The classical expression is given by Eqs. (4.13) and (4.14)

$$\begin{aligned} Q(T,V,\mathbf{N}) &= \frac{1}{h^{3N}} \prod_{k=1}^{n} \frac{1}{N_k!} \int d\mathbf{p}_1 \cdots d\mathbf{p}_N \int_V d\mathbf{r}_1 \cdots d\mathbf{r}_N \\ &\quad \times \exp(-\beta \mathcal{H}(\mathbf{p}_1,\ldots,\mathbf{p}_N,\mathbf{r}_1,\ldots,\mathbf{r}_N)) \\ &= \prod_{k=1}^{n} \frac{1}{N_k! \Lambda_k^{3N_k}} \int_V d\mathbf{r}_1 \cdots d\mathbf{r}_N \exp(-\beta \mathcal{V}(\mathbf{r}_1,\ldots,\mathbf{r}_N)) \end{aligned} \tag{9.8}$$

For the Helmholtz energy of a single replica, we obtain as equation of state

$$F^{\mathrm{C}}(T,V,\mathbf{N}) = -k_{\mathrm{B}} T \ln Q(T,V,\mathbf{N}) \tag{9.9}$$

By introducing this into Eq. (3.24), we obtain

$$p^{\mathrm{C}}(T,V,\mathbf{N}) = -\left(\frac{\partial F^{\mathrm{C}}}{\partial V}\right)_{T,\mathbf{N}} = k_{\mathrm{B}} T \left(\frac{\partial \ln Q(T,V,\mathbf{N})}{\partial V}\right)_{T,\mathbf{N}} \tag{9.10}$$

$$S^{\mathrm{C}}(T,V,\mathbf{N}) = -\left(\frac{\partial F^{\mathrm{C}}}{\partial T}\right)_{V,\mathbf{N}} = k_{\mathrm{B}} \left(\frac{\partial T \ln Q(T,V,\mathbf{N})}{\partial T}\right)_{V,\mathbf{N}} \tag{9.11}$$

$$\mu_j^{\mathrm{C}}(T,V,\mathbf{N}) = \left(\frac{\partial F^{\mathrm{C}}}{\partial N_j}\right)_{T,V,\mathbf{N}_j} = -k_{\mathrm{B}} T \left(\frac{\partial \ln Q(T,V,\mathbf{N})}{\partial N_j}\right)_{T,V,\mathbf{N}_j} \tag{9.12}$$

The internal energy is given by the average of the Hamiltonian in the canonical ensemble. From Eqs. (9.9) and (9.11), we obtain the internal energy

$$U^{\mathrm{C}}(T,V,\mathbf{N}) = F^{\mathrm{C}} - TS^{\mathrm{C}} = -\left(\frac{\partial \ln Q(T,V,\mathbf{N})}{\partial \beta}\right)_{V,\mathbf{N}} = \langle \mathcal{H} \rangle^{\mathrm{C}} \tag{9.13}$$

where $\langle \cdots \rangle^{\mathrm{C}}$ is the average in the canonical ensemble. The link between the internal energy and the average of $\mathcal{H}$ is provided by the partition function.

By reformulating the expression, we have for a quantum mechanical application

$$U^{\mathrm{C}}(T,V,\mathbf{N}) = \langle \mathcal{H} \rangle^{\mathrm{C}} = \frac{1}{Q(T,V,\mathbf{N})} \sum_k \epsilon_k(V,\mathbf{N}) \exp(-\beta \epsilon_k(V,\mathbf{N})) \tag{9.14}$$

and for a classical mechanical application

$$U^{\mathrm{C}}(T,V,\mathbf{N}) = \langle \mathcal{H} \rangle^{\mathrm{C}} \tag{9.15}$$

$$= \frac{1}{Q(T,V,\mathbf{N})} \frac{1}{h^{3N}} \prod_{k=1}^{n} \frac{1}{N_k!} \int d\mathbf{p}_1 \ldots d\mathbf{p}_N \times \int_V d\mathbf{r}_1 \ldots d\mathbf{r}_N \mathcal{H} \exp(-\beta \mathcal{H}) \tag{9.16}$$

In the last expression, we suppressed the momentum and position dependence of the Hamiltonian for ease of notation.

The heat capacity of the small system at constant volume and component particle numbers is linked to the fluctuations in $\mathcal{H}$ by

$$\begin{aligned} C_V^{\mathrm{C}}(T,V,\mathbf{N}) &\equiv \left(\frac{\partial U^{\mathrm{C}}}{\partial T}\right)_{V,\mathbf{N}} = -\frac{1}{k_{\mathrm{B}}T^2}\left(\frac{\partial U^{\mathrm{C}}}{\partial \beta}\right)_{V,\mathbf{N}} \\ &= \frac{1}{k_{\mathrm{B}}T^2}\left(\frac{\partial^2 \ln Q(T,V,\mathbf{N})}{\partial \beta^2}\right)_{V,\mathbf{N}} \\ &= \frac{1}{k_{\mathrm{B}}T^2}\left[\langle \mathcal{H}^2 \rangle^{\mathrm{C}} - \left(\langle \mathcal{H} \rangle^{\mathrm{C}}\right)^2\right] \end{aligned} \tag{9.17}$$

The expression explains how we can calculate the heat-capacity from simulations, by recording the fluctuations in $\mathcal{H}$. From the heat capacity at constant volume, we can obtain the internal energy. The Small System Method, described in Chapter 5, can be used to find values for the small and the large scale.

9.3 Isobaric Ensemble: T, p, N

The quantum mechanical isobaric partition function is given by Eq. (4.30)

$$\Delta(T,p,\mathbf{N}) = \beta p \int_0^\infty dV \exp(-\beta p V) \sum_k \exp(-\beta \epsilon_k(V,\mathbf{N})) \quad (9.18)$$

The classical isobaric partition function is given by Eqs. (4.31) and (4.32)

$$\begin{aligned}\Delta(T,p,\mathbf{N}) &= \frac{\beta p}{h^{3N}} \left(\prod_{k=1}^{n} \frac{1}{N_k!} \right) \int_0^\infty dV \exp(-\beta p V) \int d\mathbf{p}_1 \cdots d\mathbf{p}_N \\ &\quad \times \int_V d\mathbf{r}_1 \cdots d\mathbf{r}_N \exp(-\beta \mathcal{H}(\mathbf{p}_1, \ldots, \mathbf{p}_N, \mathbf{r}_1, \ldots, \mathbf{r}_N)) \\ &= \left(\prod_{k=1}^{n} \frac{1}{N_k! \Lambda_k^{3N_k}} \right) \beta p \int_0^\infty dV \exp(-\beta p V) \\ &\quad \times \int_V d\mathbf{r}_1 \cdots d\mathbf{r}_N \exp(-\beta \mathcal{V}(\mathbf{r}_1, \ldots, \mathbf{r}_N)) \end{aligned} \quad (9.19)$$

For the Gibbs energy of a single replica G^{P}, Eq. (4.38) gives as equation of state

$$G^{\mathrm{P}}(T,p,\mathbf{N}) = -k_{\mathrm{B}} T \ln \Delta(T,p,\mathbf{N}) \quad (9.20)$$

Together with Eq. (3.38), this gives

$$S^{\mathrm{P}}(T,p,\mathbf{N}) = -\left(\frac{\partial G^{\mathrm{P}}}{\partial T} \right)_{p,\mathbf{N}} = k_{\mathrm{B}} \left(\frac{\partial (T \ln \Delta(T,p,\mathbf{N}))}{\partial T} \right)_{p,\mathbf{N}} \quad (9.21)$$

$$V^{\mathrm{P}}(T,p,\mathbf{N}) = \left(\frac{\partial G^{\mathrm{P}}}{\partial p} \right)_{T,\mathbf{N}} = -k_{\mathrm{B}} T \left(\frac{\partial \ln \Delta(T,p,\mathbf{N})}{\partial p} \right)_{T,\mathbf{N}} \quad (9.22)$$

$$\mu_j^{\mathrm{P}}(T,p,\mathbf{N}) = \left(\frac{\partial G^{\mathrm{P}}}{\partial N_j}\right)_{T,p,\mathbf{N}_j} = -k_{\mathrm{B}}T\left(\frac{\partial \ln \Delta(T,p,\mathbf{N})}{\partial N_j}\right)_{T,p,\mathbf{N}_j} \tag{9.23}$$

A small system is not Euler homogeneous in V, so G^{P} is not equal to $\sum_{j=1}^{n} \mu_j^{\mathrm{P}} N_j$. It follows from Eqs. (9.22) and (9.18) that

$$V^{\mathrm{P}}(T,p,\mathbf{N}) = \langle V \rangle^{\mathrm{P}} \tag{9.24}$$

where $\langle \ldots \rangle^{\mathrm{P}}$ is the average in the isobaric ensemble. In a quantum mechanical application, the volume becomes

$$\begin{aligned} V^{\mathrm{P}}(T,p,\mathbf{N}) &= \langle V \rangle^{\mathrm{P}} \\ &= \frac{\beta p}{\Delta(T,p,\mathbf{N})} \int_0^\infty dV\ V \exp(-\beta p V) \sum_k \exp(-\beta \epsilon_k(V,\mathbf{N})) \end{aligned} \tag{9.25}$$

For a classical application, the volume is

$$\begin{aligned} V^{\mathrm{P}}(T,p,\mathbf{N}) = \langle V \rangle^{\mathrm{P}} &= \frac{\beta p}{\Delta(T,p,\mathbf{N}) h^{3N}} \prod_{k=1}^{n} \frac{1}{N_k!} \\ &\times \int_0^\infty dV\ V \exp(-\beta p V) \int d\mathbf{p}_1 \cdots d\mathbf{p}_N \\ &\times \int_V d\mathbf{r}_1 \cdots d\mathbf{r}_N \exp(-\beta \mathcal{H}) \end{aligned} \tag{9.26}$$

In the last expression, we suppressed the momentum- and position-dependence of the Hamiltonian for ease of notation.

The isothermal compressibility is

$$\begin{aligned} \frac{1}{\kappa_T^{\mathrm{P}}(T,p,\mathbf{N})} &\equiv -\frac{1}{V^{\mathrm{P}}}\left(\frac{\partial V^{\mathrm{P}}}{\partial p}\right)_{T,\mathbf{N}} = \frac{k_{\mathrm{B}}T}{V^{\mathrm{P}}}\left(\frac{\partial^2 \ln \Delta(T,p,\mathbf{N})}{\partial p^2}\right)_{T,\mathbf{N}} \\ &= -k_{\mathrm{B}}T\,\frac{\langle V^2 \rangle^{\mathrm{P}} - (\langle V \rangle^{\mathrm{P}})^2}{\langle V \rangle^{\mathrm{P}}} \end{aligned} \tag{9.27}$$

This equation gives the link between the molecular energies, as described by the partition function and the fluctuations, and the overall property which is the isothermal compressibility.

It follows from Eqs. (9.18), (3.32) and (9.21) that

$$H^{\mathrm{P}} \equiv U^{\mathrm{P}} + pV^{\mathrm{P}} = TS^{\mathrm{P}} + G^{\mathrm{P}} = \langle \mathcal{H} \rangle^{\mathrm{P}} + p \langle V \rangle^{\mathrm{P}} \tag{9.28}$$

where H^{P} is the enthalpy of a single replica. We note that

$$U^{\mathrm{P}}(T, p, \mathbf{N}) = \langle \mathcal{H} \rangle^{\mathrm{P}} \tag{9.29}$$

Another identity follows from Eqs. (9.18) and (9.19) with constant βp:

$$U^{\mathrm{P}}(T, p, \mathbf{N}) = \langle \mathcal{H} \rangle^{\mathrm{P}} = -\left(\frac{\partial \ln \Delta(T, p, \mathbf{N})}{\partial \beta} \right)_{\beta p, \mathbf{N}} \tag{9.30}$$

9.4 Grand Canonical Ensemble: T, V, $\boldsymbol{\mu}$

The quantum mechanical expression for the grand canonical partition function is given by Eq. (4.51)

$$\Xi(T, V, \boldsymbol{\mu}) = \sum_{N_1=0}^{\infty} \cdots \sum_{N_n=0}^{\infty} \exp(\beta \boldsymbol{\mu} \cdot \mathbf{N}) \sum_k \exp(-\beta \epsilon_k(V, \mathbf{N})) \tag{9.31}$$

where $\boldsymbol{\mu} \cdot \mathbf{N} \equiv \sum \mu_i N_i$. The classical expression is given by Eqs. (4.52) and (4.53)

$$\begin{aligned} \Xi(T, V, \boldsymbol{\mu}) &= \sum_{N_1=0}^{\infty} \cdots \sum_{N_n=0}^{\infty} \frac{\exp(\beta \boldsymbol{\mu} \cdot \mathbf{N})}{h^{3N} N_1! \dots N_n!} \\ &\quad \times \int d\mathbf{p}_1 \dots d\mathbf{p}_N \\ &\quad \times \int_V d\mathbf{r}_1 \cdots d\mathbf{r}_N \exp(-\beta \mathcal{H}(\mathbf{p}_1, \dots, \mathbf{p}_N, \mathbf{r}_1, \dots, \mathbf{r}_N)) \\ &= \sum_{N_1=0}^{\infty} \cdots \sum_{N_n=0}^{\infty} \frac{\exp(\beta \boldsymbol{\mu} \cdot \mathbf{N})}{\Lambda_1^{3N_1} N_1! \dots \Lambda_n^{3N_n} N_n!} \\ &\quad \times \int_V d\mathbf{r}_1 \cdots d\mathbf{r}_N \exp(-\beta \mathcal{V}(\mathbf{r}_1, \dots, \mathbf{r}_N)) \end{aligned} \tag{9.32}$$

where the total number of particles is $N \equiv \sum_{i=1}^{n} N_i$.

For a single replica, the equation of state for the integral pressure becomes

$$\widehat{p}(T, V, \boldsymbol{\mu})V = k_{\mathrm{B}}T \ln \Xi(T, V, \boldsymbol{\mu}) \tag{9.33}$$

The logarithm of the grand canonical partition function of a single replica gives the integral pressure $\widehat{p}$, *not* the differential pressure p^{GC}! In combination with Eq. (3.52), this gives

$$p^{\mathrm{GC}}(T, V, \boldsymbol{\mu}) = \left(\frac{\partial \widehat{p}V}{\partial V}\right)_{T,\boldsymbol{\mu}} = k_{\mathrm{B}}T\left(\frac{\partial \ln \Xi(T, V, \boldsymbol{\mu})}{\partial V}\right)_{T,\boldsymbol{\mu}} \tag{9.34}$$

$$S^{\mathrm{GC}}(T, V, \boldsymbol{\mu}) = \left(\frac{\partial \widehat{p}V}{\partial T}\right)_{V,\boldsymbol{\mu}} = k_{\mathrm{B}}\left(\frac{\partial (T\ln \Xi(T, V, \boldsymbol{\mu}))}{\partial T}\right)_{V,\boldsymbol{\mu}} \tag{9.35}$$

$$N_j^{\mathrm{GC}}(T, V, \boldsymbol{\mu}) = \left(\frac{\partial \widehat{p}V}{\partial \mu_j}\right)_{T,V,\boldsymbol{\mu}_j} = k_{\mathrm{B}}T\left(\frac{\partial \ln \Xi(T, V, \boldsymbol{\mu})}{\partial \mu_j}\right)_{T,V,\boldsymbol{\mu}_j} \tag{9.36}$$

It follows from Eqs. (9.31), (9.32) and (9.36) that

$$N_j^{\mathrm{GC}}(T, V, \boldsymbol{\mu}) = \langle N_j \rangle^{\mathrm{GC}} \tag{9.37}$$

where $\langle \cdots \rangle^{\mathrm{GC}}$ is the average in the grand canonical ensemble. For the number of particles of component j, we have for a quantum mechanical application that

$$N_j^{\mathrm{GC}}(T, V, \boldsymbol{\mu}) = \langle N_j \rangle^{\mathrm{GC}} = \frac{1}{\Xi(T, V, \boldsymbol{\mu})} \sum_{N_1=0}^{\infty} \cdots \sum_{N_n=0}^{\infty} \times \exp(\beta \boldsymbol{\mu} \cdot \mathbf{N})\, N_j \sum_k \exp(-\beta \epsilon_k(V, \mathbf{N})) \tag{9.38}$$

The equivalent classical expression is

$$N_j^{\mathrm{GC}}(T, V, \boldsymbol{\mu}) = \langle N_j \rangle^{\mathrm{GC}} = \frac{1}{\Xi(T, V, \boldsymbol{\mu})} \sum_{N_1=0}^{\infty} \cdots \sum_{N_n=0}^{\infty} \frac{\exp(\beta \boldsymbol{\mu} \cdot \mathbf{N})}{h^{3N} N_1! \ldots N_n!} \times N_j \int d\mathbf{p}_1 \cdots d\mathbf{p}_N \int_V d\mathbf{r}_1 \cdots d\mathbf{r}_N \exp(-\beta \mathcal{H}) \tag{9.39}$$

In the last expression we suppressed the momentum and position dependence of the Hamiltonian for ease of notation.

The matrix of inverse thermodynamic factors is given by

$$c_j^{\mathrm{GC}}\left[\boldsymbol{\Gamma}^{-1}\right]_{jk}^{\mathrm{GC}}(T,V,\boldsymbol{\mu}) \equiv B_{jk}^{\mathrm{GC}}(T,V,\boldsymbol{\mu}) \tag{9.40}$$

where the particle correlation density B_{jk}^{GC} can be given in terms of fluctuating variables by

$$\begin{aligned} B_{jk}^{\mathrm{GC}}(T,V,\boldsymbol{\mu}) &\equiv \frac{k_{\mathrm{B}}T}{V}\left(\frac{\partial N_k^{\mathrm{GC}}}{\partial \mu_j}\right)_{T,V,\boldsymbol{\mu}_j} \\ &= \frac{(k_{\mathrm{B}}T)^2}{V}\left(\frac{\partial^2 \ln \Xi(T,V,\boldsymbol{\mu})}{\partial \mu_j \partial \mu_k}\right)_{T,V,\boldsymbol{\mu}_{j,k}} \\ &= \frac{\langle N_j N_k\rangle^{\mathrm{GC}} - N_j^{\mathrm{GC}} N_k^{\mathrm{GC}}}{V} \end{aligned} \tag{9.41}$$

Here $\boldsymbol{\mu}_{j,k} \equiv (\mu_1,\ldots,\mu_{j-1},\mu_{j+1},\ldots,\mu_{k-1},\mu_{k+1},\ldots,\mu_n)$. In Chapter 5, we discussed how to obtain all matrix elements, B_{jk}^{GC}, as a function of the characteristic length by using the small system method. The matrix of thermodynamic factors is obtained by inverting the matrix $\left[\boldsymbol{\Gamma}^{-1}\right]_{jk}^{\mathrm{GC}}$. It follows from Eqs. (9.40) and (9.41) that the matrix of thermodynamic factors satisfies the symmetry relation

$$N_j^{\mathrm{GC}}\Gamma_{jk}^{\mathrm{GC}}(T,V,\boldsymbol{\mu}) = N_k^{\mathrm{GC}}\Gamma_{kj}^{\mathrm{GC}}(T,V,\boldsymbol{\mu}) \tag{9.42}$$

There are two pressures, the integral and the differential pressure, so there are also two isothermal compressibilities

$$\begin{aligned} \widehat{\kappa}_T(T,V,\boldsymbol{\mu}) &\equiv -V\frac{\partial \widehat{p}}{\partial V} = \widehat{p}(T,V,\boldsymbol{\mu}) - p^{\mathrm{GC}}(T,V,\boldsymbol{\mu}) \\ &= -k_{\mathrm{B}}T\left[\left(\frac{\partial \ln \Xi(T,V,\boldsymbol{\mu})}{\partial V}\right)_{T,\boldsymbol{\mu}} - \frac{\ln \Xi(T,V,\boldsymbol{\mu})}{V}\right] \end{aligned} \tag{9.43}$$

and

$$\kappa_T^{\mathrm{GC}}(T,V,\boldsymbol{\mu}) \equiv -V\frac{\partial p^{\mathrm{GC}}}{\partial V} = -k_{\mathrm{B}}TV\left(\frac{\partial^2 \ln \Xi(T,V,\boldsymbol{\mu})}{\partial V^2}\right)_{T,\boldsymbol{\mu}} \tag{9.44}$$

For small systems, both these compressibilities differ from the compressibility in the isobaric ensemble. It follows from Eqs. (3.80) and (9.43) that

$$\widehat{\kappa}_T(T,V,\boldsymbol{\mu}) = \widehat{p}(T,V,\boldsymbol{\mu}) - p^{\mathrm{GC}}(T,V,\boldsymbol{\mu}) = -\frac{1}{V}\varepsilon^{\mathrm{GC}}(T,V,\boldsymbol{\mu}) \quad (9.45)$$

In the thermodynamic limit, the subdivision potential $\varepsilon^{\mathrm{GC}}$, and therefore $\widehat{\kappa}_T$, approach zero, while κ_T^{GC} approaches the thermodynamic limit value of the isothermal compressibility. The subdivision potential $\varepsilon^{\mathrm{GC}}$ is additive so that $\varepsilon^{\mathrm{GC}}/V = \widehat{\kappa}_T$ can be expanded in the inverse size of the system, see Section 5.1. Also κ_T^{GC} can be expanded in this way.

For the internal energy, using Eq. (9.32) we have

$$U^{\mathrm{GC}}(T,V,\boldsymbol{\mu}) = \langle\mathcal{H}\rangle^{\mathrm{GC}} = -\left(\frac{\partial \ln \Xi(T,V,\boldsymbol{\mu})}{\partial \beta}\right)_{V,\beta\boldsymbol{\mu}} \quad (9.46)$$

The heat capacity of the small system at constant volume and chemical potentials is

$$\begin{aligned} C_V^{\mathrm{GC}}(T,V,\boldsymbol{\mu}) &\equiv \left(\frac{\partial U^{\mathrm{GC}}}{\partial T}\right)_{V,\boldsymbol{\mu}} = -\frac{1}{k_{\mathrm{B}}T^2}\left(\frac{\partial U^{\mathrm{GC}}}{\partial \beta}\right)_{V,\boldsymbol{\mu}} \\ &= \frac{1}{k_{\mathrm{B}}T^2}\left[\frac{\partial}{\partial\beta}\left(\frac{\partial \ln \Xi(T,V,\boldsymbol{\mu})}{\partial \beta}\right)_{V,\beta\boldsymbol{\mu}}\right]_{V,\boldsymbol{\mu}} \\ &= \frac{1}{k_{\mathrm{B}}T^2}\left\{\left(\frac{\partial^2 \ln \Xi}{\partial \beta^2}\right)_{V,\beta\boldsymbol{\mu}} + \boldsymbol{\mu}\cdot\left[\frac{\partial}{\partial \beta\boldsymbol{\mu}}\left(\frac{\partial \ln \Xi}{\partial \beta}\right)_{V,\beta\boldsymbol{\mu}}\right]_{\beta}\right\} \\ &= \frac{1}{k_{\mathrm{B}}T^2}\{[\langle\mathcal{H}^2\rangle^{\mathrm{GC}} - (\langle\mathcal{H}\rangle^{\mathrm{GC}})^2] \\ &\quad + \boldsymbol{\mu}\cdot[\langle\mathbf{N}\mathcal{H}\rangle^{\mathrm{GC}} - \langle\mathbf{N}\rangle^{\mathrm{GC}}\langle\mathcal{H}\rangle^{\mathrm{GC}}]\} \end{aligned} \quad (9.47)$$

In the grand canonical ensemble, the heat capacity has two contributions. The first is due to the fluctuations in the energy in the small volume

$$C_{V,\mathcal{H}}^{\mathrm{GC}}(T,V,\boldsymbol{\mu}) \equiv \frac{1}{k_{\mathrm{B}}T^2}[\langle\mathcal{H}^2\rangle^{\mathrm{GC}} - (\langle\mathcal{H}\rangle^{\mathrm{GC}})^2] \quad (9.48)$$

In Chapter 5, we discussed how to obtain $C_{V,\mathcal{H}}^{\mathrm{GC}}$ as a function of the characteristic length by using the small system method. The second is due to correlations of the fluctuations in the energy and the number of particles in the small volume

$$C_{V,N}^{\mathrm{GC}}(T,V,\boldsymbol{\mu}) \equiv \frac{1}{k_{\mathrm{B}}T^2}\boldsymbol{\mu}\cdot\left[\langle\mathbf{N}\mathcal{H}\rangle^{\mathrm{GC}} - \langle\mathbf{N}\rangle^{\mathrm{GC}}\langle\mathcal{H}\rangle^{\mathrm{GC}}\right] \tag{9.49}$$

The sum of the two is the total heat capacity

$$C_V^{\mathrm{GC}}(T,V,\boldsymbol{\mu}) = C_{V,\mathcal{H}}^{\mathrm{GC}}(T,V,\boldsymbol{\mu}) + C_{V,N}^{\mathrm{GC}}(T,V,\boldsymbol{\mu}) \tag{9.50}$$

It will be interesting to compare the relative size of the two contributions.

9.5 Osmotic Ensemble: T, p, N_1, μ_1

A quantum mechanical expression for the osmotic partition function is given by Eq. (4.63)

$$\begin{aligned}\Upsilon(T,p,N_1,\boldsymbol{\mu}_1) &= \beta p \sum_{N_2=0}^{\infty}\cdots\sum_{N_n=0}^{\infty}\exp(\beta(\mu_2 N_2 + \cdots + \mu_n N_n))\\ &\quad\times\int_0^{\infty} dV\exp(-\beta pV)\sum_k \exp(-\beta\epsilon_k(V,\mathbf{N}))\end{aligned} \tag{9.51}$$

The corresponding classical expression is given by Eqs. (4.64) and (4.65)

$$\begin{aligned}&\Upsilon(T,p,N_1,\boldsymbol{\mu}_1)\\ &= \frac{\beta p}{N_1!h^{3N_1}}\sum_{N_2=0}^{\infty}\cdots\sum_{N_n=0}^{\infty}\left[\left(\prod_{k=2}^{n}\frac{\exp(\beta\mu_k N_k)}{N_k!h^{3N_k}}\right)\right.\\ &\quad\times\int_0^{\infty} dV\exp(-\beta pV)\int d\mathbf{p}_1\ldots d\mathbf{p}_N\\ &\quad\left.\times\int_V d\mathbf{r}_1\ldots d\mathbf{r}_N\exp(-\beta\mathcal{H}(\mathbf{p}_1,\ldots,\mathbf{p}_N,\mathbf{r}_1,\ldots,\mathbf{r}_N))\right]\\ &= \frac{\beta p}{N_1!\Lambda_1^{3N_1}}\sum_{N_2=0}^{\infty}\cdots\sum_{N_n=0}^{\infty}\left[\left(\prod_{k=2}^{n}\frac{\exp(\beta\mu_k N_k)}{N_k!\Lambda_k^{3N_k}}\right)\right.\\ &\quad\left.\times\int_0^{\infty} dV\exp(-\beta pV)\int_V d\mathbf{r}_1\ldots d\mathbf{r}_N\exp(-\beta\mathcal{V}(\mathbf{r}_1,\ldots,\mathbf{r}_N))\right]\end{aligned} \tag{9.52}$$

For a single replica, the equation of state of the integral chemical potential of the first component becomes

$$\widehat{\mu}_1^{\mathrm{O}}(T,p,N_1,\boldsymbol{\mu}_1)N_1 = -k_{\mathrm{B}}T\ln\Upsilon(T,p,N_1,\boldsymbol{\mu}_1) \tag{9.53}$$

The logarithm of the osmotic partition function of a single replica gives the integral chemical potential $\widehat{\mu}_1^{\mathrm{O}}$ and *not* the differential chemical potential of component one, μ_1^{O}. Together with Eq. (3.67) this gives

$$S^{\mathrm{O}}\left(T,p,N_1,\boldsymbol{\mu}_1\right) = -\left(\frac{\partial\widehat{\mu}_1^{\mathrm{O}}N_1}{\partial T}\right)_{p,N_1,\boldsymbol{\mu}_1} = k_{\mathrm{B}}\left(\frac{\partial(T\ln\Upsilon)}{\partial T}\right)_{p,N_1,\boldsymbol{\mu}_1} \tag{9.54}$$

$$V^{\mathrm{O}}\left(T,p,N_1,\boldsymbol{\mu}_1\right) = \left(\frac{\partial\widehat{\mu}_1^{\mathrm{O}}N_1}{\partial p}\right)_{T,N_1,\boldsymbol{\mu}_1} = -k_{\mathrm{B}}T\left(\frac{\partial\ln\Upsilon}{\partial p}\right)_{T,N_1,\boldsymbol{\mu}_1} \tag{9.55}$$

$$\mu_1^{\mathrm{O}}\left(T,p,N_1,\boldsymbol{\mu}_1\right) = \left(\frac{\partial\widehat{\mu}_1^{\mathrm{O}}N_1}{\partial N_1}\right)_{T,p,\boldsymbol{\mu}_1} = -k_{\mathrm{B}}T\left(\frac{\partial\ln\Upsilon}{\partial N_1}\right)_{T,p,\boldsymbol{\mu}_1} \tag{9.56}$$

$$N_j^{\mathrm{O}}\left(T,p,N_1,\boldsymbol{\mu}_1\right) = \left(\frac{\partial\widehat{\mu}_1^{\mathrm{O}}N_1}{\partial \mu_j}\right)_{T,p,N_1,\boldsymbol{\mu}_{1,j}} = -k_{\mathrm{B}}T\left(\frac{\partial\ln\Upsilon}{\partial \mu_j}\right)_{T,p,N_1,\boldsymbol{\mu}_{1,j}}$$
$$\text{for} j = 2,\ldots,n \tag{9.57}$$

Not surprisingly, it follows from Eqs. (9.51), (9.52) and (9.57) that particle numbers are average particle numbers;

$$N_j^{\mathrm{O}}(T,V,N_1,\boldsymbol{\mu}_1) = \langle N_j\rangle^{\mathrm{O}} \quad \text{for } j = 2,\ldots,n \tag{9.58}$$

where $\langle\cdots\rangle^{\mathrm{O}}$ is the average in the osmotic ensemble. We also obtain

$$N_j^{\mathrm{O}}(T,p,N_1,\boldsymbol{\mu}_1) = \langle N_j\rangle^{\mathrm{O}} = \frac{\beta p}{\Upsilon(T,p,N_1,\boldsymbol{\mu}_1)}\sum_{N_2=0}^{\infty}\cdots\sum_{N_n=0}^{\infty}N_j$$
$$\times\ \exp(\beta(\mu_2N_2+\cdots+\mu_nN_n))$$
$$\times\int_0^{\infty}dV\exp(-\beta pV)\sum_k\exp(-\beta\epsilon_k(V,\mathbf{N})) \tag{9.59}$$

for a quantum mechanical system, and

$$N_j^{\mathrm{O}}(T, p, N_1, \boldsymbol{\mu}_1) = \langle N_j \rangle^{\mathrm{O}}$$

$$= \frac{\beta p}{\Upsilon(T, p, N_1, \boldsymbol{\mu}_1) N_1! \Lambda_1^{3N_1}} \sum_{N_2=0}^{\infty} \cdots \sum_{N_n=0}^{\infty} N_j \left(\prod_{k=2}^{n} \frac{\exp(\beta \mu_k N_k)}{N_k! \Lambda_k^{3N_k}} \right)$$

$$\times \int_0^{\infty} dV \exp(-\beta p V) \int_V d\mathbf{r}_1 \ldots d\mathbf{r}_N \exp(-\beta \mathcal{V}(\mathbf{r}_1, \ldots, \mathbf{r}_N)) \tag{9.60}$$

for a classical system.

9.6 Nanocanonical Ensemble: T, p, $\boldsymbol{\mu}$

The quantum mechanical expression Eq. (4.76) for the nanocanonical partition function is

$$\Psi(T, p, \boldsymbol{\mu}) = \beta p \sum_{N_1=0}^{\infty} \cdots \sum_{N_n=0}^{\infty} \exp(\beta \boldsymbol{\mu} \cdot \mathbf{N})$$
$$\times \int_0^{\infty} dV \exp(-\beta p V) \sum_k \exp(-\beta \epsilon_k(V, \mathbf{N})) \tag{9.61}$$

The classical mechanical expression of Eqs. (4.77) and (4.78) give

$$\Psi(T, p, \boldsymbol{\mu}) = \beta p \sum_{N_1=0}^{\infty} \cdots \sum_{N_n=0}^{\infty} \left[\left(\prod_{k=1}^{n} \frac{\exp(\beta \mu_k N_k)}{N_k! h^{3N_k}} \right) \right.$$
$$\times \int_0^{\infty} dV \exp(-\beta p V) \int d\mathbf{p}_1 \cdots d\mathbf{p}_N$$
$$\left. \times \int_V d\mathbf{r}_1 \cdots d\mathbf{r}_N \exp(-\beta \mathcal{H}(\mathbf{p}_1, \ldots, \mathbf{p}_N, \mathbf{r}_1, \ldots, \mathbf{r}_N)) \right]$$
$$= \beta p \sum_{N_1=0}^{\infty} \cdots \sum_{N_n=0}^{\infty} \left[\left(\prod_{k=1}^{n} \frac{\exp(\beta \mu_k N_k)}{N_k! \Lambda_k^{3N_k}} \right) \int_0^{\infty} dV \exp(-\beta p V) \right.$$
$$\left. \times \int_V d\mathbf{r}_1 \cdots d\mathbf{r}_N \exp(-\beta \mathcal{V}(\mathbf{r}_1, \ldots, \mathbf{r}_N)) \right] \tag{9.62}$$

The integrals over the volume V above are over the size of the volume. This replaces the size of the volume V by the pressure p as control variable. There is a dependence of Ψ, ϵ_k and $\mathcal{H}$ on the shape of the volume, but this is not explicitly indicated.

From the nanocanonical partition function, we find as equation of state for the replica energy of a single replica

$$X^{\mathrm{NC}}(T,p,\boldsymbol{\mu}) = -k_{\mathrm{B}}T\ln\Psi(T,p,\boldsymbol{\mu}) \tag{9.63}$$

For the nanocanonical ensemble, the replica energy is equal to the subdivision potential $\varepsilon^{\mathrm{NC}}(T,p,\boldsymbol{\mu})$. By using Eq. (3.79), we find that S^{NC}, V^{NC}, $\mathbf{N}^{\mathrm{NC}}$ are given by

$$\begin{aligned}
S^{\mathrm{NC}}(T,p,\boldsymbol{\mu}) &= -\left(\frac{\partial X^{\mathrm{NC}}}{\partial T}\right)_{p,\boldsymbol{\mu}} = k_{\mathrm{B}}\left(\frac{\partial\left(T\ln\Psi(T,p,\boldsymbol{\mu})\right)}{\partial T}\right)_{p,\boldsymbol{\mu}}\\
&= k_{\mathrm{B}}\Psi(T,p,\boldsymbol{\mu}) + k_{\mathrm{B}}T\left(\frac{\partial\ln\Psi(T,p,\boldsymbol{\mu})}{\partial T}\right)_{p,\boldsymbol{\mu}}\\
V^{\mathrm{NC}}(T,p,\boldsymbol{\mu}) &= \left(\frac{\partial X^{\mathrm{NC}}}{\partial p}\right)_{T,\boldsymbol{\mu}} = -k_{\mathrm{B}}T\left(\frac{\partial\ln\Psi(T,p,\boldsymbol{\mu})}{\partial p}\right)_{T,\boldsymbol{\mu}}\\
N_j^{\mathrm{NC}}(T,p,\boldsymbol{\mu}) &= -\left(\frac{\partial X^{\mathrm{NC}}}{\partial \mu_j}\right)_{T,p,\boldsymbol{\mu}_j} = k_{\mathrm{B}}T\left(\frac{\partial\ln\Psi(T,p,\boldsymbol{\mu})}{\partial \mu_j}\right)_{T,p,\boldsymbol{\mu}_j}
\end{aligned} \tag{9.64}$$

For a small system, the replica energy X^{NC} is unequal to zero. This equation then makes it possible to calculate properties that are proportional to the size of the system V^{NC} in the thermodynamic limit. In the thermodynamic limit, X^{NC} is equal to zero and these equations no longer make sense.

It follows from Eqs. (9.61) and (9.62) that the internal energy per replica is given by

$$U^{\mathrm{NC}}(T,V,\boldsymbol{\mu}) = \langle\mathcal{H}\rangle^{\mathrm{NC}} = -\left(\frac{\partial\ln\Psi(T,p,\boldsymbol{\mu})}{\partial\beta}\right)_{\beta p,\beta\boldsymbol{\mu}} \tag{9.65}$$

9.7 Concluding Remarks

In this chapter, we have shown that fluctuations in energy, volume and particle numbers contain all the thermodynamic information needed to describe a small system completely. There is a well-defined path provided by the statistical mechanical apparatus, and energy distribution functions, that gives the properties that we can also measure. The path is particular for each set of environmental variables. Our list of ensembles was not an exhaustive set.

Fluctuations in energies, volume and particle numbers are accessible, say, from molecular dynamics simulations, see Chapter 5. With the realistic particle interaction potentials that are available nowadays, we expect that the formulas can be used to generate a wide range of physical–chemical properties. This is further described in the chapters on the internal energy, the pressure and the enthalpy, as well as in Chapter 10 on Kirkwood–Buff integrals.

Chapter 10

Kirkwood–Buff Integrals

We define Kirkwood–Buff (KB) integrals for small systems; i.e., finite-volume KB integrals. On the nanoscale, these integrals obtain contributions from the surface, so they depend on the characteristic size and shape of the small system. The size-dependence can be used to find the properties of the small system as well as the properties in the thermodynamic limit.

Kirkwood–Buff (KB)-integrals were formulated in 1951 [70] in order to characterize the thermodynamic behavior of the components of a mixture. The integrals were volume integrals over pair correlation functions minus one. Ben-Naim [71] gave all the thermodynamic mixture properties in terms of such integrals, for instance, solute and solvent thermodynamic factors, chemical potentials, partial molar entropies and enthalpies. Knowledge of thermodynamic factors is needed for many purposes, i.e., to relate Fick's and Maxwell–Stefan's diffusion coefficients [24, 72], see Chapter 13. In this chapter, we examine the integrals from a small system perspective.

The original KB-integrals were given for systems in the thermodynamic limit. But attempts to compute the integrals from simulations were met with problems of convergence [33]. This made the integrals at best difficult to use, at worst, unreliable. Convergence problems can, however, be circumvented using the well-defined finite volume KB-integrals and their scaling properties. We define the integral for a small system directly following Krüger *et al.* [33], rather than following the original idea to define the integral for a large system. The small system method (SSM), cf. Chapter 5, can be applied to find both the thermodynamic limit value as well as the small system value. Both values are essential. We shall explain how to obtain them in a reliable way.

The solution theory of Ben-Naim [71] gives thermodynamic information of mixtures in terms of pair correlation functions for components i and j, $g_{ij}(\mathbf{r}_1, \mathbf{r}_2)$, see Chapter 8. When the chemical potentials are controlled, and $|\mathbf{r}_1 - \mathbf{r}_2|$ is large enough, correlations disappear and $g_{ij} = 1$. When the size of a system approaches the nanoscale, however, the correlation length in the mixture becomes comparable with the size of the system, and $g_{ij} \neq 1$ for all values of $\mathbf{r}_1$ and $\mathbf{r}_2$. Both positions are needed to compute g_{ij} correctly.

The pair correlation function $g_{ij}(\mathbf{r}_1, \mathbf{r}_2)$ depends not only on the volume V, but also on the shape of the volume. For systems in which the number of particles of a component is controlled, $g_{ij}(\mathbf{r}_1, \mathbf{r}_2) - 1$ is proportional to one divided by the number of particles for large distances [73]. This is the origin of the convergence problem of the volume integral. We shall avoid such problems, focusing on the small system properties of the finite-volume KB-integral [33].

10.1 Finite-Volume KB-Integrals

Characteristic to small systems is that the pair correlation function depends on the position coordinates of both particles in a mixture, here $\mathbf{r}_1$ and $\mathbf{r}_2$, and not only on their difference. The difference between two positions is most often not sufficient to describe system energies. We therefore define the finite-volume KB-integral as follows:

$$G_{ij}^V \equiv \frac{1}{V c_i c_j} \int_V \int_V c_i(\mathbf{r}_1) c_j(\mathbf{r}_2)[g_{ij}(\mathbf{r}_1, \mathbf{r}_2) - 1] d\mathbf{r}_1 d\mathbf{r}_2 \qquad (10.1)$$

in agreement with Krüger *et al.* [33]. Here, $c_i(\mathbf{r}_1)$ is the average density of component i in $\mathbf{r}_1$, and $c_i \equiv \frac{1}{V}\int_V c_i(\mathbf{r}) d\mathbf{r} = \langle N_i \rangle / V$ is the average density of component i in V. The brackets $\langle \cdots \rangle$ indicate an average over the ensemble. Equation 10.1 applies to any system of volume, V, of arbitrary size and shape. It also applies to any ensemble. It is in this context completely general. We shall therefore always start with Eq. (10.1) and define the KB-integral directly for the finite volume in question, well aware of the fact that the pair correlation function in general differs from unity. The finite-volume KB-integral will not only depend on the size and shape of the actual volume, but also on the ensemble.

In the next section we give an expression for G_{ij}^V in terms of fluctuations in the number of particles in V. This expression circumvents the use of the pair correlation functions and provides a convenient direct route to G_{ij}^V, and also its value in the thermodynamic limit. After this, we will clarify the traditional method and the use of the pair correlation function for finite volumes.

10.2 Finite-Volume KB-Integrals from Fluctuations

The KB-integrals have a statistical basis, which is the same regardless of the size and shape of the volume. This provides us with a direct path between any small system property and its KB-integrals. It remains to use this to connect all thermodynamic properties to KB-integrals. The route can be explained as follows.

The finite-volume KB-integral is first obtained from the density fluctuations as a function of $(T, V, \boldsymbol{\mu}, \text{shape})$. In the grand canonical ensemble [70], we obtain

$$G_{ij}^V(T, V, \boldsymbol{\mu}, \text{shape}) = V\frac{\langle N_i N_j\rangle - \langle N_i\rangle \langle N_j\rangle}{\langle N_i\rangle \langle N_j\rangle} - \frac{\delta_{ij}}{c_i^{\text{GC}}} \tag{10.2}$$

where $c_i^{\text{GC}}(T, V, \boldsymbol{\mu}, \text{shape}) = N_i^{\text{GC}}(T, V, \boldsymbol{\mu}, \text{shape})/V \equiv \langle N_i\rangle /V$ is the average number density of component i. The brackets $\langle \cdots \rangle$ denote the ensemble average in the grand canonical ensemble. Equation (10.2) as well as its proof for the grand canonical ensemble are the same for small and large volumes. Equation (10.2) and further equations in this section are completely general. In particular, they do not include any assumptions about the pair correlation function and the density distributions in V. From now on, we restrict most of the discussion to systems controlled in the grand canonical ensemble.

Remark 10.1. In order to compute a KB-integral in a system of a finite volume, it is most appropriate to use Eq. (10.2). To compute the pair correlation functions to obtain them serves no purpose.

Following Kirkwood and Buff, we introduce the auxiliary matrix, B_{ij}^{GC}, for particle number correlation densities. In the grand canonical ensemble, it is defined by

$$\begin{aligned} B_{ij}^{\mathrm{GC}}(T,V,\boldsymbol{\mu},\mathrm{shape}) &\equiv \frac{1}{V}\left[\langle N_i N_j\rangle - \langle N_i\rangle\langle N_j\rangle\right] \\ &= c_i^{\mathrm{GC}} c_j^{\mathrm{GC}} G_{ij}^{V}(T,V,\boldsymbol{\mu},\mathrm{shape}) + c_i^{\mathrm{GC}}\delta_{ij} \end{aligned} \tag{10.3}$$

This results in

$$B_{ij}^{\mathrm{GC}}(T,V,\boldsymbol{\mu},\mathrm{shape}) = \frac{k_{\mathrm{B}}T}{V}\frac{\partial\langle N_i\rangle}{\partial\mu_j} = k_{\mathrm{B}}T\frac{\partial c_i^{\mathrm{GC}}}{\partial\mu_j} \tag{10.4}$$

All elements B_{ij}^{GC} are functions of $(T,V,\boldsymbol{\mu},\mathrm{shape})$. Both matrices G_{ij}^{V} and B_{ij}^{GC} are symmetric. The equation applies to small systems which are not isotropic. Kirkwood and Buff [70] also defined the auxiliary matrix A_{ij}^{GC}. For environmental variables $\left(T,V,\mathbf{N}^{\mathrm{GC}},\mathrm{shape}\right)$, we deviate from Kirkwood and Buff's definition, adding a factor V in the definition of $\mathbf{A}^{\mathrm{GC}}$ and dividing by V in the definition of $\mathbf{B}^{\mathrm{GC}}$. This gives

$$A_{ij}^{\mathrm{GC}}\left(T,V,\mathbf{N}^{\mathrm{GC}},\mathrm{shape}\right) = \beta V\left(\frac{\partial\mu_j}{\partial N_i^{\mathrm{GC}}}\right)_{T,V,\mathbf{N}_i^{\mathrm{GC}}} = \beta\left(\frac{\partial\mu_j}{\partial c_i^{\mathrm{GC}}}\right)_{T,V,\mathbf{c}_i^{\mathrm{GC}}} \tag{10.5}$$

where

$$\begin{aligned} \mathbf{N}_i^{\mathrm{GC}} &\equiv (N_1,\ldots,N_{i-1},N_{i+1},\ldots,N_k)^{\mathrm{GC}} \\ &\equiv V\left(c_1,\ldots,c_{i-1},c_{i+1},\ldots,c_k\right)^{\mathrm{GC}} \equiv V\mathbf{c}_i^{\mathrm{GC}} \end{aligned} \tag{10.6}$$

The **A** and **B** matrices are each other's inverse, meaning that

$$\sum_j B_{ij}^{\mathrm{GC}} A_{jk}^{\mathrm{GC}} = \delta_{ik} \tag{10.7}$$

From the symmetry of $\mathbf{B}^{\mathrm{GC}}$, we conclude that $\mathbf{A}^{\mathrm{GC}}$ is also symmetric. The matrix of thermodynamic factors $\boldsymbol{\Gamma}^{\mathrm{GC}}$ can now be obtained from

$\mathbf{A}^{\mathrm{GC}}$ by

$$\Gamma_{ij}^{\mathrm{GC}}\left(T,V,\mathbf{N}^{\mathrm{GC}},\mathrm{shape}\right) \equiv c_i^{\mathrm{GC}} A_{ij}^{\mathrm{GC}}\left(T,V,\mathbf{N}^{\mathrm{GC}},\mathrm{shape}\right)$$
$$= \beta N_i^{\mathrm{GC}}\left(\frac{\partial \mu_j}{\partial N_i^{\mathrm{GC}}}\right)_{T,V,\mathbf{N}_i^{\mathrm{GC}}} = \beta c_i^{\mathrm{GC}}\left(\frac{\partial \mu_j}{\partial c_i^{\mathrm{GC}}}\right)_{T,V,\mathbf{c}_i^{\mathrm{GC}}} \tag{10.8}$$

We have thus obtained the thermodynamic factors as functions of $(T, V, \mathbf{N}^{\mathrm{GC}}$, shape) in a few steps directly from fluctuations. Expressions for one- and two-component small systems are given as remarks.

Remark 10.2. The formulas above can be written for a one-component system [71] as

$$B_{11} = c_1^2 G_{11} + c_1 \tag{10.9}$$

$$\kappa_T = \frac{1 + c_1 G_{11}}{k_{\mathrm{B}} T c_1} \tag{10.10}$$

$$V_1 = \left(\frac{\partial V}{\partial N_1}\right)_T = 1/c_1 \tag{10.11}$$

$$\left(\frac{\partial \mu_1}{\partial N_1}\right)_{T,V} = \frac{k_{\mathrm{B}} T}{N_1\left(1 + c_1 G_{11}\right)} \tag{10.12}$$

Here, κ_T is the isothermal compressibility, V_1 is the molar volume and c_1 is the density. Ben-Naim gives these relations for the thermodynamic limit only. They are also valid for a finite volume.

Remark 10.3. For two-component systems, we introduce the expressions

$$\eta \equiv c_1 + c_2 + c_1 c_2\left(G_{11} + G_{22} - 2G_{12}\right) \tag{10.13}$$

$$\zeta \equiv 1 + c_1 G_{11} + c_2 G_{22} + c_1 c_2\left(G_{11} G_{22} - G_{12}^2\right) \tag{10.14}$$

In term of these quantities, we have for Eq. (10.3)

$$B_{ij} = c_i c_j G_{ij} + \delta_{ij} c_{ij} \tag{10.15}$$

and furthermore

$$\kappa_T = \frac{\zeta}{k_\mathrm{B} T \eta} \tag{10.16}$$

$$V_1 = \frac{1 + c_2 \left(G_{22} - G_{12}\right)}{\eta} \tag{10.17}$$

$$V_2 = \frac{1 + c_1 \left(G_{11} - G_{12}\right)}{\eta} \tag{10.18}$$

$$\mu_{11} = \frac{c_2 k_\mathrm{B} T}{N_1 \eta}, \quad \mu_{22} = \frac{c_1 k_\mathrm{B} T}{N_2 \eta}, \quad \mu_{12} = \mu_{21} = \frac{k_\mathrm{B} T}{V \eta} \tag{10.19}$$

Also these relations apply to a small system with a finite volume. Ben-Naim gives the inverse relation

$$G_{ij} = k_\mathrm{B} T \kappa_T - \frac{\delta_{ij}}{c_i} + c k_\mathrm{B} T \frac{\left(1 - c_i V_i\right)\left(1 - c_j V_j\right)}{c_i c_j \mu_{ij}} \tag{10.20}$$

where $c \equiv c_1 + c_2$ and

$$\mu_{ij} \equiv \left(\frac{\partial \mu_i}{\partial N_j}\right)_{p,T,N_i} \tag{10.21}$$

Thermodynamic properties can be thus computed from KB-integrals, but KB-integrals can also be computed from thermodynamic properties.

From the symmetry of the $\mathbf{A}^{\mathrm{GC}}$-matrix, it follows that $\Gamma_{ij}^{\mathrm{GC}} c_j^{\mathrm{GC}} = \Gamma_{ji}^{\mathrm{GC}} c_i^{\mathrm{GC}}$. The factors can be used to understand solute and solvent properties of small volumes of arbitrary shape, i.e., fluid mixtures in confinement. From these factors, the remaining properties are derived.

For convenience, we give the corresponding results for the microcanonical and the canonical ensembles:

$$\begin{aligned} \Gamma_{ij}^{\mathrm{MC}} \left(U, V, \mathbf{N}, \text{shape}\right) &\equiv c_i A_{ij}^{\mathrm{MC}} \left(U, V, \mathbf{N}, \text{shape}\right) \\ &= \frac{N_i}{k_\mathrm{B} T^{\mathrm{MC}}} \left(\frac{\partial \mu_j^{\mathrm{MC}}}{\partial N_i}\right)_{U,V,\mathbf{N}_i} = \frac{c_i}{k_\mathrm{B} T^{\mathrm{MC}}} \left(\frac{\partial \mu_j^{\mathrm{MC}}}{\partial c_i}\right)_{U,V,\mathbf{c}_i} \end{aligned} \tag{10.22}$$

and

$$\Gamma_{ij}^{\mathrm{C}}(T,V,\mathbf{N},\text{shape}) \equiv c_i A_{ij}^{\mathrm{C}}(T,V,\mathbf{N},\text{shape})$$
$$= \beta N_i \left(\frac{\partial \mu_j^{\mathrm{C}}}{\partial N_i}\right)_{T,V,\mathbf{N}_i} = \beta c_i \left(\frac{\partial \mu_j^{\mathrm{C}}}{\partial c_i}\right)_{T,V,\mathbf{c}_i} \tag{10.23}$$

Inversions of the $\mathbf{A}^{\mathrm{MC}}$ and the $\mathbf{A}^{\mathrm{C}}$ matrices give the $\mathbf{B}^{\mathrm{MC}}(U,V,\boldsymbol{\mu}^{\mathrm{MC}},\text{shape})$ and the $\mathbf{B}^{\mathrm{C}}\left(T,V,\boldsymbol{\mu}^{\mathrm{C}},\text{shape}\right)$ matrices. The above procedure is required, as these matrices are not directly obtainable from fluctuations in particle numbers, unlike in the grand canonical ensemble, cf. Eq. (10.3).

10.3 The Small System Method

We can now use the small system method (SSM) to find system properties at any scale, away from the thermodynamic limit. In order to do so, we expand the property in question in the inverse characteristic length $L \equiv \sqrt[3]{V}$ of the system, as explained in Section 5.1, Eq. (5.1).

From the particle number fluctuations, cf. Eq. (10.3), we obtain the matrix $\mathbf{B}^{\mathrm{GC}}$ of particle correlations, which is expanded as follows;

$$\mathbf{B}^{\mathrm{GC}}(T,V,\boldsymbol{\mu},\text{shape}) = \mathbf{B}^{\infty} + \mathbf{B}^{s}\frac{1}{L} + \mathbf{B}^{se}\frac{1}{L^2} + \mathbf{B}^{sc}\frac{1}{L^3} + \cdots \tag{10.24}$$

All expansion coefficients depend on $(T,V,\boldsymbol{\mu})$ and the shape of the volume as well. The number of terms needed in the expansion depends on the system considered, see Chapter 5. The reason to use the expansion of $\mathbf{B}^{\mathrm{GC}}$ in $1/L$ is that the particle number fluctuations are additive. This improves the convergence in the thermodynamic limit. The thermodynamic limit-values of the KB-integrals G_{ij}^{∞} and the B_{ij}^{∞} are obtained from this expansion in $1/L$. They depend on the environmental variables $(T,V,\boldsymbol{\mu})$, but no longer on the shape. The $\mathbf{A}^{\infty}$- matrix and the corresponding thermodynamic factors $\boldsymbol{\Gamma}^{\infty}$ are found by inversion of the $\mathbf{B}^{\infty}$ matrix. Kirkwood and Buff [70] explained how we can find all thermodynamic properties of a mixture in the large system limit. In this limit (only), we can change from one to another set of environmental variables using Legendre transforms.

If we prefer to expand in the inverse characteristic length $L_\Omega \equiv V/\Omega$, we obtain, similar to Eq. (5.2)

$$\mathbf{B}^{\mathrm{GC}}(T, V, \boldsymbol{\mu}, \text{shape}) = \mathbf{B}^{\infty} + \mathbf{B}^{s}_{\Omega}\frac{1}{L_\Omega} + \mathbf{B}^{se}_{\Omega}\frac{1}{L_\Omega^2} + \mathbf{B}^{sc}_{\Omega}\frac{1}{L_\Omega^3} + \cdots \tag{10.25}$$

In this case, both $\mathbf{B}^{\infty}$ and $\mathbf{B}^{s}_{\Omega}$ are independent of the shape of the small system. $\mathbf{B}^{s}_{\Omega}$ is therefore a property of the planar surface. The expansion matrices are related by

$$\mathbf{B}^{s} = \mathbf{B}^{s}_{\Omega}\frac{L}{L_\Omega} = \mathbf{B}^{s}_{\Omega}\frac{\Omega}{V^{2/3}}, \quad \mathbf{B}^{se} = \mathbf{B}^{se}_{\Omega}\left(\frac{L}{L_\Omega}\right)^2 = \mathbf{B}^{se}_{\Omega}\left(\frac{\Omega}{V^{2/3}}\right)^2$$
$$\mathbf{B}^{sc} = \mathbf{B}^{sc}_{\Omega}\left(\frac{L}{L_\Omega}\right)^3 = \mathbf{B}^{sc}_{\Omega}\left(\frac{\Omega}{V^{2/3}}\right)^3 \tag{10.26}$$

The terms in the expansion, the matrices $\mathbf{B}^{s}, \mathbf{B}^{se}, \ldots$ can be obtained by fitting simulation results to an $1/L$ expansion. These matrices not only depend on $(T, V, \boldsymbol{\mu})$ but also on the shape. Similarly the matrices $\mathbf{B}^{s}_{\Omega}, \mathbf{B}^{se}_{\Omega}, \ldots$ can be obtained from fitting to an $1/L_\Omega$ expansion. Now $\mathbf{B}^{s}_{\Omega}$ does not depend on the shape. The expansion gives thermodynamic information on the small system.

10.4 The Traditional Method

How do we compare the expression for the finite volume to the expression for the system in the thermodynamic limit? Unlike the original definition [70] of Eq. (10.27), the definition 10.1 does not use the large volume limit. Through $g_{ij}(\mathbf{r}_1, \mathbf{r}_2)$, the integral G^V_{ij} depends not only on the volume V but also on the shape of the volume. This is not explicitly indicated in Eq. (10.1), but is implicitly assumed. Definition (10.1) for finite volumes can be shown to contain the original KB-integrals under certain conditions. When the pair correlation function approaches 1 for large distances and depends only on $r_{12} \equiv |\mathbf{r}_1 - \mathbf{r}_2|$, the double integral can be separated into two integrals. One integral can be performed over $r_{12} \equiv |\mathbf{r}_1 - \mathbf{r}_2|$, while the other one results in V. Equation (10.1) reduces then to

$$G_{ij} = 4\pi \int_0^\infty [g_{ij}(r) - 1]r^2 dr \tag{10.27}$$

This definition applies only to infinitely large, homogeneous systems. It is also for infinitely large systems only that the pair correlation functions $g_{ij}(\mathbf{r}_1, \mathbf{r}_2)$ can be reduced for all ensembles to $g_{ij}(r_{12})$, and makes the distance between particles into the sole position variable, $r = r_{12} = |\mathbf{r}_1 - \mathbf{r}_2|$. For ensembles, in which the number of particles is controlled, $g_{ij}(r) - 1$ is proportional to one divided by the number of particles for large values of r [73]. The ratio becomes negligible only for infinitely large systems.

A weight function for size and shape-dependence

In order to be able to integrate Eq. (10.1), we now make the assumption that the system is isotropic and homogeneous, so that the pair correlation function only depends on the distance r_{12}, and $c_i(\mathbf{r}_1) = c_i$. This enables us to write

$$\begin{aligned} G_{ij}^V &= \frac{1}{V} \int_V \int_V [g_{ij}(r_{12}) - 1]\, d\mathbf{r}_1 d\mathbf{r}_2 \\ &= \frac{1}{V} \int_0^\infty \delta(r - r_{12}) \int_V \int_V [g_{ij}(r_{12}) - 1]\, d\mathbf{r}_1 d\mathbf{r}_2 dr \\ &= \int_0^\infty \left[\frac{1}{V} \int_V \int_V \delta(r - r_{12}) d\mathbf{r}_1 d\mathbf{r}_2 \right] [g_{ij}(r) - 1]\, dr \end{aligned} \tag{10.28}$$

Equation (10.28) can now be written in the form

$$G_{ij}^V = \int_0^\infty w(r; V, \text{shape})[g_{ij}(r) - 1]dr \tag{10.29}$$

where the weight function

$$w(r; V, \text{shape}) \equiv \frac{1}{V} \int_V \int_V \delta\,(r - r_{12})\, d\mathbf{r}_1 d\mathbf{r}_2 \tag{10.30}$$

has been introduced to take care of the size- and shape-dependence.

Remark 10.4. When applied to systems of finite volumes, it has been common practice to compute the integral until a distance $2R$, or to approximate the KB-integral (10.27) by

$$\widetilde{G}_{ij}^{R} \equiv 4\pi \int_0^{2R} [g_{ij}(r) - 1] r^2 dr \tag{10.31}$$

where the upper limit $2R$ is the largest distance between two particles in the volume, and this distance is large compared to the correlation length. For a fluid away from the critical point, this length is typically of the order of the particle diameter. The use of larger and larger volumes to increase R, in order to obtain thermodynamic limit values of Eq. (10.27), still leads to convergence problems [33, 71].

The w has the dimensionality of a surface area. Krüger *et al.* [33][a] gave analytical expressions for spheres, see Table 10.1. Results for large systems can be obtained by setting $x \equiv r/2R = 0$. Krüger and Vlugt [34, 35] gave also the weight function for a cube with rib $L = \sqrt[3]{V}$:

$$w(r; V, \text{cube})$$
$$= \begin{cases} 4\pi r^2 \left(1 - \frac{3x}{2} + \frac{2x^2}{\pi} - \frac{x^3}{4\pi}\right) & \text{for } x < 1 \\ r^2 \left[-8\pi + 6x + 2x^3 + \frac{6\pi - 1}{x} \right. & \\ \quad \left. + 24x \arccos \frac{1}{x} + 8\left(2x^2 + 1\right)\sqrt{1 - 1/x^2}\right] & \text{for } 1 < x < \sqrt{2} \\ \simeq 0 & \text{for } \sqrt{2} < x \end{cases} \tag{10.32}$$

where $x \equiv r/L$. The fact that the weight function is negligible for $\sqrt{2} < x < \sqrt{3}$ was verified numerically. For $x > \sqrt{3}$, it is exactly zero.[b]

The integrals of $g_{ij}(r) - 1$ for the binary mixture example with the Weeks–Chandler–Anderson potential, given in Section 5.6, can be

[a]The expression for the one-dimensional case in [33] misses the factor 2, which is due to the fact that there are two endpoints.
[b]Krüger and Vlugt [34] used $x = r/(L\sqrt{3})$.

Table 10.1. The weight function $w(x)$ for $d = 1, 2, 3$, and $x \equiv r/2R$, and $w(x)$ equals 0 for $x > 1$.

d	$w(x)$
1	$2\,(1-x)$
2	$4r\left[\arccos(x) - x\left(1-x^2\right)^{1/2}\right]$
3	$2\pi r^2\left(1 - 3x/2 + x^3/2\right)$

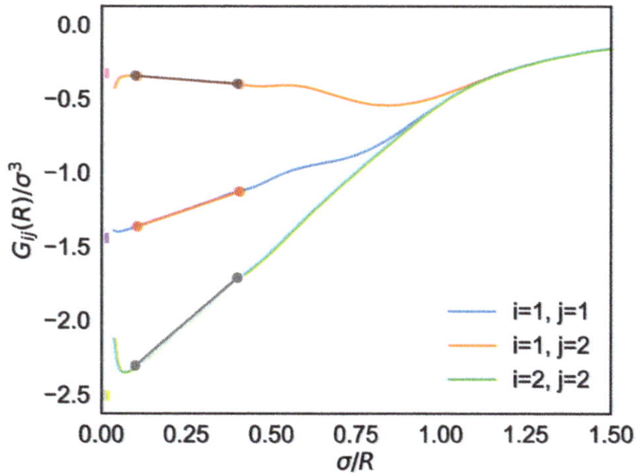

Figure 10.1. The integral over the radial distribution function, G_{ij}^V, as a function of σ/R. By extrapolation to the thermodynamic limit, we find the limit-values indicated by dots on the y-axis. We have used molecular dynamics simulations to sample the radial-distribution functions using the OCTP module [74]. The functions were corrected for finite-size effects using the method described by van der Vegt *et al.* [43, 44].

found in Fig. 10.1. The numerical values of $g_{ij}(r)$ were sampled throughout the simulation, using the OCTP-module of LAMMPS [74]. The pair correlation functions were corrected for finite-size effects using the method of van der Vegt *et al.* [43, 44]. The integration was done using the integration method implemented in the *pykbi* python module [35].

10.5 The Thermodynamic Limit Value

Finite-volume KB-integrals describe properties for small systems exactly, for any size and shape. They can be obtained from molecular dynamics simulations.

Ben-Naim's solution theory concerns the thermodynamic limit-value of the KB-integral. These values can be obtained by extrapolating the linear part of the curve of the finite-volume KB-integrals. The extrapolated values are shown in Fig. 10.1 as points on the y-axis. The linear range of the G_{ij}'s is contained within two demarcation points on each curve in this figure.

The convergence of the computed KB-integrals in this figure was improved by correcting the pair correlation function directly for finite-size effects. The method was introduced by van der Vegt *et al.* [43, 44]. While an improved convergence is beneficial, the resulting finite-volume KB-integral no longer refers to the smaller volumes. Finite-size corrections of pair correlation functions have also been proposed, cf. Cortes-Huerto *et al.* [8] and Heidari *et al.* [75]. For a discussion on how to best extrapolate the value to the thermodynamic limit, see Dawass *et al.* [76].

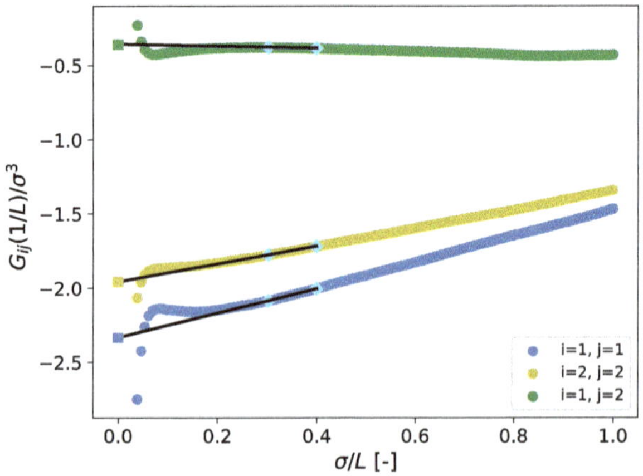

Figure 10.2. Extrapolation of the small system fluctuations to the thermodynamic limit, as a function of $1/L$, with L being the radius of the small embedded system (see Chapter 5).

10.6 Concluding Remarks

We have proposed to replace the integrals originally introduced by Kirkwood and Buff [70] by so-called finite-volume KB-integrals [33]. From the finite-volume KB-integrals, we can extract thermodynamic limit values, as well as small system property values, using the small system method (SSM).

By using the KB-integral for a finite volume, we avoid a well-known integral convergence problem in the literature [71]. For small systems, the most convenient way to obtain the KB-integrals is to use the number fluctuations, cf. Eq. (10.3). KB-integrals for finite, small volumes, Eq. (10.1), contain integrals over the pair correlation functions minus 1, which depend on two vectorial positions, $\mathbf{r}_1, \mathbf{r}_2$. This dependence is typical for small systems. It has not yet been systematically investigated. As the KB-integrals for finite volumes can be found using the number fluctuations, the use of $g_{ij}(\mathbf{r}_1, \mathbf{r}_2)$ seems unpractical. In the case that pair correlation functions only depend on the distance, $r_{12} = |\mathbf{r}_1 - \mathbf{r}_2|$, we find the expression given by Eq. (10.29), which has now been used successfully by several groups [6–9, 31, 32, 68, 72, 76, 77], see also Chapter 13.

Chapter 11

Pressures and Partial Molar Enthalpies

A small system with environmental variables T, V, $\boldsymbol{\mu}$ in the grand canonical ensemble has two pressures: the integral pressure and the differential pressure. It follows that the small system also has two enthalpies. We show how the Small System Method can be used to determine partial molar enthalpies. An integral relation is presented for the differential pressure in terms of the small system pair correlation function.

Partial molar enthalpies are essential for description of mixtures. They can be used to reveal attractive or repulsive forces between the components, and are needed to describe mixing processes in industrial contexts [11]. The properties are therefore central in chemical engineering. Systems that are small, e.g., droplets, bubbles, nuclei, are also relevant. Small system thermodynamics may be needed to describe confined fluids, micro- and nanofluidics [78] or heterogeneous catalysis.

We shall see in this chapter how the small system method (SSM) of Chapter 5 can be used to determine partial molar enthalpies of components in a mixture. A binary mixture is discussed in detail as an example. The SSM provides data for the grand canonical ensemble. Other sets of environmental variables are more common in chemical engineering. We shall therefore explain how data sets in the thermodynamic limit can be transformed from one set into a set with other environmental control variables. The chapter follows Chapter 7 on the internal energy. In addition, the computation of the integral and the differential pressure becomes central.

The SSM applies to the grand canonical (GC) ensemble. The integral pressure, $\widehat{p}$, is a characteristic property for this ensemble, see Section 3.4. This pressure gives the differential pressure p^{GC} from

$$p^{\mathrm{GC}} \equiv \left[\frac{\partial\left(\widehat{p}V\right)}{\partial V}\right]_{T,\boldsymbol{\mu}} \tag{11.1}$$

The name "differential pressure" of a small system arises from this expression, cf. Chapter 3. Both pressures can be related to work done on the system. The expression $p^{\mathrm{GC}}dV$ is the work done by each replica when we change the volume of all replicas. The integral pressure, on the other hand, describes the work done by adding one replica of volume V.

The two pressures of the small system give rise to two definitions of the enthalpy of the GC ensemble

$$\begin{aligned} H^{\mathrm{GC}} &= U^{\mathrm{GC}} + p^{\mathrm{GC}}V \\ \widehat{H} &= U^{\mathrm{GC}} + \widehat{p}V \end{aligned} \tag{11.2}$$

where the classical expression is the differential enthalpy, while $\widehat{H}$ is called the integral enthalpy. Partial molar enthalpies can be defined with both starting points. The pressures are related by Eq. (11.1). For the differential pressure of a small system, we find an integral relation (Section 11.2).

11.1 The Integral Pressure and the Integral Enthalpy Contributions

The formulas for the enthalpies, Eq. (11.2), contain two terms on the right-hand side. The internal energy was given in Chapter 7. To the internal energy, we must add the term $p^{\mathrm{GC}}V$ or the term $\widehat{p}V$ in order to obtain the respective enthalpies. We proceed to explain how these terms are computed.

In the grand canonical ensemble, the integral pressure of a single replica is given by the equation of state, Eq. (4.56),

$$\widehat{p}(T, V, \boldsymbol{\mu})V = k_{\mathrm{B}}T \ln \Xi(T, V, \boldsymbol{\mu}) \tag{11.3}$$

This expression is correct not only for large systems but also for small systems [4]. The integral pressure $\widehat{p}$ is in general not equal to the differential pressure, see Eq. (11.1). Only in the thermodynamic limit is $\widehat{p}$ independent of the volume, and the two pressures are the same.

Equation (3.50) gives

$$N_i^{\mathrm{GC}} = \left(\frac{\partial \widehat{p}V}{\partial \mu_i}\right)_{T,V,\boldsymbol{\mu}_i} \tag{11.4}$$

where $\boldsymbol{\mu}_i \equiv (\mu_1, \ldots, \mu_{i-1}, \mu_{i+1}, \ldots, \mu_n)$. From Eq. (9.41), we know that

$$\left(\frac{\partial N_i^{\mathrm{GC}}}{\partial \mu_j}\right)_{T,V,\boldsymbol{\mu}_j} = \beta\left(\langle N_i N_j\rangle^{\mathrm{GC}} - N_i^{\mathrm{GC}} N_j^{\mathrm{GC}}\right) \tag{11.5}$$

The quantities $\widehat{p}V$, $N_i^{\mathrm{GC}} \equiv \langle N_i\rangle^{\mathrm{GC}}$ and $\partial N_i^{\mathrm{GC}}/\partial \mu_j$ are additive, and can thus be expanded in the inverse characteristic size of the small volume, cf. Chapter 5. The product $\widehat{p}V$ is minus the compressional energy of the system, as follows:

$$\begin{aligned}\left(\frac{\partial \widehat{p}V}{\partial N_i^{\mathrm{GC}}}\right)_{T,V,\boldsymbol{\mu}_i} &= \left(\frac{\partial \widehat{p}V}{\partial \mu_i}\right)_{T,V,\boldsymbol{\mu}_i}\left(\frac{\partial \mu_i}{\partial N_i^{\mathrm{GC}}}\right)_{T,V,\boldsymbol{\mu}_i} \\ &= N_i^{\mathrm{GC}}\left(\frac{\partial \mu_i}{\partial N_i^{\mathrm{GC}}}\right)_{T,V,\boldsymbol{\mu}_i} = \frac{k_{\mathrm{B}}T\, N_i^{\mathrm{GC}}}{\langle N_i^2\rangle^{\mathrm{GC}} - (N_i^{\mathrm{GC}})^2}\end{aligned} \tag{11.6}$$

where we used Eqs. (11.4) and (11.5). By also using Eq. (7.6), we find the partial molar integral enthalpy:

$$\left(\frac{\partial \widehat{H}^{\mathrm{GC}}}{\partial N_i^{\mathrm{GC}}}\right)_{T,V,\boldsymbol{\mu}_i} = \frac{\langle \mathcal{H} N_i\rangle^{\mathrm{GC}} - U^{\mathrm{GC}} N_i^{\mathrm{GC}} + k_{\mathrm{B}}T\, N_i^{\mathrm{GC}}}{\langle N_i^2\rangle^{\mathrm{GC}} - (N_i^{\mathrm{GC}})^2} \tag{11.7}$$

where $\langle \mathcal{H}\rangle^{\mathrm{GC}} = U^{\mathrm{GC}}$. These expressions relate thermodynamic properties of a small system to fluctuations on the particle level. The numerator as well as the denominator are both additive in Eqs. (11.6) and (11.7). The word additive is used in the sense that bulk- (surface- or line-) properties can be added. We can accordingly make use of this property to find data for the thermodynamic limit. This will be done as follows:

1. We apply the SSM, cf. Chapter 5, to each additive term, $\langle \mathcal{H} N_i \rangle^{\rm GC} - U^{\rm GC} N_i^{\rm GC}$, $\langle N_i^2 \rangle^{\rm GC} - (N_i^{\rm GC})^2$ and $N_i^{\rm GC}$, separately.
2. We find the thermodynamic limit-value of each term of Eq. (11.7) in the GC ensemble by extrapolation.
3. From the set of limit-values, we find the partial molar properties for other ensembles using Legendre transforms. Formulas for transformation from the GC to the $T, p, \mathbf{N}$-ensemble in the thermodynamic limit are given in Appendix C.

11.2 Integral Relation for the Differential Pressure

For large systems, we can compute the pressure from an integral over the products of the moment of the inter-particle force and the pair correlation function. We show now that an analogous relation exists for the differential pressure of a small system. We restrict ourselves in this section to a one-component small system. The generalization to a mixture is straightforward.

Consider the grand canonical ensemble where the temperature, volume and chemical potential are environmental variables. The integral pressure is given by Eq. (11.3) where the grand canonical partition function is given by

$$\Xi(T, V, \mu) = \sum_{N=0}^{\infty} \frac{\exp(\beta \mu N)}{h^{3N} N!} \int d\mathbf{p}_1 \cdots d\mathbf{p}_N \int_V d\mathbf{r}_1 \cdots d\mathbf{r}_N \exp(-\beta \mathcal{H}) \tag{11.8}$$

We do not indicate the dependence of $\mathcal{H}$ on $\mathbf{p}_1, \ldots, \mathbf{p}_N, \mathbf{r}_1, \ldots, \mathbf{r}_N$ for ease of notation. After introducing Eq. (4.1), we integrate over the momenta and obtain for a single replica

$$\Xi(T, V, \mu) = \sum_{N=0}^{\infty} \frac{\exp(\beta \mu N)}{\Lambda^{3N} N!} \int_V d\mathbf{r}_1 \cdots d\mathbf{r}_N \exp(-\beta \mathcal{V}) \tag{11.9}$$

We do not show the dependence of $\mathcal{V}$ on $\mathbf{r}_1, \ldots, \mathbf{r}_N$ for ease of notation. The differential pressure is equal to

$$p^{\mathrm{GC}}(T,V,\mu) = \frac{\partial \widehat{p}(T,V,\mu)V}{\partial V} = \frac{k_B T}{\Xi(T,V,\mu)} \frac{\partial \Xi(T,V,\mu)}{\partial V}$$
$$= \frac{k_B T}{\Xi(T,V,\mu)} \sum_{N=0}^{\infty} \frac{\exp(\beta\mu N)}{\Lambda^{3N} N!} \frac{\partial}{\partial V} \int_V d\mathbf{r}_1 \cdots d\mathbf{r}_N \exp(-\beta\mathcal{V}) \tag{11.10}$$

Before differentiating the integral with respect to V, we make all coordinates dimensionless by dividing them by $L = V^{1/3}$. The subsequent differentiation then leads to

$$p^{\mathrm{GC}}(T,V,\mu) = \frac{k_B T}{\Xi(T,V,\mu)} \sum_{N=0}^{\infty} \frac{\exp(\beta\mu N)}{\Lambda^{3N} N!} \frac{N}{V}$$
$$\times \int_V d\mathbf{r}_1 \ldots d\mathbf{r}_N \left[1 - \frac{\beta}{3} \sum_{j=1}^{N} \mathbf{r}_j \cdot \frac{\partial \mathcal{V}}{\partial \mathbf{r}_j}\right] \exp(-\beta\mathcal{V})$$
$$= k_B T c^{\mathrm{GC}} - \frac{1}{3\Xi(T,V,\mu)V} \sum_{N=0}^{\infty} \frac{\exp(\beta\mu N)}{\Lambda^{3N} N!} N$$
$$\times \int_V d\mathbf{r}_1 \ldots d\mathbf{r}_N \mathbf{r}_1 \cdot \frac{\partial \mathcal{V}}{\partial \mathbf{r}_1} \exp(-\beta\mathcal{V}) \tag{11.11}$$

We used the fact that the same integral is obtained by interchanging particles. Furthermore, $c^{\mathrm{GC}} \equiv N^{\mathrm{GC}}/V = \langle N \rangle^{\mathrm{GC}}/V$. The potential is a sum of pair potentials $\phi(\mathbf{r}_i - \mathbf{r}_j)$, see Eq. (4.2). We use a pair potential which only depends on the distance $r_{ij} = |\mathbf{r}_i - \mathbf{r}_j|$, and write the pressure as

$$p^{\mathrm{GC}}(T,V,\mu) = k_B T c^{\mathrm{GC}} - \frac{1}{3\Xi(T,V,\mu)V} \sum_{N=0}^{\infty} \frac{\exp(\beta\mu N)}{\Lambda^{3N} N!} N(N-1)$$
$$\times \int_V d\mathbf{r}_1 \ldots d\mathbf{r}_N r_{12} \frac{\partial \phi(r_{12})}{\partial r_{12}} \exp(-\beta\mathcal{V}) \tag{11.12}$$

The pair correlation function for an isotropic small system in the grand canonical ensemble is given by, cf. Eq. (8.12),

$$g_2(\mathbf{r}_1, \mathbf{r}_2; T, V, \mu) = \frac{V^2}{\Xi(T,V,\mu)N^{\mathrm{GC}}(N^{\mathrm{GC}}-1)} \sum_{N=0}^{\infty} \frac{\exp(\beta\mu N)}{\Lambda^{3N} N!} N(N-1)$$
$$\times \int_V d\mathbf{r}_3 \dots d\mathbf{r}_N \exp(-\beta\mathcal{V}) \tag{11.13}$$

This gives the integral relation for the differential pressure.

$$p^{\mathrm{GC}}(T,V,\mu) = k_B T c^{\mathrm{GC}} - \frac{c^{\mathrm{GC}}(c^{\mathrm{GC}} - 1/V)}{3V}$$
$$\times \int_V d\mathbf{r}_1 d\mathbf{r}_2 r_{12} \frac{\partial\phi\,(r_{12})}{\partial r_{12}} g_2(\mathbf{r}_1, \mathbf{r}_2; T, V, \mu) \tag{11.14}$$

In the thermodynamic limit, the pair correlation function only depends on the distance $r_{12} = |\mathbf{r}_1 - \mathbf{r}_2|$. Equation (11.14) then reduces to the familiar one.

11.3 Partial Molar Enthalpies

The SSM of Chapter 5 can be used to find system extensive properties in the thermodynamic limit. The simulation method yields data for the grand canonical ensemble, where the chemical potential, μ, temperature, T, and volume, V, are controlled. A transformation of data from one ensemble to another is often needed. Here, we demonstrate the transformation of partial molar enthalpies from the grand canonical ensemble to the more practical isothermal–isobaric ensemble for a binary mixture in the thermodynamic limit. The procedure is applied to a simple reaction in Chapter 14, see also Schnell *et al.* [47] and Skorpa *et al.* [46]. In order to illustrate the sampling method and the consecutive transforms, we consider a mixture of WCA-particles, cf. Appendix D.

The total differential of $H = H(p, T, N_i)$ for the large system is

$$dH = \left(\frac{\partial H}{\partial T}\right)_{p,N_1,N_2} dT + \left(\frac{\partial H}{\partial p}\right)_{T,N_1,N_2} dp$$
$$+ \left(\frac{\partial H}{\partial N_1}\right)_{T,p,N_2} dN_1 + \left(\frac{\partial H}{\partial N_2}\right)_{T,p,N_1} dN_2 \quad (11.15)$$

This enthalpy is an Euler homogeneous function of the first order.

$$H = H_1 N_1 + H_2 N_2 \quad (11.16)$$

At constant p, T we find the partial molar enthalpies. They are defined by

$$H_1 = \left(\frac{\partial H}{\partial N_1}\right)_{T,p,N_2} \quad (11.17)$$

$$H_2 = \left(\frac{\partial H}{\partial N_2}\right)_{T,p,N_1} \quad (11.18)$$

The name *partial molar enthalpy* of component i is used for derivatives with respect to N_i, at constant conditions; here constant p, T (isothermal–isobaric ensemble).

The partial molar enthalpy is obtained from SSM by extrapolating a small system value to the thermodynamic limit, see Chapter 5. The variables controlled in SSM are V, T and μ_j, i.e., we are in the grand canonical ensemble. The total differential of the enthalpy in the thermodynamic limit is then

$$dH = \left(\frac{\partial H}{\partial T}\right)_{V,\mu_1,\mu_2} dT + \left(\frac{\partial H}{\partial V}\right)_{T,\mu_1,\mu_2} dV$$
$$+ \left(\frac{\partial H}{\partial \mu_1}\right)_{V,T,\mu_2} d\mu_1 + \left(\frac{\partial H}{\partial \mu_2}\right)_{V,T,\mu_1} d\mu_2 \quad (11.19)$$

The problem now is to convert the partial differentials of Eq. (11.19) into the partial molar enthalpies defined in Eq. (11.15). The conversion is done in several steps, where the controlled variables are changed in each step:

$$\left(\frac{\partial H}{\partial N_i}\right)_{T,V,\mu_j} \Rightarrow \left(\frac{\partial H}{\partial N_i}\right)_{T,V,N_j} \Rightarrow \left(\frac{\partial H}{\partial N_i}\right)_{T,P,N_j} \tag{11.20}$$

Transformations of properties from one ensemble to another can only be carried out in the thermodynamic limit. There is no simple relation between thermodynamic quantities of different ensembles when the system is small [4], see Section 4.9 and Chapter 21 on Legendre-Fenchel transforms. The partial molar enthalpies, the derivatives of Eq. (11.15), can be found by a series of steps. We shall illustrate the procedure for a mixture of WCA particles.

11.4 Transformation in the Thermodynamic Limit

As explained above, there is more practical interest in $(\partial H/\partial N_i)_{T,P,N_j}$ than in $(\partial H/\partial N_i)_{T,V,\mu j}$. Here, we show the final equations for property transformation between ensembles. The full derivation is shown in Appendix C, see also Kirkwood and Buff (see Eqs. (20) and (21)) [70], Liu *et al.* [68] (see Eqs. (19)–(21)) and Chapter 4 (and Appendix A) of Ben-Naim's book [71].

The starting point for the transformations is the total differential of the enthalpy:

$$dH = \left(\frac{\partial H}{\partial T}\right)_{V,N_1,N_2} dT + \left(\frac{\partial H}{\partial V}\right)_{T,N_1,N_2} dV + \left(\frac{\partial H}{\partial N_1}\right)_{T,V,N_2} dN_1 + \left(\frac{\partial H}{\partial N_2}\right)_{T,V,N_1} dN_2 \tag{11.21}$$

As explained in Appendix C, the partial molar enthalpy in the canonical ensemble is

$$\left(\frac{\partial H}{\partial N_1}\right)_{T,V,N_2} = B_{22}\left[A_{22}\left(\frac{\partial H}{\partial N_1}\right)_{T,V,\mu_2} + A_{12}\left(\frac{\partial H}{\partial N_2}\right)_{T,V,\mu_1}\right] \tag{11.22}$$

$$\left(\frac{\partial H}{\partial N_2}\right)_{T,V,N_1} = B_{11}\left[A_{11}\left(\frac{\partial H}{\partial N_2}\right)_{T,V,\mu_1} + A_{12}\left(\frac{\partial H}{\partial N_1}\right)_{T,V,\mu_2}\right] \tag{11.23}$$

The matrices A and B follow from Kirkwood and Buff [70], or more transparently from Ben-Naim's [71] Eqs. (4.23) and (4.24):

$$B_{12} = B_{21} = \frac{k_B T}{V}\left(\frac{\partial N_2}{\partial \mu_1}\right)_{T,V,\mu_2} = k_B T\left(\frac{\partial c_2}{\partial \mu_1}\right)_{T,V,\mu_2}$$
$$= c_1 c_2 G_{12} + c_1 \delta_{12} = \frac{\langle N_1 N_2\rangle - N_1 N_2}{V} \tag{11.24}$$

where G_{ij} are Kirkwood–Buff integrals, see Chapter 10. Furthermore,

$$A_{12} = A_{21} = \beta V\left(\frac{\partial \mu_2}{\partial N_1}\right)_{T,V,N_2} = \beta\left(\frac{\partial \mu_2}{\partial c_1}\right)_{T,V,c_2} \tag{11.25}$$

The B- and the A-matrices are each other's inverse:

$$\sum_j B_{ij} A_{jk} = \delta_{ik} \tag{11.26}$$

The final step to obtain partial molar enthalpies of the binary system is

$$V\left(\frac{\partial H}{\partial N_1}\right)_{T,P,N_2} = V_1 H + N_2\left[V_2\left(\frac{\partial H}{\partial N_1}\right)_{T,V,N_2} - V_1\left(\frac{\partial H}{\partial N_2}\right)_{T,V,N_1}\right] \tag{11.27}$$

$$V\left(\frac{\partial H}{\partial N_2}\right)_{T,P,N_1} = V_2 H + N_1\left[V_1\left(\frac{\partial H}{\partial N_2}\right)_{T,V,N_1} - V_2\left(\frac{\partial H}{\partial N_1}\right)_{T,V,N_2}\right] \tag{11.28}$$

where V_1 and V_2 are the partial molar volumes. These can be calculated using Kirkwood–Buff integrals [71], or equivalently using Eq. (14.14). The partial molar enthalpies can thus be calculated from Eqs. (11.27) and (11.28) when we know the values of H, V, V_1, V_2, N_1, N_2. We demonstrate such calculations for a mixture in what follows.

So far, we have seen above how we can compute the partial molar enthalpy from the partial enthalpy at constant V, T, μ_j. The transformation applies only to the thermodynamic limit. The partial quantity that we obtain for the small system, is *not* the normal partial molar enthalpy.

11.5 Determining Partial Molar Enthalpies

The complete procedure can now be applied to the chosen example. In the first step, we obtain the partial molar internal energies $U_i = (\partial U/\partial N_i)_{V,T,\mu_j}$ from SSM in the thermodynamic limit [47]. In the following steps, we transform the results to the conditions of interest:

Step I: SSM-determination of $(\partial U/\partial N_i)_{T,V,\mu_j}$

We obtain the fluctuation results for $(\langle \mathcal{H} N_i \rangle - UN_i)/V$ and $(\langle N_i^2 \rangle - N_i^2)/V$ as a function of the small system size. With the SSM, these values are extrapolated to the thermodynamic limit, see Chapter 5. In Fig. 11.1, we show how well the extrapolation works when we compare results for a binary mixture over various compositions. Results of simulations for large systems in the grand canonical ensemble are matched with these data-points.

Step II: Intermediate results: Matrices A_{ij} and B_{ij}

In order to be able to do the necessary transforms, we need the matrix elements A_{ij} and B_{ij}. In Fig. 11.2, we show values of G_{ij}, A_{ij}

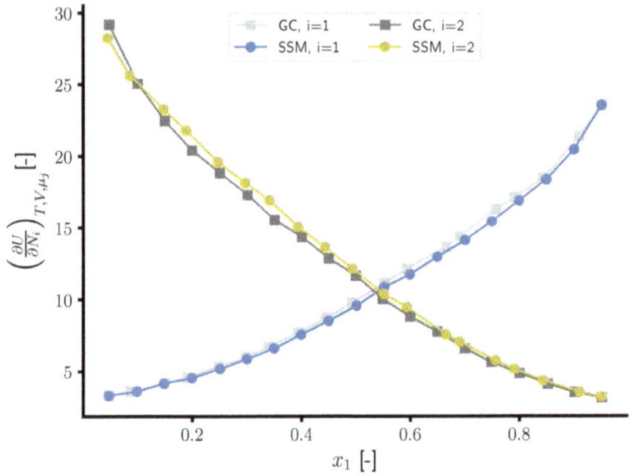

Figure 11.1. The partial molar internal energy in the grand canonical ensemble. Values labeled SSM are obtained by extrapolating Eq. (7.6) to the thermodynamic limit using the Small System Method. Data-points labeled GC are obtained directly from the grand-canonical ensemble for comparison.

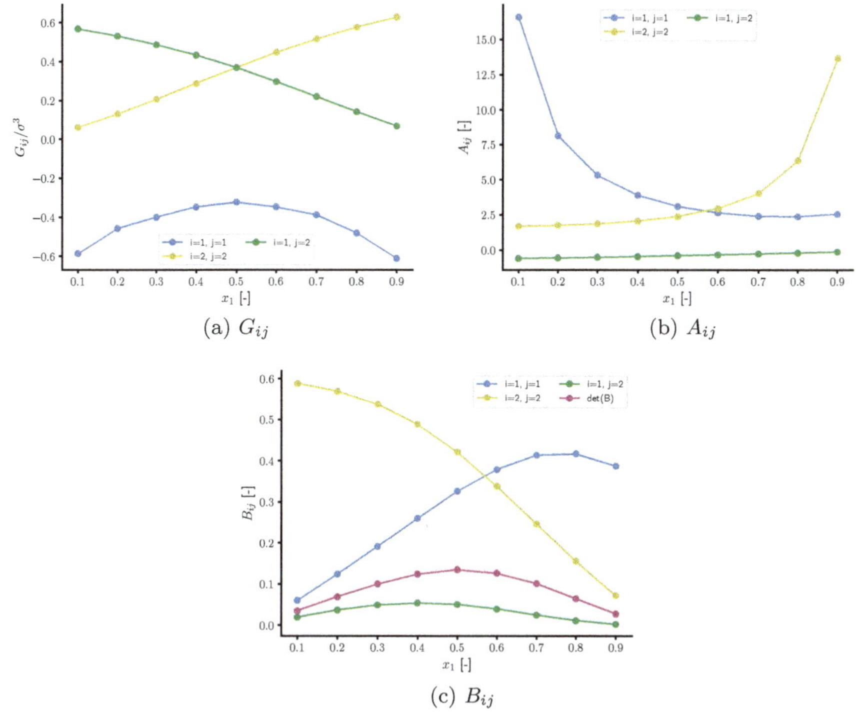

Figure 11.2. The values of (a) G_{ij}, (b) A_{ij} and (c) B_{ij} for the binary mixture of WCA particles described in the text.

and B_{ij} as well as the determinant of the matrix $\mathbf{B}$. The elements B_{ij} were determined from Eq. (11.24) and the restriction $\mathbf{AB} = \mathbf{I}$. In Fig. 11.2(a), we see how the determinant approaches zero as $x_1 \downarrow 0.1$. The small determinant, in addition to significantly larger values of B_{ij}, result in diverging values of A_{11} and A_{22} when $x_1 \to 0$ and $x_1 \to 1$. This calls for care in the choice of conditions.

Step III: Intermediate results: Differential $(\partial U/\partial N_i)_{V,T,N_j}$

We present the differential $(\partial U/\partial N_i)_{V,T,N_j}$ for $i = 1, 2$ in Fig. 11.3. The values were computed from Eqs. (11.22) and (11.23), where $(\partial H/\partial N_i)_{T,V,\mu_j}$ was taken from the fluctuation equation (cf. Fig. 11.1), and A and B were calculated as described above. The

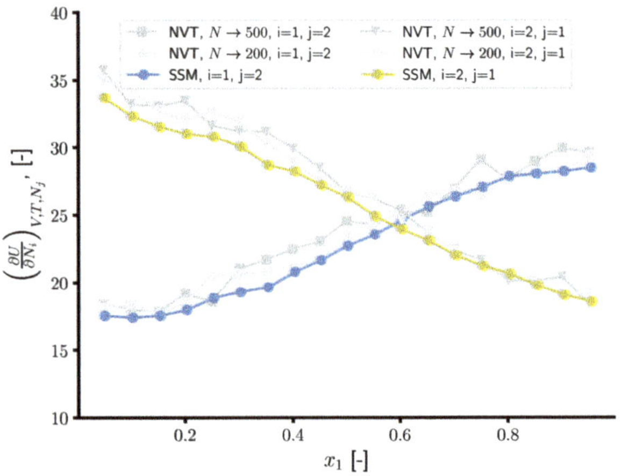

Figure 11.3. Change in partial molar enthalpy of the N, V, T ensemble. The N, V, T simulation results are obtained by inserting an extra particle in the N, V, T ensemble, while the yellow and blue lines are obtained by converting the partial molar enthalpies from Fig. 11.1, using Kirkwood–Buff coefficients to calculate **A** and **B**, see Fig. 11.2.

differential $(\partial H/\partial N_i)_{V,T,N_j}$ was also calculated directly from N, V, T-simulations, as explained in the previous section. No significant difference was seen by increasing the particle number from 200 to 500 in the numerical approach.

Step IV: Partial molar enthalpies

The partial molar enthalpies using the SSM are shown in Fig. 11.4. The enthalpy variation with composition in the isothermal–isobaric ensemble is numerically differentiated, and the results are also shown in the figure. The methods result in the same value of the partial molar enthalpy of both components over most of the composition range. The SSM-based method becomes less reliable as the mixtures become dilute. This behavior is expected. The fluctuations tend to diverge slowly as the ratio between particle numbers becomes very large or very small.

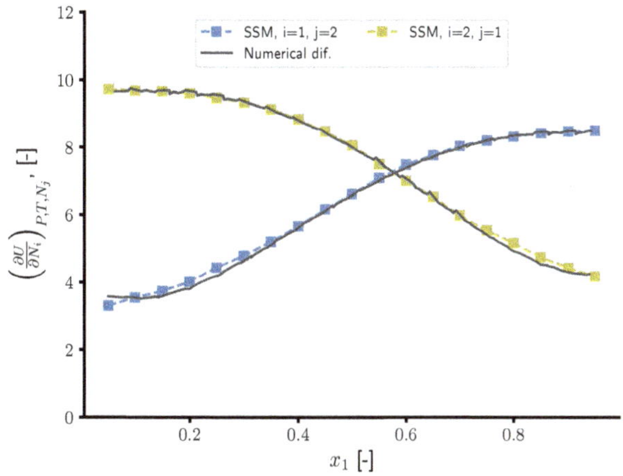

Figure 11.4. Partial molar enthalpies of components 1 and 2 as calculated from SSM (points) and from direct numerical differentiation (line).

11.6 Concluding Remarks

We have shown how to obtain partial molar enthalpies of small systems in the grand canonical ensemble, applying the SSM of Chapter 5. The SSM-results for the thermodynamic limit, which we derive from the GC ensemble, were transformed by several steps into values for the isothermal, isobaric ensemble. We have given a stepwise procedure to do so, hoping to facilitate such determinations in the future.

For a small system, there are two characteristic pressures, and therefore two enthalpies. An integral relation to compute the differential pressure in the grand canonical ensemble was presented for a single-component system. Integral relations for multi-component mixtures can be obtained by the same procedure.

Chapter 12

Hill's versus Gibbs' Thermodynamics

Gibbs and followers have given a thermodynamic theory of curved surfaces. Here we show how this theory is contained in Hill's thermodynamic analysis of spheres and cylinders in a one-component system. The surface pressure can be understood as the negative surface tension. This and other relations can be helpful in studies of more complex surfaces.

We have seen throughout this book how properties of small systems deviate from those of large ones, because their surface-to-volume ratio is so large. Thermodynamic properties of small systems are not proportional to their volume. Gibbs and followers [20, 52, 53] already accounted long ago for surface energy contributions to the internal energy; and also contributions due to surface curvature [20] (see Eq. (493), page 225 of his collected works, volume 1). Their expressions for the surface excess energy and the surface energy density were

$$
\begin{aligned}
dU^s &= TdS^s + \mu dN^s + \gamma d\Omega + C_1 dc_1 + C_2 dc_2 \\
u^s &= Ts^s + \mu n^s + \gamma \qquad (12.1)
\end{aligned}
$$

Superscript s indicates that we are dealing with a property of a dividing surface, see e.g. [25]. The symbols $U, S, N, \Omega, \mu, T, \gamma$ have their usual meanings and C_1, C_2 will be defined below. The variables c_1 and c_2 are the principal curvatures of the surface. They are by definition the inverse of the principle radii R_1 and R_2, giving $c_1 \equiv 1/R_1$ and $c_2 \equiv 1/R_2$. We shall see in this chapter how the variables

C_1 and C_2 can be related to variables in Hill's description, building on the work of Bedeaux and Kjelstrup [36] and Dong [50].

Hill's method contains also terms that describe curvature energies. In short, Hill's method is relevant when variables other than S, N and V are needed. In Eq. (12.1), such variables are known, used and well defined. In a porous medium, the proper choice of the variables is less clear. This will be discussed in Chapters 17–20.

In general, it is an advantage to have two methods to use in the study of any property. For curved surfaces, Gibbs' method is direct and therefore practical. Equation (12.1) gives an explicit relation to the surface-dependent variables. Hill's method is more systematic in its way to account for environmental variables, cf. Chapter 3. Thereby, it can provide a link between the macro- and nano-scale, called a scaling law, see Chapter 5. With a scaling law at hand, it is sufficient to simulate one small system, in order to find thermodynamic properties of another. In this aspect, Hill's description can be regarded as being more general than Gibbs' description.

In Gibbs' approach, the size-, curvature- and shape-effects are included step by step: first the flat surface, then the Tolman length, then the rigidities and shape effects (edge- and corner effects). This is possible in a well-defined system. Hill's description contains all possible size-, curvature- and shape-effects in the outset. Complex size-effects can be described directly from Hill's procedure, i.e., from the partition function of a small system. An example was given for an ideal gas with surface adsorption in Chapter 6.

The essential new variable in Hill's method is the subdivision potential ε. This is the variable that contains all effects due to size, curvature and shape. In an intermediate situation, it is however possible to separate out the surface area as a thermodynamic variable. The corresponding part of ε then disappears, as shown by Galteland *et al.* [51], see Chapter 17. By introducing furthermore the curvature as a variable, epsilon is further reduced. In this manner, we can obtain Gibbs' description from Hill's method. Hills' method, though less specific, is in this sense more general.

We show here that Hill's and Gibbs' descriptions are equivalent for the simple case of spherical or cylindrical one-component systems. The systems are both controlled by the grand canonical (GC) ensemble. All formulas in the chapter refer to this ensemble, so we suppress GC as a superscript. The derivations will be done for a one-component system. Generalizations to mixtures are possible.

12.1 Hill's Description of Small, Curved Systems

Let us recapitulate the equations that govern a small system in the GC ensemble. The system has volume V and is in contact with a heat- and particle-bath with temperature T and chemical potential μ. The system was described in Section 3.4, cf. also Table 3.1.

The system is illustrated in Fig. 12.1. Following Hill, we construct an ensemble of $\mathcal{N}$ identical replicas of the system. Only two replicas are shown in contact with a heat- and particle-bath in Fig. 12.1.

The average energy, the energy density, the Gibbs equation and the Hill–Gibbs–Duhem equation of the single small one-component system were derived in Section 3.4. We repeat

$$\begin{aligned} U(T,V,\mu) &= TS(T,V,\mu) + \mu N(T,V,\mu) - \widehat{p}(T,V,\mu)V \\ u &= Ts + \mu n - \widehat{p} \\ dU &= TdS - pdV + \mu dN \\ d(\widehat{p}V) &= SdT + pdV + Nd\mu \end{aligned} \tag{12.2}$$

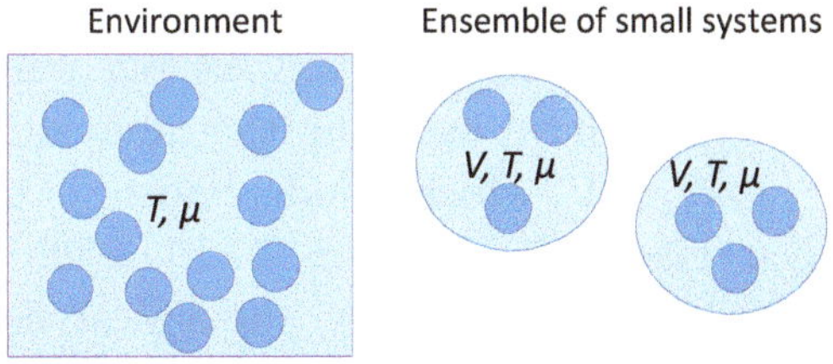

Figure 12.1. An example of an ensemble of small systems. Only two replicas of the ensemble with volume V are pictured on the right-hand side. The ensemble exchanges energy and mass with the environment which serves as a heat- and particle-bath, which controls T and μ. Reproduced from [36] with permission from Elsevier B.V.

The symbols have their usual meaning. For ease of notation we do not show the functional dependence on the ensemble variables T, V, μ after the first line. The left-hand side of the last equation expresses that the system is not extensive. The replica energy is equal to the grand potential or minus the compressional energy in the GC case, cf. Table 3.1:

$$X = -\widehat{p}V \tag{12.3}$$

The differential (p) and integral ($\widehat{p}$) pressures are related by

$$p = \left(\frac{\partial \widehat{p}V}{\partial V}\right)_{T,\mu} = \widehat{p} + V\left(\frac{\partial \widehat{p}}{\partial V}\right)_{T,\mu} \tag{12.4}$$

In the thermodynamic limit, $\widehat{p}$ is independent of V so that $p = \widehat{p}$, and we obtain the classical expressions.

12.2 Volume- and Surface-Contributions

Consider next one of the extensive properties, for instance, the entropy or the compressional energy. Small systems differ from large ones, because of their large surface-to-volume ratio. We may therefore find it useful to separate volume and surface effects and write any extensive property as the sum of the large system value plus a correction due to size:

$$A(T, V, \mu) \equiv a(T, V, \mu)V \equiv a^{\infty}(T, \mu)V + a^{s}(T, \mu, c_1, c_2)\Omega \tag{12.5}$$

Here, $a(T, V, \mu)$ is the density of $A(T, V, \mu)$ per unit of volume. In the second equality, we have used the thermodynamic limit value, $a^{\infty}(T, \mu) \equiv A^{\infty}(T, V, \mu)/V$, which is independent of the volume. The symbol Ω is the surface area. Equation (12.5) defines $a^{s}(T, \mu, c_1, c_2)$.

The properties a and A depend on the size of the small system, V, but also on the shape of this volume. We did not explicitly indicate this dependence earlier. But now we introduce the shape-dependence through the last term, $a^{s}(T, \mu, c_1, c_2)$, through the principle curvatures c_1 and c_2 of the small volume.

For a sphere, these principal curvatures are identical. For a cylinder, one curvature is zero, while the other is finite. The thermodynamic limit value is uniquely determined, meaning that Eq. (12.5) also determines a^s uniquely.

The curvature of interest changes when V changes. For a sphere, we have

$$\frac{\partial c_1}{\partial V} = -\frac{c_1}{3V} \quad \text{and} \quad \frac{\partial c_2}{\partial V} = -\frac{c_2}{3V} \tag{12.6}$$

And for a cylinder, we have

$$\frac{\partial c_1}{\partial V} = -\frac{c_1}{2V} \quad \text{and} \quad c_2 = 0 \tag{12.7}$$

In general, the principle curvatures vary along the surface of the small volume. There are local values everywhere on the surface. Here, the curvatures c_1 (and c_2) are constants, since we are dealing with a sphere or a cylinder only.

Equation (12.5) applies to U, S, N, pV and $\widehat{p}V$ and gives

$$\begin{aligned} uV &= u^{\infty}(T,\mu)V + u^{s}(T,\mu,c_1,c_2)\Omega \\ \cdots &= \cdots \\ p(T,V,\mu)V &= p^{\infty}(T,\mu)V - \gamma(T,\mu,c_1,c_2)\Omega \\ \widehat{p}(T,V,\mu)V &= p^{\infty}(T,\mu)V - \widehat{\gamma}(T,\mu,c_1,c_2)\Omega \end{aligned} \tag{12.8}$$

where we used that $\widehat{p^{\infty}} = p^{\infty}$.[a]

In order to make a link to the surface properties used in Eq. (12.1), we introduce a separation of variables in all of the terms explained in Eq. (12.2), and obtain a set of relations for the thermodynamic limit (denoted by ∞) as well as for the contribution of the curved surface (denoted by s). The equations for the thermodynamic limit

---

[a]Bedeaux and Kjelstrup [36] used $-p^s$ instead of γ and $-\widehat{p}^s$ instead of $\widehat{\gamma}$. They also identified $-\widehat{p}^s$ with γ, rather than with $\widehat{\gamma}$.

values are

$$\begin{aligned} U^{\infty} &= TS^{\infty} + \mu N^{\infty} - p^{\infty} V \\ u^{\infty} &= Ts^{\infty} + \mu n^{\infty} - p^{\infty} \\ dU^{\infty} &= TdS^{\infty} + \mu dN^{\infty} - p^{\infty} dV \\ V dp^{\infty} &= S^{\infty} dT + N^{\infty} d\mu \end{aligned} \tag{12.9}$$

The corresponding set for the surface contributions is

$$\begin{aligned} U^{s} &= TS^{s} + \mu N^{s} + \widehat{\gamma}\Omega \\ u^{s} &= Ts^{s} + \mu n^{s} + \widehat{\gamma} \\ dU^{s} &= TdS^{s} + \mu dN^{s} + \gamma \frac{\Omega}{V} dV \\ d(\widehat{p}^{s}\Omega) &= S^{s} dT + N^{s} d\mu + \gamma \frac{\Omega}{V} dV \end{aligned} \tag{12.10}$$

In the thermodynamic limit, $\widehat{p}^{\infty} = p^{\infty} = p$. For a sphere, we replace $(\Omega/V)dV = (3/2)d\Omega$, and for a cylinder, $(\Omega/V)dV = 2d\Omega$.

12.3 Comparing Hill's and Gibbs' Expressions

We are now in a position to compare Eq. (12.1) to Hill's formulations (12.9) and (12.10) for bulk and surface energies. We split the contributions to p and $\widehat{p}$, like we do in Eq. (12.8), and obtain

$$p = p^{\infty} - \gamma \frac{\Omega}{V} = \frac{\partial \widehat{p} V}{\partial V} = \frac{\partial \left(\widehat{p}^{\infty} V\right)}{\partial V} - \frac{\partial \left(\widehat{\gamma}(T, \mu, c_1, c_2)\Omega\right)}{\partial V} \tag{12.11}$$

This results in the equality

$$\begin{aligned} \gamma\left(T, \mu, c_1, c_2\right) \frac{\Omega}{V} = \widehat{\gamma}(T, \mu, c_1, c_2) \frac{\partial \Omega}{\partial V} &+ \Omega \frac{\partial \widehat{\gamma}(T, \mu, c_1, c_2)}{\partial c_1} \frac{\partial c_1}{\partial V} \\ &+ \Omega \frac{\partial \widehat{\gamma}(T, \mu, c_1, c_2)}{\partial c_2} \frac{\partial c_2}{\partial V} \end{aligned} \tag{12.12}$$

The curvatures change when V changes.

By introducing Eq. (12.6) and $(\Omega/V)dV = (3/2)d\Omega$ into Eq. (12.12), we obtain for a sphere that

$$\gamma(T,\mu,c_1,c_2) = \frac{2}{3}\widehat{\gamma}(T,\mu,c_1,c_2) - \frac{c_1}{3}\frac{\partial\widehat{\gamma}(T,\mu,c_1,c_2)}{\partial c_1} - \frac{c_2}{3}\frac{\partial\widehat{\gamma}(T,\mu,c_1,c_2)}{\partial c_2} \tag{12.13}$$

For the cylinder, we obtain

$$\gamma(T,\mu,c_1) = +\frac{1}{2}\widehat{\gamma}(T,\mu,c_1) - \frac{c_1}{2}\frac{\partial\widehat{\gamma}(T,\mu,c_1)}{\partial c_1} \tag{12.14}$$

By introducing these identities into Eq. (12.10), we can write for the sphere and the cylinder in compact form

$$dU^s = TdS^s + \widehat{\gamma}d\Omega - \frac{1}{2}\frac{\partial\widehat{\gamma}}{\partial c_1}c_1 d\Omega - \frac{1}{2}\frac{\partial\widehat{\gamma}}{\partial c_2}c_2 d\Omega + \mu dN^s \tag{12.15}$$

It follows from $(\Omega/V)dV = (3/2)d\Omega$ and Eq. (12.6) that

$$c_1 d\Omega = -2\Omega dc_1 \quad \text{and} \quad c_2 d\Omega = -2\Omega dc_2 \tag{12.16}$$

for the sphere. For the cylinder ($c_2 = 0$), we obtain

$$c_1 d\Omega = -\Omega dc_1 \tag{12.17}$$

By introducing this into Gibbs' equation for the surface contribution, we can write for both geometries

$$dU^s(T,\Omega,\mu,c_1,c_2) = TdS^s(T,\Omega,\mu,c_1,c_2) + \widehat{\gamma}(T,\mu,c_1,c_2)d\Omega + \frac{\partial\widehat{\gamma}(T,\mu,c_1,c_2)}{\partial c_1}\Omega dc_1 + \frac{\partial\widehat{\gamma}(T,\mu,c_1,c_2)}{\partial c_2}\Omega dc_2 + \mu dN^s(T,\Omega,\mu,c_1,c_2) \tag{12.18}$$

where for the cylinder c_2 and dc_2 are zero. The corresponding Euler relation is

$$U^s(T,\Omega,\mu,c_1,c_2) = TS^s(T,\Omega,\mu,c_1,c_2) + \widehat{\gamma}(T,\mu,c_1,c_2)\Omega + \mu dN^s(T,\Omega,\mu,c_1,c_2) \tag{12.19}$$

and the Hill–Gibbs–Duhem equation becomes

$$d\widehat{\gamma}(T,\mu,c_1,c_2) = -s^s(T,\mu,c_1,c_2)dT + \frac{\partial\widehat{\gamma}(T,\mu,c_1,c_2)}{\partial c_1}dc_1 + \frac{\partial\widehat{\gamma}(T,\mu,c_1,c_2)}{\partial c_2}dc_2 - n^s(T,\mu,c_1,c_2)d\mu \quad (12.20)$$

We can now consider the accomplishment. Equation (12.18) is exactly the equation given by Gibbs [20] (see Eq. (493) on page 225 of his collected works, volume 1; see also Eq. (12.1)) for surfaces in heterogeneous systems. The measured surface tension is Hill's $\widehat{\gamma}$ rather than his γ. The coefficients C_1 and C_2 in his expression can be identified as

$$C_1 = \frac{\partial\widehat{\gamma}(T,\mu,c_1,c_2)}{\partial c_1}\Omega \quad \text{and} \quad C_2 = \frac{\partial\widehat{\gamma}(T,\mu,c_1,c_2)}{\partial c_2}\Omega \quad (12.21)$$

which are identities which follow from Eq. (12.18) as Maxwell relations.

12.4 Concluding Remarks

We have seen above that Gibbs' [20] surface description of small cylinders and spheres is contained in Hill's thermodynamic description of the same small systems [4].

Hill extended the treatment of small systems beyond Eq. (12.1). His method allows for a broader, more general study of curved surfaces. For instance, his method sets the study of spheres, tetrahedra and dodecahedra on the same footing [48]. Dong related the volume dependence of the integral surface tension to the disjoining pressure [50]. The emerging theoretical relations can be used to decompose data from molecular dynamics simulations. We demonstrated in Chapter 5 how this can be done, using the so-called Small System Method — a method that relates surface properties to properties of the thermodynamic limit [1, 48]. We expect that there is always an equivalence in the methods, when all variables that are particular to the small system, can be specified.

Chapter 13

Maxwell–Stefan's and Fick's Diffusivities

Fick's law is often used to describe experimental results on diffusion. The average molar or solvent frame of reference is then convenient. Maxwell–Stefan diffusion coefficients are reference-free and directly obtainable from molecular dynamics simulations. In order to transform Maxwell–Stefan coefficients into Fick's coefficients and compare the results, we need to know the thermodynamic factors of the mixture where the diffusion takes place. We describe here how these factors enter in the transformation, and how they can be determined from the Small System Method.

Inter-diffusion is the collective movement of molecules due to gradients in chemical potential. Diffusion can be slow and seriously hamper other processes, such as chemical reactions. Diffusion can be rate-limiting for the overall process [79–81]. This makes it important to predict.

Self-diffusion, in contrast to diffusion, takes place in a uniform solution. A molecule in a uniform solution will move in a random manner, and we can observe a mean square displacement. This movement is relatively easy to compute, and is often used to estimate diffusion. Better models are needed, however, and for their development, it is important to be able to compare simulated data with experimental data. The aim of the present chapter is to show how this can be done for two sets of diffusion coefficients or diffusivities. The Maxwell–Stefan (MS) diffusivities are mostly obtained from molecular dynamics simulations, while Fick's diffusivities are commonly obtained from experiments. Sets of coefficients are equivalent if they describe the same entropy production [24]. This condition of invariance allows us to find relations that can be used to transform one set into another. The procedure that predicts the last set from the first was worked out by Liu and coworkers [6, 31, 32, 68, 77, 82, 83]. It will be recapitulated here as an example of application of nanothermodynamic procedures.

The fundamental starting point for a description of diffusion is always the flux–force relations of Onsager [84, 85]. These are derived directly from the entropy production, and have the form

$$\mathbf{J}_i = \sum_{j=1}^{n} \Lambda_{ij}\mathbf{X}_j = -\frac{c}{RT}\sum_{j=1}^{n} \Lambda_{ij}\nabla\mu_j \tag{13.1}$$

Here, R and T are the gas constant and the absolute temperature, c is the total concentration in $\mathrm{mol\,m^{-3}}$ and μ_j, the chemical potential of component j in $\mathrm{J\,mol^{-1}}$. There are n molar fluxes, $\mathbf{J}_i$. Each flux has a conjugate driving force, $\mathbf{X}_i$. At constant temperature, $\mathbf{X}_i = -(c/RT)\nabla\mu_i$. Thermal diffusion, driven by a temperature gradient, is not considered here.

The matrix of Onsager coefficients in $\mathrm{m^2\,s^{-1}}$, Λ_{ij}, is symmetric, so that $\Lambda_{ij} = \Lambda_{ji}$ [25]. Because of isotropy, the Onsager coefficient matrix, Λ_{ij}, is furthermore scalar rather than tensorial. The coefficients can be computed in terms of average equilibrium time correlations of the component velocities [72, 83];

$$\Lambda_{ij} = \frac{1}{3N}\int_0^\infty dt \sum_{l=1}^{N_i}\sum_{k=1}^{N_k} \langle \mathbf{v}_{l,i}(0)\cdot\mathbf{v}_{k,j}(t)\rangle \tag{13.2}$$

where N is the total number of particles in the mixture. The symbol $\cdot$ indicates the contraction of two vectors. In the system there are N_j particles of component j. The velocity of particle l of component j is $\mathbf{v}_{l,j}$. In equilibrium, the average of this velocity is zero. In molecular dynamics simulations, we compute velocities of all molecules as a function of time. Fluxes and velocities are in the mean molar frame of reference.

In practice, other formulations are used. Fick's and MS's formulations are the most common ones. A thermodynamically consistent model must however always obey the symmetry requirement of Eq. (13.2). The aim of this chapter is to describe how one set of coefficients can be transformed into another, taking advantage of the Small System Method (SSM), outlined in Chapter 5. We present first the two common formulations, and the relations that follow from entropy production invariance. We proceed to show how they can be used to derive diffusivities, and compare these to simple models.

In terms of the above variables, the entropy production in $\mathrm{J\,K^{-1}\,m^{-3}\,s^{-1}}$ is given by [25]

$$\sigma = \frac{R}{c}\sum_{j=1}^{n} \mathbf{J}_j \cdot \mathbf{X}_j = -\frac{1}{T}\sum_{j=1}^{n} \mathbf{J}_j \cdot \nabla\mu_j \tag{13.3}$$

The reason to use $(R/c)\,\mathbf{X}_j$ as a driving force in this equation is that, as a consequence, the Λ_{ij} matrix obtains the dimension of Fick's diffusion coefficients, $\mathrm{m^2\,s^{-1}}$.

13.1 Fick's and Maxwell–Stefan's Equations

Fick's law describes diffusion as a linear combination of composition gradients. For an n-component system, the molar flux $\mathbf{J}_i$ of component i (in $\mathrm{mol\,m^{-2}\,s^{-1}}$) follows from

$$\mathbf{J}_i = -c\sum_{j=1}^{n-1} D_{ij}\,\nabla x_j \tag{13.4}$$

in which D_{ij} are Fick's diffusivities, and $x_j \equiv c_j/c$ is the mole fraction of component j [80]. The coefficients depend on the frame of reference used for the fluxes, on the concentrations, c_j, and the temperature, but not on the magnitude of the driving forces.

The set of Eq. (13.4) is often used to describe experimental data. Gases have diffusivities typically around $10^{-5}\mathrm{m^2\,s^{-1}}$. In liquids, the diffusivities are about $10^{-9}\mathrm{m^2\,s^{-1}}$. In solids, diffusion is even slower [80, 81, 86]. Diffusion is slowed down by confinement. The frame of reference for the fluxes in Eq. (13.4), is the average molar velocity. The sum of the diffusion fluxes is therefore zero

$$\sum_{j=1}^{n} \mathbf{J}_i = 0 \tag{13.5}$$

The number of Fick diffusivities is $(n-1)^2$. The matrix of Fick diffusivities is not symmetric.

Other frames of reference, like the barycentric, the mean volume, or the solvent frame of reference, are common [25, 87]. The experimental conditions decide which frame of reference is most convenient.

The MS equations are convenient for data reduction of molecular dynamics simulations [79, 80, 88]. The driving force for diffusion, i.e., minus the chemical potential gradient, $-\nabla\mu_i$, at constant temperature and pressure, are regarded in this approach as being balanced by a sum of frictional forces:

$$-\frac{1}{RT}\nabla\mu_i = \sum_{j=1, j\neq i}^{n} \frac{x_j(\mathbf{u}_i - \mathbf{u}_j)}{\mathcal{D}_{ij}} \tag{13.6}$$

The frictional force between components i and j is proportional to the difference in average relative velocities of the components, $(\mathbf{u}_i - \mathbf{u}_j)$. The MS-diffusivity $\mathcal{D}_{ij}$ is an inverse friction coefficient, which describes the magnitude of friction between components i and j. It can be shown, using the Onsager coefficient matrix, that the matrix of MS-coefficients is also symmetric. These diffusivities depend usually less strongly on concentration than the Fick diffusivities do [79, 80].

By comparing Fick's and MS's laws of diffusion [25], we observe that

- MS diffusivities do not depend on any frame of reference, while Fick's diffusivities depend on the frame of reference.
- The matrix of MS diffusivities is symmetric, i.e. $\mathcal{D}_{ij} = \mathcal{D}_{ji}$, while the matrix of Fick's diffusivities is not, i.e., $D_{ij} \neq D_{ji}$.
- In a system of n components, we need $n(n-1)/2$ MS diffusivities to describe mass transport, while $(n-1)^2$ Fick diffusivities are needed.

The form of Eq. (13.6) suggests that MS diffusivities can be estimated from values of binary mixtures. Such estimates often give the correct order of magnitude [79, 80]. It remains difficult to correctly predict multicomponent Fick diffusivities using knowledge of binary systems alone. The MS diffusivities cannot be directly obtained from experiments, as the chemical potential gradient cannot be measured directly. In order to assist model development, we need to compare experiments and simulations. The transformation of one set into another is therefore of interest.

13.2 Transformation Relations

The entropy production is absolute [79, 80] and invariant under transformation from one frame of reference to another. When we compute the entropy production from Eqs. (13.4) and (13.6), and set them equal to each other [80, 88, 89], we find the relation

$$[D] = [B]^{-1} \cdot [\Gamma] \tag{13.7}$$

in which $[D]$ is the $(n-1) \times (n-1)$ matrix of Fick diffusivities. The elements of the matrix $[B]$ are given by [80, 89, 90]

$$\begin{aligned} B_{ii} &= \frac{x_i}{\mathcal{D}_{in}} + \sum_{j=1, j \neq i}^{n} \frac{x_j}{\mathcal{D}_{ij}}, \quad \text{with} \ \ i = 1, \ldots, (n-1) \\ B_{ij} &= -x_i \left(\frac{1}{\mathcal{D}_{ij}} - \frac{1}{\mathcal{D}_{in}} \right), \quad \text{with} \ \ i, j = 1, \ldots, (n-1) \text{ and } i \neq j \end{aligned} \tag{13.8}$$

The elements of the matrix of thermodynamic factors $[\Gamma]$ are defined by [68, 80, 88]

$$\Gamma_{ij} \equiv \frac{x_i}{RT} \left(\frac{\partial \mu_i}{\partial x_j} \right)_{T,p,\Sigma} = \delta_{ij} + x_i \left(\frac{\partial \ln \gamma_i}{\partial x_j} \right)_{T,p,\Sigma} \quad \text{with} \ \ i, j = 1, \ldots, (n-1) \tag{13.9}$$

where we used the chemical potential $\mu_i = \mu_i^0(T) + RT \ln(x_i \gamma_i)$. Furthermore, δ_{ij} is the Kronecker delta, γ_i is the activity coefficient of component i and $\mu_i^0(T)$ is the standard state chemical potential. The symbol Σ indicates that the partial differentiation with respect to the mole fraction x_j is carried out at constant mole fraction of all other components except the nth one, so that $\sum_{i=1}^{n} x_i = 1$ during the differentiation [80].

Equation (13.7) simplifies the binary mixtures to

$$D = \Gamma \times \mathcal{D}_{12} \tag{13.10}$$

in which D is the binary Fick diffusivity, $\mathcal{D}_{12}$ is the MS diffusivity and Γ is the thermodynamic factor given by [80]

$$\Gamma = 1 + x_1 \left(\frac{\partial \ln \gamma_1}{\partial x_1} \right)_{T,p,\sum} = 1 + x_2 \left(\frac{\partial \ln \gamma_2}{\partial x_2} \right)_{T,p,\sum} \tag{13.11}$$

The latter equality follows from the Gibbs–Duhem equation.

13.3 Diffusion Models

From the set of correlation functions, Eq. (13.2), we can define also a self-diffusion coefficient as follows:

$$D_i^{\text{self}} = \frac{1}{3N_i} \int_0^{\infty} dt \sum_{l=1}^{N_i} \langle \mathbf{v}_{l,i}(0) \cdot \mathbf{v}_{l,i}(t) \rangle = \frac{\Lambda_{ii}}{x_i} \tag{13.12}$$

Self-diffusivities of acetone–methanol mixtures according to [68] are shown in Fig. 13.1(a). We see that there is a dependence of the self-diffusion coefficient on the concentration, stronger for methanol than for acetone. The points, computed from Eq. (13.12), can be reasonably fitted to a straight line. The line in the figure is predicted by the Darken approximation, Eq. (13.14), together with Eq. (13.15), both given in what follows. A similar picture is obtained for an acetone–carbon tetrachloride mixture from Fig. 13.1(b).

In terms of Onsager coefficients, Eq. (13.2), the MS diffusivities of a binary mixture are

$$\mathcal{D}_{12} = \frac{x_2}{x_1} \Lambda_{11} + \frac{x_1}{x_2} \Lambda_{22} - 2\Lambda_{12} \tag{13.13}$$

The Darken approximation for binary mixtures follows from this equation by neglecting the term containing Λ_{12}:

$$\mathcal{D}_{12}^{\text{Darken}} = x_2 D_1^{\text{self}} + x_1 D_2^{\text{self}} \tag{13.14}$$

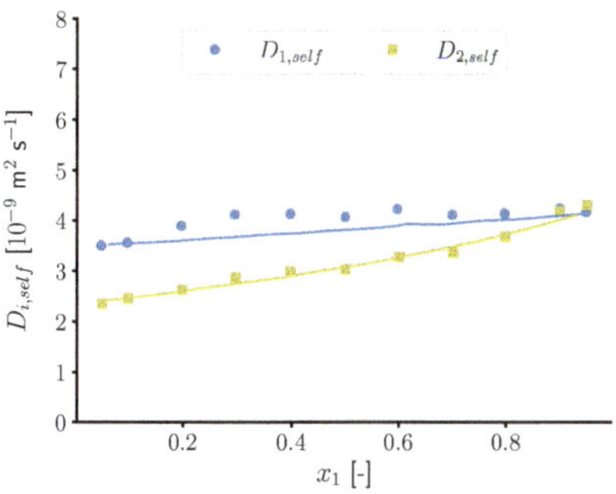

(a) Acetone–methanol

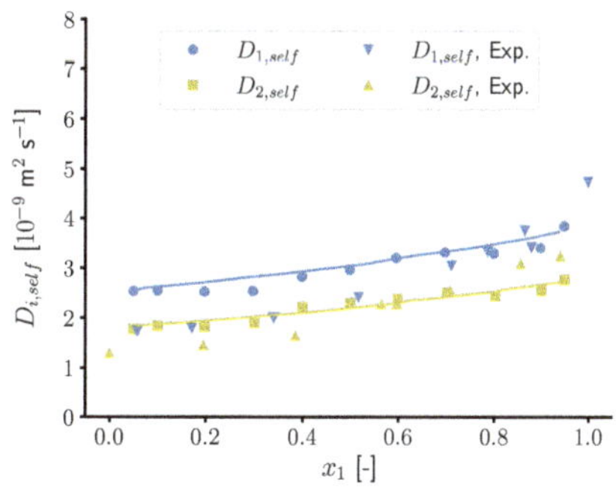

(b) Acetone–carbon tetrachloride

Figure 13.1. Self diffusivities in acetone–methanol (a) and acetone–carbon tetrachloride (b) mixtures as a function of mole fraction at 298 K, 1 atm according to [68]. Experimental values taken from [91]. Solid lines are predictions from Eq. (13.14) together with (13.15). Reproduced from [68].

The contributions to the right-hand side of this equation are easy to compute. The Darken relation has often been used as a starting point for further model development [68, 72, 82]. An approximation for the self-diffusion coefficient, to be used in Eq. (13.14) was proposed by Liu *et al.* [31] as follows:

$$\frac{1}{D_i^{\text{self}}} = \sum_{j=1}^{n} \frac{x_j}{D_i^{\text{self},x_j \to 1}} \tag{13.15}$$

This models the self-diffusivity of species i in the mixture as the inverse mole-fraction weighted sum of its pure component value and its values at infinite dilution. We refer to [31] for a more thorough discussion of the Darken approximation and Eq. (13.15).

13.4 Maxwell–Stefan Diffusivities

The MS diffusivities $\mathcal{D}_{ij}$ were computed from the Onsager coefficients Λ_{ij}, and these were calculated from molecular dynamics simulations of mixtures at equilibrium. Figure 13.2 shows the MS diffusivities of acetone–methanol and acetone–carbon tetrachloride [72], as a function of mixture composition. Both data sets apply for 298 K

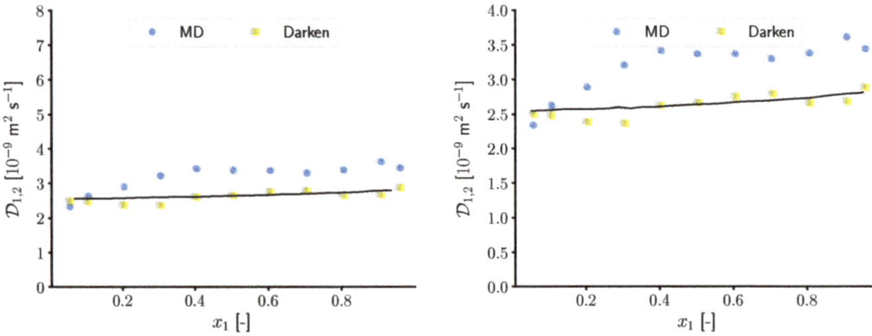

(a) Acetone–methanol (b) Acetone–carbon tetrachloride

Figure 13.2. MS diffusivities in acetone–methanol (a), and acetone–carbon tetrachloride (b) mixtures at 298 K, and 1 atm according to [68]. Filled triangles are computed MS diffusivities using MD simulations. Squares are predictions using the Darken equation (13.14) with self-diffusivities taken from the MD simulations, Eq. (13.12). The solid line represents predictions using the Darken equation (13.14) with self-diffusivities predicted from Eq. (13.12).

and 1 atm. Filled triangles are computed MS diffusivities using MD simulations [68]. Open symbols are predictions using the Darken equation, Eq. (13.14) with self-diffusivities computed from the MD simulations, Eq. (13.12). The solid line represents predictions using the Darken equation (13.14) with self-diffusivities predicted from Eq. (13.12). A large discrepancy is shown between the MS diffusivities and the Darken equation.

13.5 The Thermodynamic Factor at Work

The matrix of thermodynamic factors relate the $[D]$ and $[B]$-matrices in Eq. (13.7). Knowing this matrix, we can transform the MS coefficients into Fick's law coefficients. The thermodynamic factor can be computed from Kirkwood–Buff (KB) integrals, G_{ij}, cf. Chapter 10. In a binary mixture, the thermodynamic factor is related to KB integrals by

$$\Gamma = [1 + x_1 c_2 (G_{11} + G_{22} - 2G_{12})]^{-1} \tag{13.16}$$

The integrals are most easily obtained from fluctuations in particle numbers as follows:

$$G_{ij} = V\frac{\langle N_i N_j\rangle - \langle N_i\rangle\langle N_j\rangle}{\langle N_i\rangle\langle N_j\rangle} - \frac{\delta_{ij}}{c_i} \tag{13.17}$$

The relation applies for the whole range of small volumes, when the system is controlled in the grand canonical ensemble. The determination of KB integrals for small systems was discussed in Chapter 10. Using the scaling law, we find the thermodynamic limit value by extrapolation. The thermodynamic limit values of the KB integrals can be used for any ensemble. In the ideal mixture limit, $\Gamma = 1$. The matrix of thermodynamic factors was obtained using the SSM on simulations in the grand canonical ensemble [94].

The SSM of Chapter 5 has an efficient algorithm, since only small systems need be sampled. In order to control the outcome of the method, the thermodynamic factors were also computed from experimentally determined equations of state [67] and excess Gibbs energies G^E [80]. Computed thermodynamic factors compared well to those predicted from excess Gibbs energy-models which were fitted to experimental vapor–liquid equilibrium data, see Fig. 13.3(a) for acetone–methanol, and Fig. 13.3(b) for acetone–carbontetrachloride.

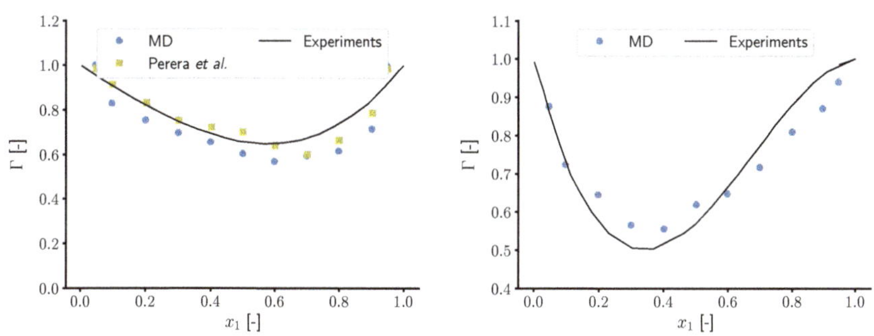

(a) Acetone–methanol (b) Acetone–carbon tetrachloride

Figure 13.3. Thermodynamic factor, Γ, of acetone–methanol (a), and acetone–carbon tetrachloride (b) mixtures at 298 K, 1 atm. Circles denote Γ computed from MD simulations (error bars are equal to the symbol size). Squares denote Γ as computed by Perera *et al.* using MD simulations [92]. The solid line is computed from experimental vapor–liquid equilibrium data [93].

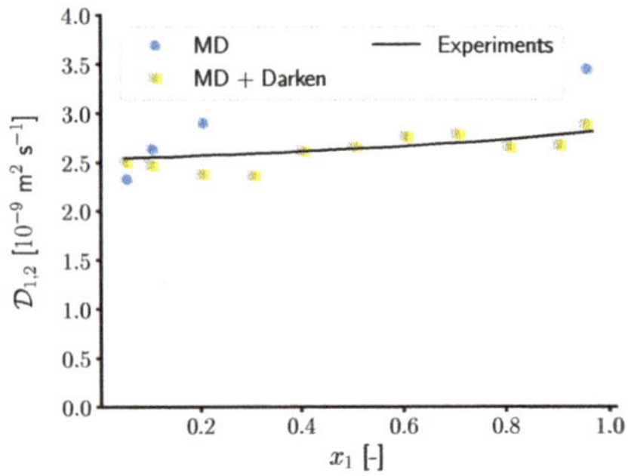

(a) Acetone–methanol

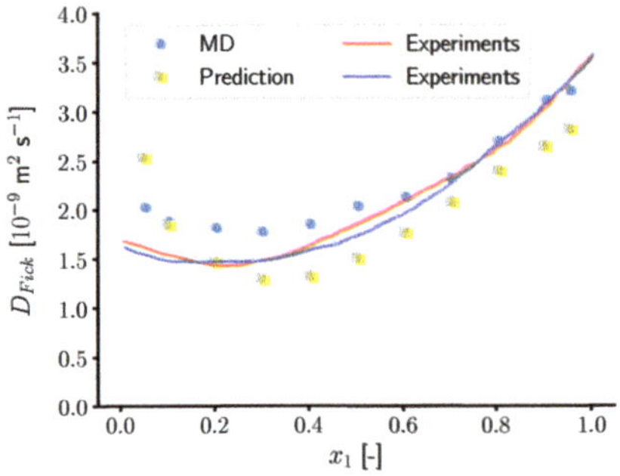

(b) Acetone–carbon tetrachloride

Figure 13.4. Fick diffusivities in mixtures of acetone and methanol (a) and acetone–carbon tetrachloride (b) mixtures at 298 K and 1 atm according to [68], as measured (solid line) and computed (filled symbols) from MS diffusivities and thermodynamic factors.

We see from Fig. 13.4 that there is excellent agreement between computed and experimental diffusivities.

13.6 Concluding Remarks

The purpose of this chapter was to show how MS's diffusivities can be used to compute Fick's diffusivities using thermodynamic factors of the mixture. The crucial variable Γ_{ij} was computed by the SSM using the scaling properties of finite-volume KB integrals presented in Chapter 10. From the simulated MS diffusivities and the matrix of thermodynamic factors, we obtained the measurable Fick diffusivities with high precision. This opens up the door to find these coefficients for ternary, even quarternary systems [31, 72, 77].

Chapter 14

Chemical Reactions

We show how we can find the thermodynamic properties of a chemical reaction using the Small System Method to sample fluctuations in a small volume element. With knowledge of scaling laws for size, we find properties in the thermodynamic limit. Thermodynamic factors, partial molar volumes, fluid compressibility, partial molar enthalpies and reaction enthalpies are found. The dissociation of hydrogen from molecule to atoms is used to illustrate the procedures. The fluid mixtures are far from ideal conditions. The accurate small system method gives access to these quantities in one simulation.

The aim of this work is to determine most prominently the reaction enthalpy of a chemical reaction, but also other thermodynamic properties. It is cumbersome to obtain reliable values of enthalpies, in the laboratory as well as by computations [95]. In particular, it is difficult to find partial molar enthalpies. With the Small System Method (SSM) [2, 67, 96] one can find partial molar properties in a single simulation, sampling fluctuations in a small volume element. We have chosen the hydrogen dissociation reaction as an example reaction to demonstrate how this can be done [56], and present results for the fluid compressibility, component partial molar volumes, thermodynamic correction factors and partial molar enthalpies. From this we find the reaction enthalpy and the thermodynamic equilibrium constant.

The hydrogen dissociation reaction is interesting in physics, e.g. in plasma modeling [97–99]. In the chemical world it is relevant for separation of hydrogen gas from a gas mixture (e.g., from the water gas shift reaction) in the presence of a palladium membrane [100–102]. At the surface of this membrane, molecular hydrogen dissociates and atomic hydrogen diffuses through the membrane, recombining to molecular hydrogen on the opposite surface [100, 101, 103]. We describe first the system; which is the chemical reaction in a box (Section 14.1). The formulas for the fluctuating variables and the size scaling laws are presented next (Section 14.2) and used to find the thermodynamic properties in Section 14.4.

14.1 The Hydrogen Dissociation Reaction

Particle interaction potentials

We have chosen to study the dissociation of hydrogen because necessary information is available for modeling. The reaction takes place under extreme conditions on the sun where the temperature is near 10^5 K. The reaction has been modeled accurately with interaction potentials derived from quantum mechanics [104–107]. Stillinger and Weber [105] showed that it was sufficient to use an interaction potential, U, that is the sum of two- and three-particle interaction contributions. They describe the essentials of a chemical reaction:

$$U(\mathbf{r}_1, \ldots, \mathbf{r}_N) = \sum_{i<j} u_{(2)}(r_{ij}) + \sum_{i<j<k} u_{(3)}(\mathbf{r}_i, \mathbf{r}_j, \mathbf{r}_k) \tag{14.1}$$

here $u_{(2)}$ and $u_{(3)}$ are the two- and three-particle potentials, given in Eqs. (14.2) and (14.3), respectively. Diedrich and Anderson [108, 109] and Kohen *et al.* derived expressions for two- and three-particle potentials [104] of hydrogen as follows:

$$u_{(2)}(r) = \begin{cases} \alpha\left(\beta_2 r^{-p} - 1\right)\exp\left[\dfrac{\gamma_2}{r - r_c}\right] & \text{if } r < r_c \\ 0 & \text{if } r > r_c \end{cases} \tag{14.2}$$

here $\alpha = 5.59 \cdot 10^{-21}\,\text{kJ}$, $\beta_2 = 0.044067\,\text{Å}^p$, $\gamma_2 = 3.902767\,\text{Å}$, $r_c = 2.8\,\text{Å}$ and $p = 4$ are constants [104]. The α is chosen such that the minimum of the potential gives the binding energy of hydrogen (432.065 kJ mol^{-1}) [104] at the bond distance between two hydrogen atoms, $r_e = 0.74\,\text{Å}$ [108]. When the distance between two atoms is larger than the cut-off distance, $r \geq r_c$, the potential is zero. Reduced units, based on the pair potential, are used throughout the paper. We have σ defined at $u_2(\sigma) = 0$, giving $\sigma = \sqrt[p]{\beta_2} = 0.458\,\text{Å}$. The value of ϵ is based on the binding energy of hydrogen, so that $-\epsilon$ is the minimum of the pair potential, giving $\epsilon/k_B = 51991$ K, where k_B is Boltzmann's constant. Relations between real and reduced units are given in Appendix D.

The three-particle potential is given as a sum over the contribution from each particle, and it's role is to prevent formation of more than one bond to each hydrogen atom.

$$u_{(3)} = h_{i,j,k}(r_{ij}, r_{jk}, \theta_{i,j,k}) + h_{j,i,k}(r_{ji}, r_{ik}, \theta_{j,i,k}) + h_{i,k,j}(r_{ik}, r_{kj}, \theta_{i,k,j}) \tag{14.3}$$

Where the h-functions are given by

$$h_{j,i,k}(r_{ji}, r_{ik}, \theta_{j,i,k}) = \begin{cases} \lambda a \exp\left[\dfrac{\gamma_3}{(r_{ji} - r_c)} + \dfrac{\gamma_3}{(r_{ik} - r_c)}\right] & \text{if } r_{ji} < r_c \text{ and } r_{ik} < r_c \\ 0 & \text{otherwise} \end{cases} \tag{14.4}$$

and

$$a = 1 + \mu\cos(\theta_{j,i,k}) + \nu\cos^2(\theta_{j,i,k}) \tag{14.5}$$

The symbols $\lambda = 2.80 \cdot 10^{-21}$ kJ, $\mu = 0.132587$, $\nu = -0.2997$ and $\gamma_3 = 1.5\,\text{Å}$ are constants [104]. The cut-off distance, r_c, is the same for both the two- and three-particle interactions (2.8 Å). The interaction potential that we use has a finite range (has no long-range dispersion attraction). A potential tail plays no role for the chemical reaction. In the triad subscript j, i, k, the middle letter i refers to the atom at the subtended angle vertex. A full description and analysis of the interaction potential was given earlier [56]. The total potential was used to determine the distance of chemically bonded particles. When the distance between two particles was shorter than $r^*_{ji} \leq 4.0$ in reduced units, they were labeled as part of a molecule, in agreement with the procedure used by Stillinger and Weber [105]. Reduced units are indicated by superscript *.

Molecular dynamics simulations

Molecular dynamics simulations were performed of the reaction for three densities; $\rho = 0.0052\,\mathrm{g\,cm^{-3}}$ (gas), $\rho = 0.0191\,\mathrm{g\,cm^{-3}}$ (compressed gas) and $\rho = 0.0695\,\mathrm{g\,cm^{-3}}$ (liquid) with temperatures in the range; $T = 3640$–$20{,}800\,\mathrm{K}$. The simulation procedure has been described elsewhere [56], so only the main properties are given here. The simulation box contained 1000 particles ($N_p = N_\mathrm{H} + 2N_{\mathrm{H}_2} = 1000$), with dimensions $V = L_x L_y L_z$, where $L_y = L_z = L_x/2$. The mass of one hydrogen atom, $m_0 = 1.67 \cdot 10^{-27}\,\mathrm{kg}$, was used to define the reduced total mass density $\rho^* = \rho\sigma^3/m_0$. This implies that the reduced total mass density equals the reduced total molar density in terms of ($N_p = N_\mathrm{H} + 2N_{\mathrm{H}_2}$).

Mixture compositions

The degree of dissociation, the mole fractions and the pressure for three densities applied are given in Tables 14.1–14.3. For the gas-like density, $\rho^* = 0.0003$ in Table 14.1, we have almost 1% dissociation at the lowest temperature ($T^* = 0.07$). For the highest temperature ($T^* = 0.4$), we have almost 80% dissociation. The total pressure and the dissociation constant $K_x = {x_\mathrm{H}}^2/x_{\mathrm{H}_2}$ are shown for all cases.

By increasing the density to that of a denser gas, say, to $\rho^* = 0.0011$ in Table 14.2, the dissociation degree decreases as expected. It is less than 1% at $T^* = 0.07$, and about 55% at $T^* = 0.4$.

Table 14.1. Number of H atoms (N_H), mole fraction (x_i), total pressure (P^*) and the dissociation constant (K_x) for the density, $\rho^* = 0.0003$.

T^*	0.07	0.15	0.2	0.25	0.3	0.4
N_H	8	254	477	574	711	761
N_{H_2}	496	373	261	213	145	120
x_{H_2}	0.984	0.595	0.353	0.270	0.169	0.136
K_x	—	0.276	1.183	1.973	4.088	5.502
$10^5 P^*$	1.138	3.045	4.672	6.347	7.983	11.148

Table 14.2. Number of H atoms (N_H), mole fraction (x_i), total pressure (P^*) and the dissociation constant (K_x) for the density, $\rho^* = 0.0011$.

T^*	0.07	0.15	0.2	0.25	0.3	0.4
N_H	6	148	285	396	469	555
N_{H_2}	497	426	358	302	266	228
x_{H_2}	0.988	0.743	0.556	0.432	0.362	0.295
K_x	—	0.089	0.354	0.745	1.126	1.689
$10^5 P^*$	5.082	12.119	17.554	23.287	29.054	40.492

Table 14.3. Number of H atoms (N_H), mole fraction (x_i), total pressure (P^*) and the dissociation constant (K_x) for the density, $\rho^* = 0.004$.

T^*	0.07	0.15	0.2	0.25	0.3	0.4
N_H	2	67	138	192	232	289
N_{H_2}	499	467	431	404	384	356
x_{H_2}	0.996	0.875	0.758	0.677	0.623	0.552
K_x	—	0.018	0.077	0.154	0.228	0.363
$10^4 P^*$	4.094	8.478	11.292	14.051	16.729	21.892

At the highest (liquid-like) density, $\rho^* = 0.004$ in Table 14.3, a further decrease is seen in the degree of dissociation. For the lowest temperature, $T^* = 0.07$, we now have 0.2% dissociation, while there is an increase to 29% for $T^* = 0.4$. Simulations were carried out for the series of composition and temperature listed in these tables, and thermodynamic data were computed from this and the equations given above. As reference point we chose the nearly ideal state, of $T = 3640$ K and $\rho = 0.0052$g cm^{-3} ($T^* = 0.07$ and $\rho^* = 0.0003$). The variations are all as expected.

14.2 Fluctuations, Scaling Laws

Thermodynamic correction factor

The thermodynamic correction factors, Γ_{ij}, are defined as

$$\Gamma_{ij}^{-1} = \frac{1}{\beta}\left(\frac{\partial \ln \langle N_i\rangle}{\partial \mu_j}\right)_{T,V,\mu_{k\neq i}} = \frac{1}{\beta\langle N_i\rangle}\left(\frac{\partial \langle N_i\rangle}{\partial \mu_j}\right)_{T,V,\mu_{k\neq i}} \tag{14.6}$$

where $\beta = 1/k_BT$, μ is the chemical potential and subscripts i and j denote H and H_2, respectively. Furthermore, Γ_{ij}^{-1} is the ij element of the inverse of the Γ-matrix. At constant (T, V, μ_j), we determine Γ_{ij} from the fluctuation theory formula derived by Reed and Ehrlich [110].

$$\Gamma_{ij}^{-1}(L) = \frac{\langle N_i N_j \rangle - \langle N_i \rangle \langle N_j \rangle}{\langle N_j \rangle} \tag{14.7}$$

The SSM gives the inverse of the thermodynamic correction factor as a (predominantly) linear function of the inverse of L [2, 67, 96] for constant (T, V, μ_j):

$$\Gamma_{ij}^{-1}(L) = \Gamma_{ij,\infty}^{-1} + \frac{A_{ij}}{L} + \cdots \tag{14.8}$$

This relation can be used to obtain $\Gamma_{ij,\infty}^{-1}$ in the thermodynamic limit as a function of T, V, μ_j. The sampling procedure for the SSM was described in Chapter 5.

To show that the method performs in the expected way, we have plotted results for the thermodynamic correction factors, obtained from particle fluctuations, versus the inverse size of the sphere used for sampling, $1/L$. The example given in Fig. 14.1 applied to density $\rho^* = 0.0011$ and temperature $T^* = 0.07$. Linear regression was done with points in the region $0.125 \leq 1/L^* \leq 0.16$, the range where the $1/L$-dependence was observed. This range corresponds to sampling spheres that include from 12 to 16 particles.

We see that the inverse correction factor varies in an expected manner. When the radius of the small system approaches zero, the value of Γ_{ii}^{-1} approaches 1. This is expected as a system of one particle is ideal. The value of $\Gamma_{\mathrm{HH}}^{-1} = 1$ obeys the ideal property already for a wide range of values, while $\Gamma_{\mathrm{H_2H_2}}^{-1}$ increases from approximately 0.9 for the biggest sphere to 1 for $1/L \approx 0.4$. The correlation between H and H_2 is near zero, explained by the dominating concentration of H_2. The results for the higher densities are similar. Only $\Gamma_{\mathrm{H_2H_2}}^{-1}$ varies with $1/L$, and the factor increases with decreasing radius. The limiting value of 1 for infinitely small spheres is reached for all densities, the approach being faster for the smallest density. The value for the thermodynamic limit decreases with increasing density

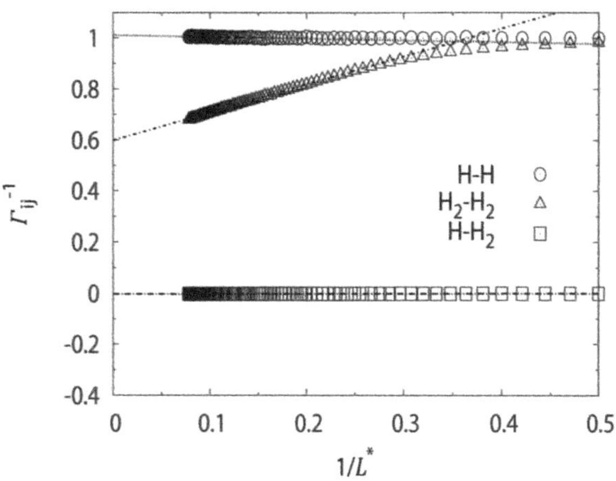

Figure 14.1. The inverse thermodynamic correction factor, Γ_{ij}^{-1}, as a function of $1/L$ for density $\rho^* = 0.0011$ at $T^* = 0.07$. At this temperature, the degree of dissociation was independent of density, within the numerical accuracy. Linear regression was done in the region $0.125 \leq 1/L^* \leq 0.16$ to find the value in the thermodynamic limit by extrapolation, cf. straight lines. Reproduced from [46] with permission from RSC publisher.

(or number of H_2). The values of $\Gamma_{\mathrm{HH}}^{-1}$ and $\Gamma_{\mathrm{HH_2}}^{-1}$ remain constant at 1 or 0, respectively, for all densities. These results on $\Gamma_{ij}^{-1}(L)$ confirm that the scaling method works in a manner expected for the range of variables chosen, see Ref. [47] for more details.

Partial molar enthalpies

The partial molar enthalpy of component i, h_i, at constant T, V, μ_j, is determined from the following formula for fluctuations of the total energy and density in the small system [47]

$$\widehat{h}_i(L) = \left(\frac{\partial \widehat{H}}{\partial N_i}\right)_{T,V,\mu_{j \neq i}} = \frac{\langle \mathcal{H} N_i \rangle - \langle \mathcal{H} \rangle \langle N_i \rangle + k_B T \langle N_i \rangle}{\langle N_i^2 \rangle - \langle N_i \rangle^2} \qquad (14.9)$$

Brackets denote time and/or ensemble averages. The internal energy is equal to $U = \langle \mathcal{H} \rangle$. According to Chapter 11, the partial molar enthalpies are (predominantly) linear functions of the inverse of the radius of the small sampling sphere, see also [2, 67, 96], L. The partial molar enthalpies $\widehat{h}_i$ are used to give

$$\widehat{h}_i(L) = h_{i,\infty} + \frac{A'_i}{L} + \cdots \tag{14.10}$$

where $h_{i,\infty} = \widehat{h}_{i,\infty}$ is the partial molar enthalpy at constant (T, V, μ_j) in the thermodynamic limit. With a spherical sampling volume, the scaling law is in good approximation a straight line, as curvature effects are small [96].

The partial molar enthalpy, $\widehat{h}_i$, was determined for small systems at constant (T, V, μ_j). The results for $\widehat{h}_i$ for H and H_2 are shown as a function of the inverse radius of the small system in Fig. 14.2 for $T^* = 0.2$. At this temperature, the mole fraction of H_2 varies from $x_{H_2} = 0.353$ ($\rho^* = 0.0003$) to $x_{H_2} = 0.758$ ($\rho^* = 0.004$) between the lowest and highest densities. Linear regression was preformed in the range $0.15 \leq 1/L^* \leq 0.35$. This range was chosen after comparing results for systems with 1000 and 4096 particles. No size effect was found in this range. At the other temperatures, results are similar.

The range which shows $1/L$ behavior is somewhat smaller for $\widehat{h}_i$ than for the inverse thermodynamic factor. The densities differ in the

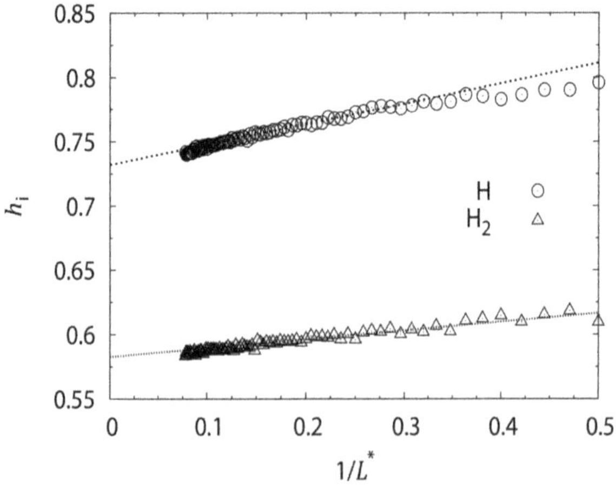

Figure 14.2. The partial molar hat-enthalpies of H (upper curve) and H_2 (lower curve) as a function og $1/L^*$. Linear regression was done in the region $0.125 \leq 1/L^* \leq 0.5$ to find the value in the thermodynamic limit by extrapolation, cf. straight lines. Reproduced from [46] with permission from RSC publisher.

two determinations, and this may offer an explanation. For liquid-like densities, there is even a cross-over of the curves for partial molar enthalpies. Nevertheless, a range of linear behavior is consistently there and can be used to find thermodynamic data for macroscopic systems. Results from extrapolations to the thermodynamic limit are described in what follows.

14.3 Thermodynamic Limit Values

Kirkwood–Buff integrals

As described in Chapter 10, the Kirkwood–Buff integrals G_{ij}, [70], can be used to relate all thermodynamic variables to the fluctuating variables. The G_{ij} is defined as the integral over the pair correlation function g_{ij} of the particle fluctuations. But we shall also use the same name for the right-hand side of the following equation:

$$G_{ij} = \int [g_{ij}(r) - 1]dV = V\frac{\langle N_i N_j\rangle - \langle N_i\rangle\langle N_j\rangle}{\langle N_i\rangle\langle N_j\rangle} - \frac{\delta_{ij}}{c_i} = \frac{1}{c_i}\Gamma_{ij}^{-1} - \frac{\delta_{ij}}{c_i} \tag{14.11}$$

Here δ is the Kronecker delta, and c_i is the concentration of i ($c_i = N_i/V$). The last equality is obtained by introducing Eq. (14.7) into the middle equality of Eq. (14.11). This Kirkwood–Buff integral can then be computed from simulation results, as described in the preceding section, at constant T, V, μ_j. [33]

The derivative of the chemical potential of component i of a binary mixture is related to G_{ij} at constant T, P, by

$$\frac{1}{k_B T}\left(\frac{\partial \mu_i}{\partial x_i}\right)_{T,P} = \frac{1}{x_i} + \frac{c_j\left(2G_{ij} - G_{ii} - G_{jj}\right)}{1 + x_i c_j\left(G_{ii} + G_{jj} - 2G_{ij}\right)} \tag{14.12}$$

where x_i is the mole fraction of i ($x_i = N_i/(N_{\mathrm{H}} + N_{\mathrm{H_2}})$; here $N_{\mathrm{H_2}}$ is the number of molecules so that $x_{\mathrm{H}} + x_{\mathrm{H_2}} = 1$). The relation between the thermodynamic correction factor and the activity coefficient, γ_i, was also given in terms of Kirkwood–Buff integrals [67]

$$\left(\frac{\partial \ln \gamma_i}{\partial \ln x_i}\right)_{T,P} = -\frac{x_i c_j (G_{ii} + G_{jj} - 2G_{ij})}{1 + c_j x_i (G_{ii} + G_{jj} - 2G_{ij})} \tag{14.13}$$

The partial molar volume, Eq. (14.14), and the isothermal compressibility, Eq. (14.15), for a binary mixture are

$$V_i = \frac{1 + (G_{jj} - G_{ij})c_j}{c_i + c_j + c_i c_j (G_{ii} + G_{jj} - 2G_{ij})} \tag{14.14}$$

and

$$\kappa k_B T = \frac{1 + G_{ii}c_i + G_{jj}c_j + \left(G_{ii}G_{jj} - G_{ij}^2\right) c_i c_j}{c_i + c_j + (G_{ii} + G_{jj} - 2G_{ij})\, c_i c_j} \tag{14.15}$$

We have shown in the preceding section how to find thermodynamic properties in the thermodynamic limit. The SSM gives access to values for the (T, V, μ_j) ensemble. These values can be transformed to conditions that are more interesting to chemists, to (T, V, N_j) or (T, P, N_j). Transformations can only be done in the thermodynamic limit, however. We transform one set of variables into a set in another ensemble, say from (T, V, μ_j) to (T, V, N_j), through Eqs. (14.19) and (14.20), using enthalpy as an example. We follow Ben-Naim [71] (his Eqs. (4.23) and (4.24)) and obtain first the symmetric B_{ij} matrix, at constant (T, V, μ_j)

$$\begin{aligned} B_{ij} = B_{ji} &= \frac{k_\mathrm{B}T}{V}\left(\frac{\partial N_j}{\partial \mu_i}\right)_{T,V,\mu_{k\neq i}} \\ &= k_\mathrm{B}T\left(\frac{\partial c_j}{\partial \mu_i}\right)_{T,V,\mu_{k\neq i}} = c_i c_j G_{ij} + c_i \delta_{ij} = \frac{1}{V}\left(\langle N_i N_j\rangle - \langle N_i\rangle\,\langle N_j\rangle\right) \end{aligned} \tag{14.16}$$

where the Kirkwood–Buff integrals always refer to the grand canonical ensemble. The matrix A_{ij} is defined at constant T, V, N_j by

$$A_{ij} = A_{ji} = \beta V\left(\frac{\partial \mu_j}{\partial N_i}\right)_{T,V,N_{k\neq i}} = \beta\left(\frac{\partial \mu_j}{\partial c_i}\right)_{T,V,c_{k\neq i}} \tag{14.17}$$

The A_{ij} and B_{ij} matrices are the inverse of one another:

$$\sum_j B_{ij}A_{jk} = \delta_{ik} \tag{14.18}$$

These properties can then be computed.

$$\left(\frac{\partial H}{\partial N_i}\right)_{T,V,N_{j\neq i}} = B_{jj}\left[A_{jj}\left(\frac{\partial H}{\partial N_i}\right)_{T,V,\mu_{j\neq i}} + A_{ij}\left(\frac{\partial H}{\partial N_j}\right)_{T,V,\mu_{i\neq j}}\right] \tag{14.19}$$

$$\left(\frac{\partial H}{\partial N_j}\right)_{T,V,N_{i\neq j}} = B_{ii}\left[A_{ii}\left(\frac{\partial H}{\partial N_j}\right)_{T,V,\mu_{i\neq j}} + A_{ij}\left(\frac{\partial H}{\partial N_i}\right)_{T,V,\mu_{j\neq i}}\right] \tag{14.20}$$

A summation is performed over all doubly appearing subscripts in Eqs. (14.19) and (14.20). The ensemble (T, P, N_j) is most commonly used in chemistry. We obtain

$$\left(\frac{\partial H}{\partial N_i}\right)_{T,P,N_{j\neq i}} = \frac{V_i H}{V} + \frac{N_j}{V}\left[V_j\left(\frac{\partial H}{\partial N_i}\right)_{T,V,N_{j\neq i}} - V_i\left(\frac{\partial H}{\partial N_j}\right)_{T,V,N_{i\neq j}}\right] \tag{14.21}$$

To summarize, we see that we can compute values for all ensembles from the thermodynamic limit values of

- Thermodynamic correction factors, Γ_{ij} at T, V, μ
- Partial molar enthalpies at constant T, V, μ
- Partial molar volumes, V_i, at constant T, V, μ
- Matrices A_{ij} and B_{ij} as defined by Kirkwood and Buff [70]

This shall be done in the following.

Properties of a chemical reaction

The enthalpy of a system is $H = U + PV$, where U is the internal energy, P is the pressure and V is the volume. The enthalpy at constant T and P can be written as

$$H = \sum_{i=1}^{n} H_i N_i \tag{14.22}$$

where H_i is the partial molar enthalpy of component H or H_2. Thus, the partial molar enthalpy is the derivative of H with respect to the number, N_i, of either H or H_2 as follows:

$$H_i = \left(\frac{\partial H}{\partial N_i}\right)_{T,P,N_j} \tag{14.23}$$

The enthalpy of the reaction is given from the partial molar enthalpies of the components

$$\Delta_r H = 2H_{\mathrm{H}} - H_{\mathrm{H}_2} \tag{14.24}$$

The thermodynamic equilibrium constant K_{th} derives from the standard reaction Gibbs energy

$$\Delta_r G^0 = -RT \ln K_{th} \tag{14.25}$$

The equilibrium constant can be found from the van't Hoff equation, at constant pressure, if the enthalpy of reaction is known at standard conditions.

$$\left[\frac{d(\ln K_{th})}{d(1/T)}\right]_P = -\frac{\Delta_r H^0}{R} \tag{14.26}$$

Superscript ^{0} denotes the standard state (1 bar). In the presence of a varying pressure, there is an additional term in this equation:

$$d \ln K_{th} = -\frac{\Delta_r H^0}{R} d\left(\frac{1}{T}\right) - \frac{\Delta_r V^0}{RT} dP \tag{14.27}$$

where the last term gives the pressure dependence, and $\Delta_r V^0$ is the reaction volume.

With the ideal gases of the pure components k as standard state at 1 bar pressure, the chemical potential is

$$\mu_i = \mu_i^0 + RT \ln f_i/P^0 = \mu_i^0 + RT \ln(x_i P \gamma_i)/P^0 \tag{14.28}$$

The fugacity of the gas component, f_i, is equal to the partial pressure of the gas, which is $x_i P$ times the activity coefficient γ_i. By introducing this into the condition $\Delta G = 0$ for the reaction in question, we obtain the equilibrium constant in terms of the dissociation constant, K_x, total pressure, and the activity coefficient ratio:

$$K_{th} = \frac{x_{\mathrm{H}}^2}{x_{\mathrm{H}_2}} \frac{P}{P^0} \frac{\gamma_{\mathrm{H}}^2}{\gamma_{\mathrm{H}_2}} = K_x \frac{P}{P^0} \frac{\gamma_{\mathrm{H}}^2}{\gamma_{\mathrm{H}_2}} = K_P \frac{\gamma_{\mathrm{H}}^2}{\gamma_{\mathrm{H}_2}} \tag{14.29}$$

where $K_P = K_x P/P^0$ is the pressure-based dissociation constant.

The partial molar heat capacities of the components and the reaction and the isothermal compressibility of the mixture are found from standard expressions as follows:

$$C_P = \left(\frac{\partial H}{\partial T}\right)_P \tag{14.30}$$

$$\kappa_T = -\frac{1}{V}\left[\frac{\partial V}{\partial P}\right]_T \tag{14.31}$$

The ideal value of the heat capacity of an atomic fluid is $3R/2$, when rotational and vibrational modes are not excited. For an ideal gas, $\kappa_T = 1/P$. The temperature T is calculated from the average kinetic energy per degree of freedom of all particles as follows:

$$T = \frac{1}{3k_B N_p} \sum_{i=1}^{N_p} m_i \mathrm{v}_i^2 \tag{14.32}$$

where $\mathrm{v}_i^2 = \mathrm{v}_{x,i}^2 + \mathrm{v}_{y,i}^2 + \mathrm{v}_{z,i}^2$ and $N_p = N_{\mathrm{H}} + 2N_{\mathrm{H}_2}$ is the total number of atoms.

From the virial theorem, the expression for the pressure in the presence of two- and three-particle interactions is

$$P = \frac{k_B T N_p}{V} - \frac{1}{3V} \sum_{i=1}^{N_p} \left[\frac{1}{2} \sum_{j \text{ pair with } i} \frac{\partial u_2(r_{ij})}{\partial r_{ij}} r_{ij} + \sum_{j<k \text{ triplet with } i} \left(\frac{\partial h_{j,i,k}}{\partial r_{ji}} r_{ji} + \frac{\partial h_{j,i,k}}{\partial r_{ik}} r_{ik} + \frac{\partial h_{j,i,k}}{\partial r_{jk}} r_{jk} \right) \right] \tag{14.33}$$

where $h_{j,i,k} = h_{j,i,k}(r_{ji}, r_{ik}, \theta_{j,i,k})$. The first term in Eq. (14.33) gives the ideal contribution to the pressure, while the second term gives the contribution from the two- and three particle potentials, respectively. For a discussion regarding the pressure calculations, see [56].

14.4 Numerical Examples

Thermodynamic correction factor

The inverse thermodynamic correction factor in the thermodynamic limit, $\Gamma^{-1}_{ij,\infty}$, is plotted in Fig. 14.3 as a function of temperature for all densities studied. Red open symbols correspond to $\Gamma^{-1}_{\mathrm{HH},\infty}$, blue filled symbols, to $\Gamma^{-1}_{\mathrm{H_2H_2},\infty}$ and black cross-shaped symbols, to $\Gamma^{-1}_{\mathrm{HH_2},\infty}$.

The figure shows that $\Gamma^{-1}_{\mathrm{HH},\infty}$ varies little with temperature and density. For all densities, $\rho^* = 0.0003$ (red triangle), $\rho^* = 0.0011$ (red circle) and $\rho^* = 0.004$ (red rectangle), we find that $\Gamma^{-1}_{\mathrm{HH},\infty} \in [0.9 : 1]$. A small decrease is found from 1 at $T^* = 0.07$ with increasing

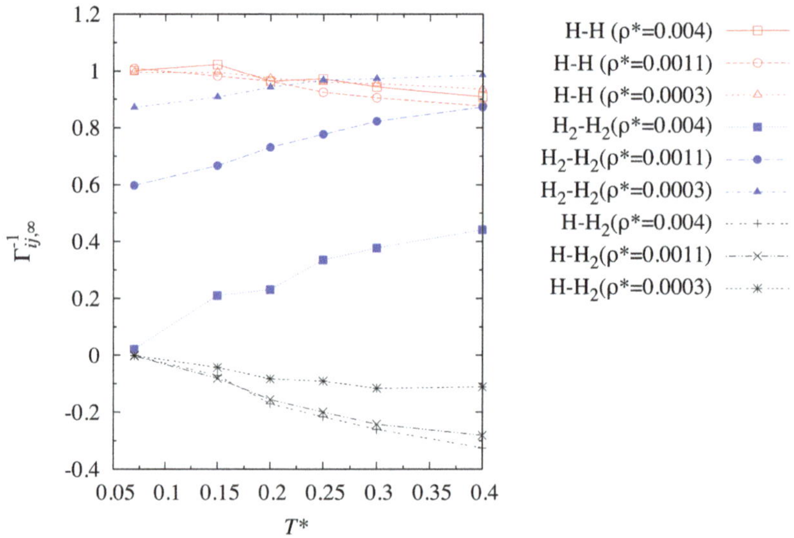

Figure 14.3. The inverse thermodynamic correction factor at the thermodynamic limit, $\Gamma^{-1}_{ij,\infty}$ for H and H_2 for all three densities under consideration ($\rho^* = 0.0003$, $\rho^* = 0.0011$ and $\rho^* = 0.004$). Red open symbols represent $\Gamma^{-1}_{\mathrm{HH},\infty}$, blue closed symbols, $\Gamma^{-1}_{\mathrm{H_2H_2},\infty}$ and black cross symbols, $\Gamma^{-1}_{\mathrm{HH_2},\infty}$. Reproduced from [46] with permission from RSC publisher.

temperature, but the value stays within the given interval. An increasing temperature leads to an increasing amount of H in the system, see Tables 14.1–14.3.

The value of $\Gamma^{-1}_{\mathrm{H_2H_2},\infty}$ (filled blue symbols) is seen to increase with temperature, as the amount of H_2 decreases. For the lowest density, $\rho^* = 0.0003$ (blue triangle), we see that it increases from approximately 0.9 to 1 for the investigated temperature interval. By increasing the density to $\rho^* = 0.0011$ (blue circle), we find a steeper increase with temperature, and this is further enhanced by increasing the density to $\rho^* = 0.004$ (blue square). If we compare $\Gamma^{-1}_{\mathrm{HH_2},\infty}$ for the different densities, we see the smallest temperature dependence for the lowest density, $\rho^* = 0.0003$ ($*$). For the two higher densities, $\rho^* = 0.0011$ ($\times$) and $\rho^* = 0.004$ ($+$), we see a bigger decrease with increasing temperature.

The value of the thermodynamic correction factor for one component interacting with itself is 1 if the system is ideal (pure component; ideal gas). The value is zero for interactions between different components in this state. We conclude that the gas-like fluid at the lowest temperature and density, $T^* = 0.07,\ \rho^* = 0.0003$, is close to an ideal mixture. This is not surprising since, only 0.8% of the hydrogen is dissociated at the temperature (8 H atoms out of a 1000 particles). This is supported by a very small contribution from the three-particle interaction on the pressure (see what follows).

The conclusion is supported by the expression for the chemical potential given by Kirkwood and Buff [70]

$$\mu_1/k_BT = \log x_1 + f_2x_2^2 + \cdots + \mu^\circ/k_BT \tag{14.34}$$

where

$$f_2 = \frac{G^\circ_{11} + G^\circ_{22} - 2G^\circ_{12}}{2v^\circ_1} \tag{14.35}$$

Here the superscript $^\circ$ indicates that the cluster integrals are evaluated in an infinitely dilute solvent (pure H_2). Component 1 represents the solvent (H_2) and 2, the solute (H). We used the values at $T^* = 0.07$ for G°_{ii} and G°_{ij}. At this temperature we do not have a pure component (0.8% dissociation), but we assume that the effect of this is negligible, and that the system still behaves approximately like a pure component system. The value of f_2 in the above equation can then be directly linked to the activity coefficient for H, γ_{H}, via

$$f_2 x_2^2 = \ln \gamma_2 \tag{14.36}$$

From this we obtained $\gamma_2 = \gamma_{\mathrm{H}} = 0.99$, confirming that we are indeed very close to ideal solution conditions.

The compressibility and the reaction volume

The isothermal compressibility of the reacting mixture, κ_T, was calculated using Eq. (14.15). The results for all densities ($\rho^* = 0.0003$, $\rho^* = 0.0011$ and $\rho^* = 0.004$) are given in Fig. 14.4. By studying Fig. 14.4, we see that κ_T varies with density. As expected, the lowest density, $\rho^* = 0.0003$, is easier to compress than the liquid, $\rho^* = 0.004$. The smaller the value of κ_T, the bigger the pressure needed to compress the mixture. This is reflected in Fig. 14.4, where we see that the compressibility of the high density mixture,

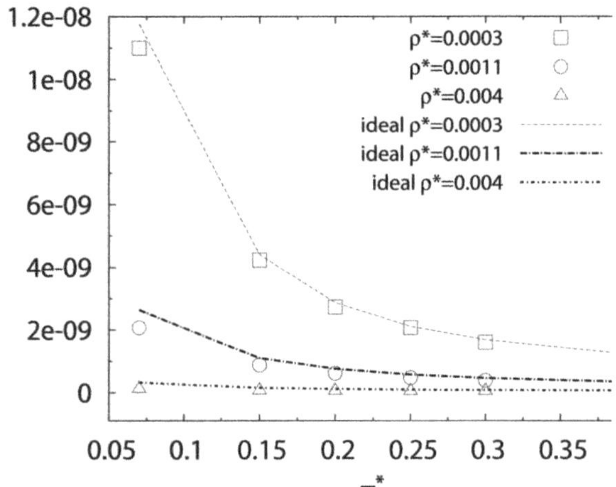

Figure 14.4. The compressibility of the reacting mixture, κ_T, as a function of temperature for densities; $\rho^* = 0.0003$, $\rho^* = 0.0011$ and $\rho^* = 0.004$. The ideal compressibility ($1/P$) is given with lines for all densities. Reproduced from [46] with permission from RSC publisher.

$\rho^* = 0.004$, is constant, and approximately zero throughout the temperature interval. For the two lower densities, $\rho^* = 0.0011$ and $\rho^* = 0.0003$, we see a decrease in the compressibility with increasing temperature. The gas-like fluid with the lowest density, $\rho^* = 0.0003$, is the easiest mixture to compress. It is also possible to compress the moderate density, $\rho^* = 0.0011$, corresponding to a compressed gas mixture, though significantly less than the lowest density. For a moderate density, the molecules are sufficiently close enough to be affected by intermolecular forces, and are thus still compressible [111]. The ideal compressibility ($1/P$) is shown for comparison in Fig. 14.4. It is fair to say that the low-density mixture at low temperature has results that are, relatively speaking, closest to ideal conditions. Relatively speaking, the other fluids are further away from the ideal result.

The partial molar volumes of H and H_2 were calculated from Eq. (14.14) and used to compute $\Delta_r V$ for all densities, see Fig. 14.5. The difference between the partial molar volumes of H and H_2 increases with decreasing density and temperature, so a higher curvature is seen in the function $\Delta_r V$ for the lowest density compared to the higher densities.

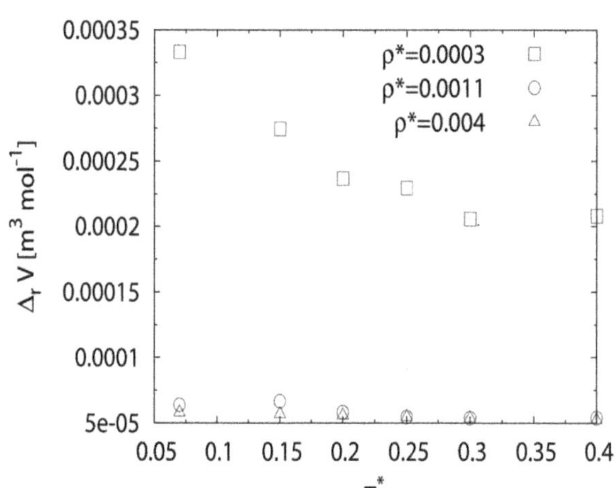

Figure 14.5. The reaction volume $\Delta_r V$ ($m^3 \ mol^{-1}$) as a function of temperature for all densities. Reproduced from [46] with permission from RSC publisher.

Partial molar enthalpies

We present results for the partial molar enthalpies in the different ensembles, the thermodynamic correction factors, the partial molar volumes and the isothermal compressibility in the thermodynamic limit. The temperature interval $T^* = 0.07 - 0.4$ was used. We shall first see how we find the sets of enthalpies at constant (T, V, μ_j), (T, V, N_j) and (T, P, N_j), where the first set is obtained by the Small System Method, and the sets for other ensembles are obtained by Lagrange transformations, see Eqs. (14.19) and (14.20). The matrices A_{ij} and B_{ij} will not be reported as such. We rather give Γ_{ij}^{-1}, which is equivalent to B_{ij}, and which gives A_{ij}.

The extrapolated partial molar enthalpies (the thermodynamic limit values) are given in Table 14.4 for all temperatures, $T^* = 0.07-0.4$, and all densities; $\rho^* = 0.0003$, $\rho^* = 0.0011$ and $\rho^* = 0.004$. All results showed a linear dependence and the linear regression was done for $0.15 \leq L^* \leq 0.3$ in all cases. The partial molar enthalpy in the table increases with increasing temperature, for H as well as H_2, for all densities.

From the partial molar enthalpies $h_{i,\infty} = (\partial H^*/\partial N_i)_{T,V,\mu_{j\neq i}}$, we find $(\partial H^*/\partial N_i)_{T,V,N_{j\neq i}}$ and the partial molar enthalpies $(\partial H^*/\partial N_i)_{T,P,N_{j\neq i}}$. The last set of partial molar enthalpies gives access to the reaction enthalpy of the system as a function of the temperature for the state of the system in question. All sets of partial

Table 14.4. Partial molar enthalpies, $h_{i,\infty}$ at $T, V, \mu_j \neq i$, in the thermodynamic limit.

	$\rho^* = 0.0003$		$\rho^* = 0.0011$		$\rho^* = 0.004$	
T^*	$h_{\mathrm{H},\infty}$	$h_{\mathrm{H}_2,\infty}$	$h_{\mathrm{H},\infty}$	$h_{\mathrm{H}_2,\infty}$	$h_{\mathrm{H},\infty}$	$h_{\mathrm{H}_2,\infty}$
0.07	0.34	−0.46	0.44	−0.42	0.59	−0.27
0.15	0.60	0.18	0.59	0.21	0.59	0.46
0.2	0.79	0.57	0.73	0.58	0.59	0.86
0.25	0.98	0.95	0.89	0.93	0.61	1.15
0.3	1.19	1.31	1.07	1.27	0.66	1.47
0.4	1.59	2.04	1.44	1.92	0.78	2.08

Note: Values are given in reduced units.

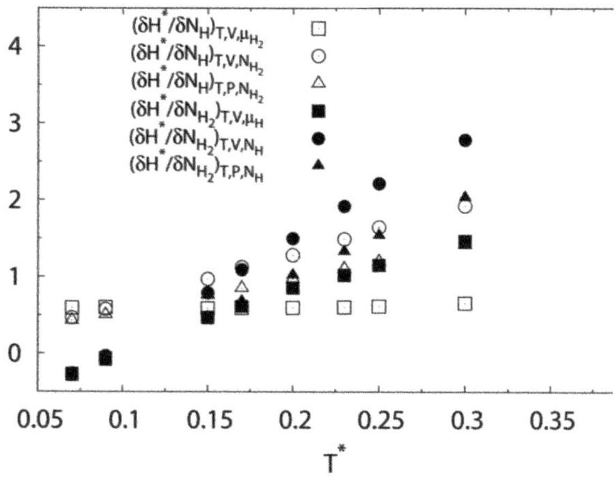

Figure 14.6. Partial molar enthalpies for all densities at the different ensembles; T, V, μ_j (*rectangle*), T, V, N_j (○) and T, P, N_j (△). Open symbols correspond to values for H while closed symbols represent H_2. Reproduced from [46] with permission from RSC publisher.

molar enthalpies, for (T, V, μ_j), (T, V, N_j) and (T, P, N_j), are shown in Fig. 14.6 for the highest density, where their differences become substantial.

At a higher temperature, the values for H_2 become more different, while they are approximately constant for H. This is in agreement with the results observed for the inverse thermodynamic correction factor. The partial enthalpy of H_2 is constant at low temperature, but varies significantly at high temperature. For H, a variation is seen over the whole temperature range, increasing with the increasing temperature. The results for the ensemble (T, P, N_j) in these figures will be used to construct the enthalpy of reaction. Before we show the outcome of that, we consider the other thermodynamic results.

The reaction enthalpy and the equilibrium constant

In a simplified study [56] of this reaction, a constant reaction enthalpy was computed for two densities, of 430 or 380 kJ mol^{-1}, respectively. Ideal conditions were assumed for the whole range of temperatures. The densities were $\rho^* = 0.004$ and $\rho^* = 0.0011$. The value was obtained from a plot of $\ln K_x$ vs $1/T$. We are now in a position to evaluate this calculation for the lowest density, $\rho^* = 0.0003$.

The standard state: Pressure variations of the reaction enthalpy

In order to determine the thermodynamic equilibrium constant from van't Hoff's equation, Eq. (14.29), we need information on the standard state, in the present case defined by an ideal gas at 1 bar. The lowest density mixture of reactant and product is close to being an ideal mixture, because the thermodynamic correction factors for the main components are close to unity at this condition, $\Gamma^{-1}_{\text{HH},\infty} = 1.00$, $\Gamma^{-1}_{\text{H}_2\text{H}_2,\infty} = 0.87$ and $\Gamma^{-1}_{\text{HH}_2,\infty} \approx 0$. In addition, the compressibility was near $1/P$, the ideal value. We find an ideal standard state from the low-density results at the lowest temperature, ($T^* = 0.07$, $\rho^* = 0.0003$), and use Eq. (14.27) to correct for the distance to 1 bar. At this temperature and density, the overall pressure of the system is approximately 850 bar. The pressure variation from 1 to 850 bar gives a contribution of 7 units to K_{th} which must be added to the normal van't Hoff term (see Eq. (14.27)) for these conditions. For the whole range of pressures and densities used, the correction term varies from 3–24 units. This gives a correction within the accuracy in the determination used otherwise.

Constant reaction enthalpy

The enthalpy of reaction does not vary much for the lowest density, so a first estimate of K_{th} as a temperature function can be obtained using the approximation $\Delta_r H(T^* = 0.07) = \Delta_r H^0 = 460$ kJ mol^{-1}. With a constant enthalpy of reaction, the integrated van't Hoff equation gives

$$\ln\left(\frac{K_{th,2}}{K_{th,1}}\right) = -\frac{\Delta_r H^0}{R}\left(\frac{1}{T_2} - \frac{1}{T_1}\right) \tag{14.37}$$

Table 14.5. K_{th} and K_P and $\gamma_{\rm H}^2/\gamma_{\rm H_2}$ as a function of the temperature, assuming that $\Delta_r H^0 = 460\,\text{kJ/mol}$ is constant, at $\rho^* = 0.0003$.

T^*	K_{th}	K_P	$\gamma_{\rm H}^2/\gamma_{\rm H_2}$
0.15	722	628	1.2
0.2	4256	4116	1.0
0.25	12335	9344	1.3
0.3	25075	24356	1.0
0.4	60867	45778	1.3

Application of this equation from $T_1 = 0.07$ to $T_2 = 0.15$, 0.2, 0.25, 0.3 or 0.4 gives $K_{th}(T_2)$. From the known value of the equilibrium constant, $K_{th,1}$, we find the ratio of the activity coefficients from Eq. (14.29). This ratio is given in Table 14.5 for the lowest density only.

Here, $K_P(T^* = 0.07) = K_{th}(T^* = 0.07)$ per definition as $T^* = 0.07$ was used as a reference point. From Table 14.5, we see that K_{th} has the same order of magnitude as K_P and that the fraction of the activity coefficients varies from 1.0 to 1.3.

Reaction enthalpy varies with temperature

The temperature dependence on the enthalpy can be included in the van't Hoff equation. The variation in $\Delta_r H$ with temperature is given in Fig. 14.7 for the lowest density. The line in the plot corresponds to the value found for ideal conditions (plot of $\ln K_x$ vs. $1/T$) [56].

The reaction enthalpy (in J mol^{-1}) was fitted to a quadratic function of the temperature, giving

$$\Delta_r H(\rho^* = 0.0003) = 0.0003T^2 - 8.9T + 4.9 \cdot 10^5 \qquad (14.38)$$

$$\Delta_r H(\rho^* = 0.0011) = 0.0006T^2 - 19.5T + 5.5 \cdot 10^5 \qquad (14.39)$$

$$\Delta_r H(\rho^* = 0.004) = 0.0007T^2 - 31.3T + 6.5 \cdot 10^5 \qquad (14.40)$$

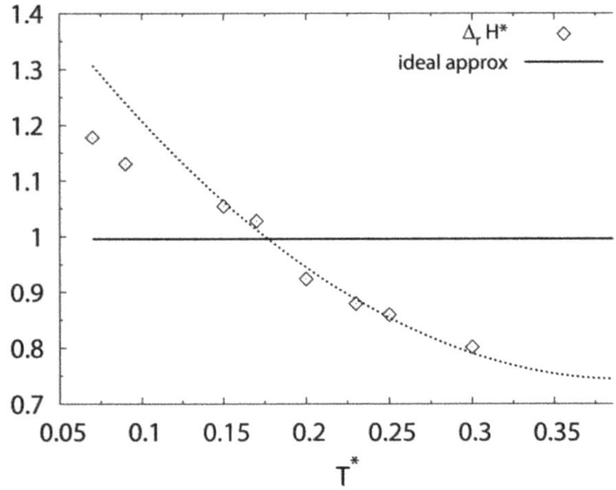

Figure 14.7. Reaction enthalpy, $\Delta_r H^*$, as a function of temperature and density $\rho^* = 0.004$. The points ($\diamond$) represent results of the simulation, where $\Delta_r H = 2H_{\rm H} - H_{\rm H_2}$. The dotted line represents the fit of the reaction enthalpy to a temperature function. The straight line represents the constant from the ideal mixture [56] found by plotting $\ln K_x$ as a function of temperature ($1/T$).

We rewrite the van't Hoff equation, using the identity $d(1/T) = -(1/T^2)dT$, and $\Delta_r H = aT^2 + bT + c$ and obtain after integration

$$\ln\left(\frac{K_{th,2}}{K_{th,1}}\right) = \frac{1}{R}\left[aT + b\ln T - \frac{c}{T}\right]\Bigg|_1^2 \tag{14.41}$$

The thermodynamic equilibrium constant and the ratio of the activity coefficients, calculated from this equation and Eq. (14.29), are shown in Table 14.6 for the lowest density.

When we include the temperature dependence in $\Delta_r H$, we see from Table 14.6 that we obtain a value of K_{th} in the same order of magnitude as K_P and K_{th} from Table 14.5 (where we used a constant $\Delta_r H$). The ratio of the activity coefficients is not constant and follow the same temperature dependence as we observe for $\Delta_r H$. The values indicate that we have repulsive forces (as expected) when $\gamma_{\rm H} \approx 1$ (Eq. (14.36)), at variance with the data in Table 14.5.

From the approximation $\Delta_r V(T^* = 0.07) = \Delta_r V^0$ in the last term in Eq. (14.27), we calculate a correction due to pressure variation in K_{th} of 3–24, for the lowest to the highest temperature.

Table 14.6. K_{th} and $\gamma_{\mathrm{H}}^2/\gamma_{\mathrm{H}_2}$ as a function of the temperature at $\rho^* = 0.0003$ assuming the enthalpy of the reaction is described by a polynomial fit.

T^*	K_{th}	K_P	$\gamma_{\mathrm{H}}^2/\gamma_{\mathrm{H}_2}$
0.15	604	628	1.0
0.2	3208	4126	0.8
0.25	8606	9344	0.9
0.3	16564	24356	0.7
0.4	37932	45778	0.8

By comparing these values to the calculated values of K_{th} in Tables 14.5 and 14.6, we see that they are negligible.

The results in Table 14.6 are therefore in agreement with results from the fluid compressibility, and molar reaction volume of the low-density mixture, which all show essentially ideal behavior. We therefore conclude that the last method, where the temperature dependence on the enthalpy is included, gives the most accurate results for K_{th}, as the repulsive forces between H and H_2 are captured.

14.5 Concluding Remarks

We have seen above how the SSM can be used to give accurate data for chemical reactions. Using the hydrogen dissociation reaction as an example, we have calculated relevant thermodynamic data for the reaction from one simulation. A range of temperatures and densities were studied. It may seem cumbersome to have to change between ensembles, but the transformation procedure is well established [47, 71]. The bonus is that information can be obtained about non-ideal behavior in mixtures with chemical reactions. This information is otherwise not easily accessible. As new force fields become available for calculation of chemical reactions using classical potentials, such as ReaxFF [112], this may turn out to be useful.

The results elucidate that the normal assumption, to take the reaction enthalpy constant in the integration of van't Hoff's equation, can be much improved, see Fig. 14.7. The biggest error is observed for the lowest density, where the ideal value is obtained when $K_{th} = K_x$, is far below the calculated reaction enthalpy. The reaction enthalpy calculated from SSM is for all densities of the order of magnitude 350–460 kJ mol^{-1} for $\rho^* = 0.0003$ to 0.004 as the enthalpy of reaction at standard state conditions (436 kJ mol^{-1}); compare also previous results obtained by us [56] of 430 and 380 kJ mol^{-1} for $\rho^* = 0.004$ and $\rho^* = 0.0011$.

Data for the thermodynamic correction factor, compressibility and partial molar volumes were all obtained from the same set of simulations. The method can provide information on the non-ideality of a reacting system, which is presently difficult to obtain in a computationally effective way.

Chapter 15

Adsorption on Flat, Large Surfaces

We show how the Small System Method can be used to find thermodynamic factors for the adsorbate on a large, flat surface. We study carbon dioxide adsorption into two layers on graphite, and compute the thermodynamic factor of each layer. The corresponding activity coefficient, entropy- and enthalpy of adsorption that apply in the thermodynamic limit, are computed.

The adsorbed state of a species can differ widely from bulk equilibrium states. An adsorbed species can, for instance, undergo a two-dimensional phase transformation [113, 114], clustering [113, 115] or chemical reactions [116]. Adsorption in pores, including pore mouths, can therefore affect transport processes [117, 118]. Easy access to thermodynamic data for adsorbed or confined materials is therefore important. Knowledge of such data can improve the modeling and understanding of equilibrium and transport in pores.

But thermodynamic data for surfaces are not easily measurable. This is where the Small System Method (SSM) of Chapter 5 can be helpful. We shall here see how thermodynamic data for adsorption can be generated using this method. The purpose of the chapter is to explain how we can use the SSM to produce information on adsorbed species in general, and on CO_2 on graphite in particular. Graphite membranes are cheap candidate membranes for CO_2-separation purposes [119]. We refer to Trinh *et al.* [37] for more details on the example and the procedure. Collell and Galliero [120] presented results for a confined slit.

The adsorption of carbon dioxide on graphite will be denoted as

$$\mathrm{CO_2(g)} + C_x \rightleftarrows \mathrm{CO_2}(\mathrm{s}, x) \tag{15.1}$$

where g means gas phase, s means adsorbed phase and C_x denotes the graphite surface with fractional coverage x. Adsorption has been found to take place in two distinct thermodynamic layers [37]. We shall see how thermodynamic data can be found for each layer. The data, which are specific to each layer, can be found from the fluctuations sampled for each layer. We shall illustrate how this can be done, using molecular simulation results, building on the work of Trinh *et al.* [37].

We describe first the film of gas molecules on graphite, and show that we can distinguish between two layers. The corresponding thermodynamic factors are obtained with the SSM. We proceed to show how the factor can be used with standard equations to provide other thermodynamic properties.

15.1 System

Classical molecular dynamics simulations can be performed with the LAMMPS package [66]. We construct the initial configuration by randomly distributing the gas molecules above the graphite surface. The system is stable after 2000 ps. A snapshot of the carbon dioxide fluid in a pore with graphite walls at 500 K is shown in Fig. 15.1. The rectangular box in the snapshot has periodic boundary conditions in all directions.

The pore in Fig. 15.1 has a diameter 70 Å. The picture shows a relatively dense gas. There is equilibrium between the fluid in the center of the pore and the adsorbed phase on the pore walls. Wall effects become negligible, and the gas can be regarded as a bulk fluid, at a distance 15 Å from the wall, see Fig. 15.2.

The graphite wall is constructed by adding five sheets of graphene. We see two sheets at the bottom and three at the top of the figure, indicating the periodic boundary conditions in the z-direction (the direction perpendicular to the surface). The graphite layers are fixed in space.

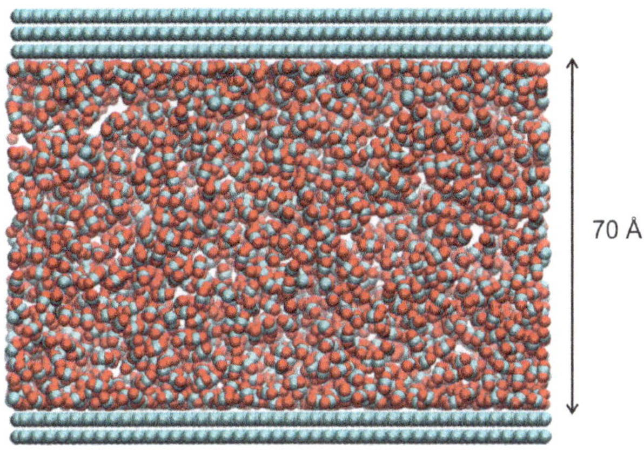

Figure 15.1. Snapshot of fluid CO_2 in a pore of diameter 70 Å. Reproduced from [37] with permission from The Royal Society of Chemistry.

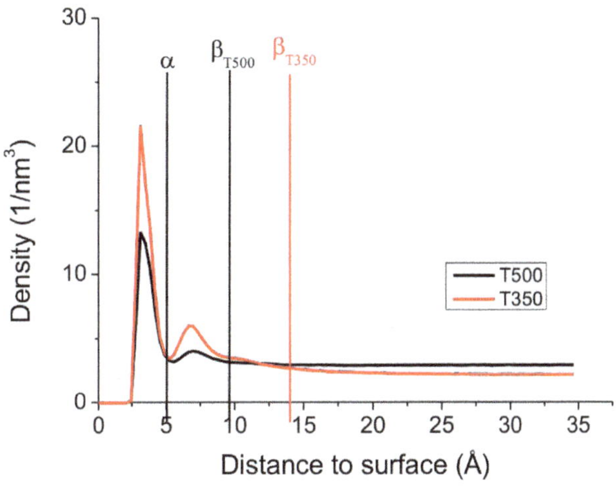

Figure 15.2. Density of CO_2 as a function of the distance from the graphite surface, and layer boundaries α and β. Results are shown for temperatures 350 and 500 K. Reproduced from [37] with permission from The Royal Society of Chemistry.

Good results for gas adsorption and surface diffusion can be obtained with the non-moving carbon atom model [121, 122], while the rigid body model, TraPPE, can be used to model CO_2 [123].

Intermolecular potentials can be modeled by shifted and truncated 12-6 Lennard-Jones (LJ) potentials, see Appendix D, with long-range Coulomb interactions [124]. The potentials are dealt with using the Ewald summation technique [124], see also [121].

The box serves as a reservoir for the small system that we shall examine. The total number of particles, volume and temperature of the reservoir ($N_{\text{tot}}, V_{\text{tot}}, T$) are constant in the procedure. Reservoir simulations are done with a typical time step of 0.001 ps. The box controls the temperature, T, and the chemical potential, μ, in the system that we shall sample.

15.2 Surface Densities

The density profile in the z-direction, $c(z)$, can be obtained by measuring the density as a function of the distance from the so-called dividing surface. Figure 15.2 gives the density variation in the present case. We see two peaks in the density for the present simulation conditions, and interpret these peaks as due to two layers, layer number 1 and 2, respectively. The first peak is located between 0 and $\alpha \approx 5$ Å. The thickness of this layer is a little more than the diameter of an oxygen ion, 3.6 Å. The linear carbon dioxide molecule in this layer must therefore lie on the graphite surface or stand on the surface in a tilted position. Most of the molecules are lying, but a few must be standing. The second layer, on the other hand, is thicker and less dense. The molecules here are more free to move along with, as well as in and out of, the surface. The position α does not depend significantly on the temperature. But the second peak, between 5 and 15 Å, is temperature-dependent. To see this, compare the values of β at 350 and 500 K.

The number of layers of CO_2 on the surface also depends on the temperature. For example, at high temperature (700 K) there is mostly a single layer. The methods presented in this chapter can deal with any number of layers in the same fashion. The densities of the surface are defined by [25]:

$$c^{\text{s}} = c_1^{\text{s}} + c_2^{\text{s}} = \int_0^{\alpha} c(z)dz + \int_{\alpha}^{\beta} c(z)dz \qquad (15.2)$$

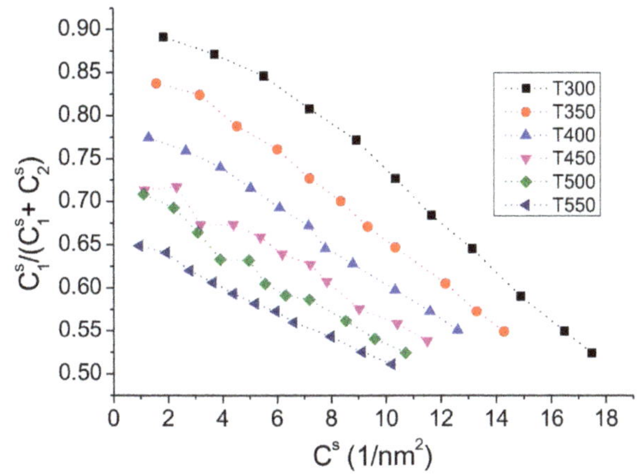

Figure 15.3. Fraction of molecules in the layer next to the graphite wall (layer 1) compared to in the adsorption in the whole layer (layers 1 and 2 together.) Reproduced from [37] with permission from The Royal Society of Chemistry.

Here $c(z)$ refers to a density in the pore at location z. This formula (15.2) will be applied to the first and the second peak in order to find the density or the adsorption in each layer and the total adsorption.

The fraction of molecules in layer 1 with respect to total number of molecules in the surface can be computed. Results are shown in Fig. 15.3. The fraction decreases as the layers are being filled up in parallel, and as the temperature rises.

15.3 Thermodynamic Factors

We are seeking a procedure for determination of thermodynamic data from knowledge of thermodynamic factors for each layer. The inverse of the thermodynamic factor is always obtained from fluctuations in particle number, N. The formula applies to any system size. When we control the temperature, T, the chemical potential, μ, and the surface area, Ω, we obtain the following from the fluctuation theory:

$$\frac{1}{\Gamma^s} = \left[\frac{\langle N^2\rangle - \langle N\rangle^2}{\langle N\rangle}\right]_{T,\mu,\Omega} \tag{15.3}$$

Superscript s indicates that the property applies to the surface. The formula will be used for layer 1 or 2 or for their combination, the total layer.

In the SSM (Chapter 5), the area of the sampling system (the two-dimensional small system) is varied, by varying the radius of the disk, L. The smallest radius is, in fact, so small that only one molecule is allowed inside the disk. The number of particles is proportional to L^2. For the radius $L = 10$ Å, the disk contains 20–30 molecules. A schematic illustration of the sampling system is shown in Fig. 15.4. We see the varying density from the wall to the gas phase, and the sampling disks. On the nanoscale, the small system is a disk in the surface with thickness δ and area Ω. The thickness of the sampling system is set equal to the thickness of the first layer, $\delta_1 = \alpha$, or the second layer, $\delta_2 = \beta - \alpha$, see Fig. 15.2 or both. By this choice, we can sample all layers at the same time.

In order to control the temperature and the chemical potential of the small system, we have to choose a reservoir that is large enough. We must first find a suitable reservoir size. A box with dimensions ($x = a = 42$ Å, $y = b = 54$ Å, $z = c = 84$ Å) or (a, b, c) was chosen for a test. Reservoir sizes of X1 $= (a, b, c)$, X4 $= (2a, 2b, c)$ and

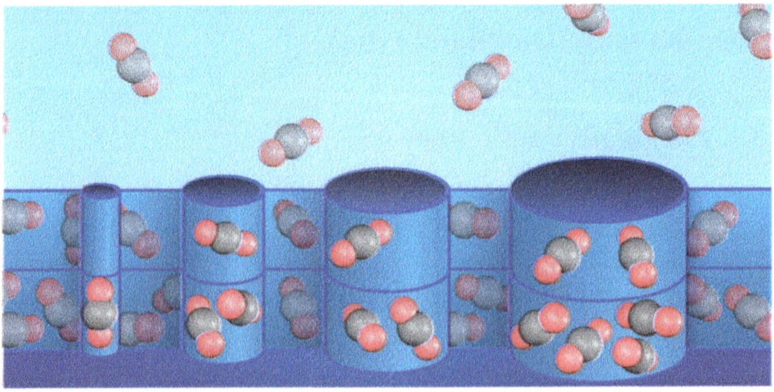

Figure 15.4. Illustration of the sampling method: Cylinders of radius L are used to sample fluctuations in particle numbers. The cylinder is composed of two disks that have the thicknesses of surface layers 1 and 2. Reproduced from [37] with permission from The Royal Society of Chemistry.

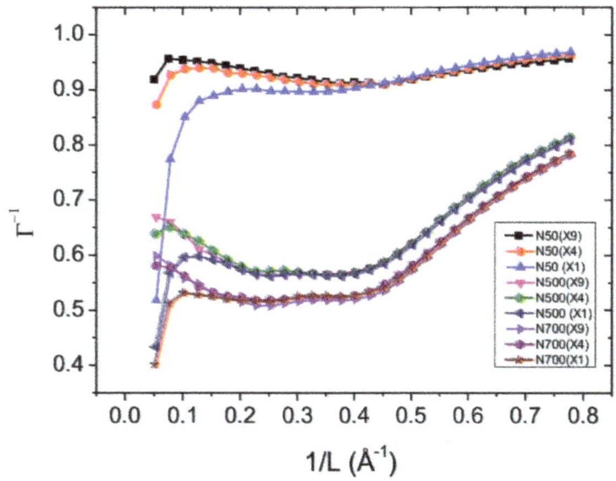

Figure 15.5. Impact on $(\Gamma^{s})^{-1}$ of reservoir sizes X1, X4 and X9 and for three fluid densities. See text for explanation of reservoirs and densities. Reproduced from [37] with permission from The Royal Society of Chemistry.

X9 $= (3a, 3b, c)$ were tested. The total number of particles was varied from 50 to 500 and 700 particles in reservoir X1, and 4 or 9 times these numbers in reservoirs X4 and X9, respectively.

The reservoir size has an impact on the inverse thermodynamic factor as illustrated in Fig. 15.5. We conclude that the X4 reservoir is big enough (disk or cylinder radius up to 20 Å) to give converging results for the two highest densities. This reservoir is therefore used in the following. For other conditions, the smallest density in particular, the mean free path is close to the size of the box.

Samplings were done for 2000 ps with the X4 reservoir at thermal equilibrium. The radius of the cylinders L varied from 1.3 to 20 Å. Thirty cylinders were put in a random position on the graphite surface. The total number of frames can amount to 10.000. Values for $\Gamma^{s}(L)$ were obtained from an average of 300.000 samples. The temperature varied from 300 to 550 K, while particle numbers varied from 200 to 2800. The value of Γ^{s} in the thermodynamic limit was obtained by plotting the left-hand side of Eq. (15.4) vs $1/L$ and extrapolating to 0

$$\frac{1}{\Gamma^{s}} = \frac{1}{\Gamma^{s,\infty}} + \frac{B}{L} \tag{15.4}$$

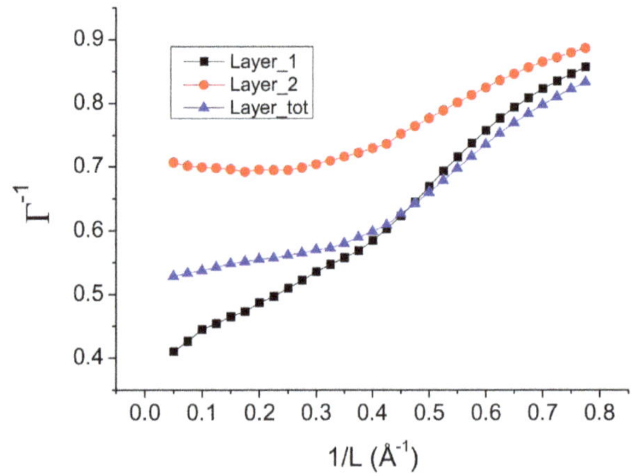

Figure 15.6. Variation in inverse thermodynamic factor with small system size. Results are shown for the first and second layers, plus the total layer. Reproduced from [37] with permission from The Royal Society of Chemistry.

Superscript ∞ means the value in the thermodynamic limit. Results for layers 1, 2 and the total layer are shown in Fig. 15.6. A linear regime is seen in the figure for values of $1/L$ between 0.1 and 0.3–0.4 Å^{-1}. This corresponds to cylinders with diameters between 2–3 and 10 Å, which is 1–3 times the length of a molecule.

The linear regime in Fig. 15.6 is specific for each layer, and is expressed by the particular value of B. By extrapolating the linear regime to $1/L \to 0$, we can obtain the wanted property $\Gamma^{s,\infty}$. In this manner, we can determine the thermodynamic factor that applies in the macroscopic limit, as well as values that apply for smaller dimensions.

We have seen above that a small system can be of molecular size. The thermodynamic equations do not break down. This seems counterintuitive, as thermodynamics normally is connected with an ensemble of molecules, but this is an outcome of Hill's nanothermodynamics. Thermodynamic equations apply, also at this small scale, because we have considered an ensemble of sampling cylinders (disks). The inverse thermodynamic factor is plotted vs the surface excess adsorption in Fig. 15.7. The plot is a straight line. Each layer contributes independently to the overall factor. The thermodynamic

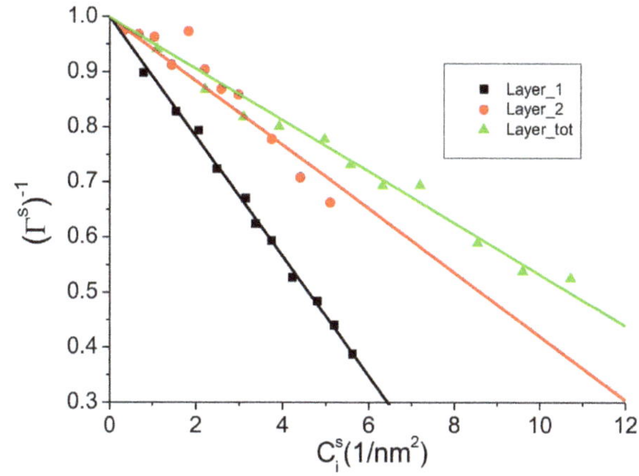

Figure 15.7. Thermodynamic factor in the macroscopic limit as a function of surface excess concentration for the first and second, plus the total layer. The low density limit gives the ideal value for Γ. Reproduced from [37] with permission from The Royal Society of Chemistry.

factor can now be calculated from its inverse value and used to derive a set of thermodynamic properties in a standard manner (see what follows).

15.4 Adsorption Thermodynamics

We present the common equations for adsorption, ending with a prescription on how such properties can be obtained from the thermodynamic factor. The chemical potential of a gas with pressure p is

$$\mu^{\mathrm{g}} = \mu^{0,\mathrm{g}} + RT \ln \left(\varphi \frac{p}{p^0} \right) \tag{15.5}$$

Here, $\mu^{0,\mathrm{g}}$ is the standard chemical potential at standard pressure $p^0 = 1$ bar. The activity coefficient φ describes how the real gas deviates from an ideal gas. For the adsorbed single component in surface layer i, we have

$$\mu_{\mathrm{i}}^{\mathrm{s}} = \mu_{\mathrm{i}}^{0,\mathrm{s}} + RT \ln a_{\mathrm{i}}^{\mathrm{s}} = \mu_{\mathrm{i}}^{0,\mathrm{s}} + RT \ln \left(\gamma_{\mathrm{i}}^{\mathrm{s}} \frac{c_{\mathrm{i}}^{\mathrm{s}}}{c_{\mathrm{i}}^{0,\mathrm{s}}} \right) \tag{15.6}$$

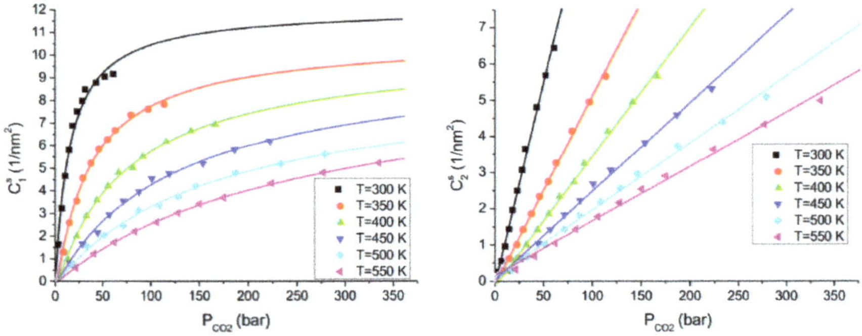

Figure 15.8. Isotherms for layers 1 and 2. The Langmuir model fits to layer 1 (left), and Henry's law applies to layer 2 (right). Reproduced from [37] with permission from The Royal Society of Chemistry.

Here, $\mu_i^{0,s}$ is the standard chemical potential for a surface with adsorption $c_i^{0,s}$, a_i^s is the (dimensionless) activity, c_i^s is the surface density or adsorption (in mol m^{-2}), and γ_i^s is the activity coefficient, all properties of the adsorbed component in layer i. A common choice for standard state is the hypothetical state given by Henry's law at temperature T. This state is found by extrapolation to zero coverage in the second layer as shown in Fig. 15.8. In this state, by definition $\gamma^s = 1$. Henry's law constant for adsorption equilibrium at various temperatures with an ideal gas is

$$K_H = \frac{c_2^s}{c^{0,s}} \frac{p^0}{p} \tag{15.7}$$

By introducing Eqs. (15.5), (15.6) and (15.7) into the equilibrium condition $\mu^g = \mu_2^s$, we obtain the surface standard chemical potential expressed by the ideal gas standard state and Henry's law's constant:

$$\mu^{0,g} = \mu^{0,s} + RT \ln K_H \tag{15.8}$$

The standard state is temperature-dependent. In order to determine the surface thermodynamic properties, we need to know $K_H(T)$. The starting point is the system's isotherms, shown in Fig. 15.8. There are isotherms for each layer, plotted left and right in Fig. 15.8. The isotherm of layer 1, to the left in the figure, shows a typical Langmuir behavior. The isotherm of layer 2, to the right in the

Table 15.1. Henry's law's constant as a function of temperature for CO_2 adsorption on graphite, in the layer next to the gas.

Temperature (K)	K_H
300	0.111
350	0.052
400	0.036
450	0.024
500	0.019

Note: Results are found from Fig. 15.8.

figure, shows straight lines, meaning that Henry's law is followed in this layer. This is therefore most convenient for standard state determination. From the slopes of the lines to the right in Fig. 15.8, we find $K_H(T)$ as tabulated in Table 15.1.

Each layer is treated separately or combined. We have $c^s = c_1^s + c_2^s$. The equilibrium condition for CO_2 in the gas phase, the total adsorbed phase and for the single layers is

$$\mu^g = \mu^s = \mu_i^s \tag{15.9}$$

Consider the total adsorbed phase. The thermodynamic factor is defined by

$$\Gamma^s = 1 + \left[\frac{\partial \ln \gamma^s}{\partial \ln c^s}\right]_{T,\Omega} \tag{15.10}$$

We use this expression first to find the activity coefficient, γ^s. We integrate from zero adsorption, where Henry's law applies and $\gamma^s = 1$, to the actual state, and obtain

$$\int_1^{\gamma^s} d\ln\gamma^s = \int_0^{c^s} (\Gamma^s - 1)\, d\ln c^s \tag{15.11}$$

Equation (15.6) gives the chemical potential at any adsorption relative to the standard state

$$\mu^s - \mu^{0,s} = RT \int_{c^{s,0}}^{c^s} d\ln(\gamma^s c^s / c^{s,0}) \tag{15.12}$$

With knowledge of Henry's law's constant $K_{\mathrm{H}}(T)$, we find the surface standard state from Eq. (15.8), and the chemical potential at any value of the adsorption and temperature from Eq. (15.12).

When the chemical potential is known from Eq. (15.12) as a function of temperature, we obtain the entropy of adsorption from its negative temperature derivative. The enthalpy of adsorption follows next from the chemical potential and the entropy. Results are presented in Figs. 15.9 and 15.10. The entropy of the surface is a linear function of the logarithm of the surface activity, as expected. The slope is near $-R$ (the activity coefficient contributions are non-negligible). The standard entropy of adsorption, $S_{\mathrm{i}}^{0,\mathrm{s}}$, is given in Table 15.2. Thermodynamic properties of layers 1 and 2, plus the total layer are given in this table. The enthalpy of adsorption is a weak function of temperature and surface adsorption.

It is interesting to consider the layer contributions to the total standard enthalpy of adsorption. Results for the enthalpy of adsorption in the first layer are more negative ($-4 \pm 1\,\mathrm{kJ/mol}$) than in the second layer ($3 \pm 1\,\mathrm{kJ/mol}$). The first layer adsorbs more strongly than the second, since the enthalpy of adsorption is negative in the

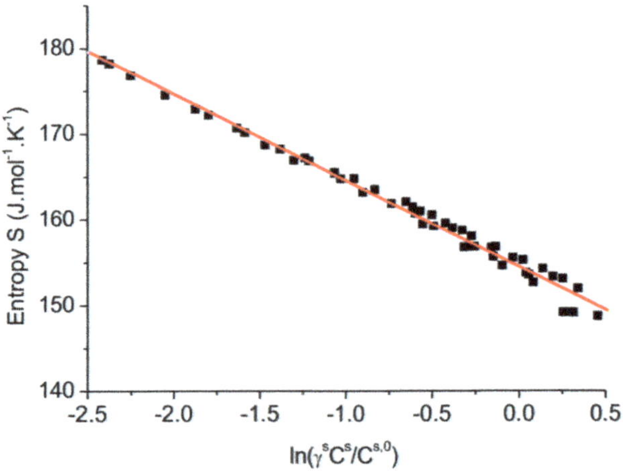

Figure 15.9. The entropy of adsorption of CO_2 on graphite as a function of the logarithm of the surface activity. Reproduced from [37] with permission from The Royal Society of Chemistry.

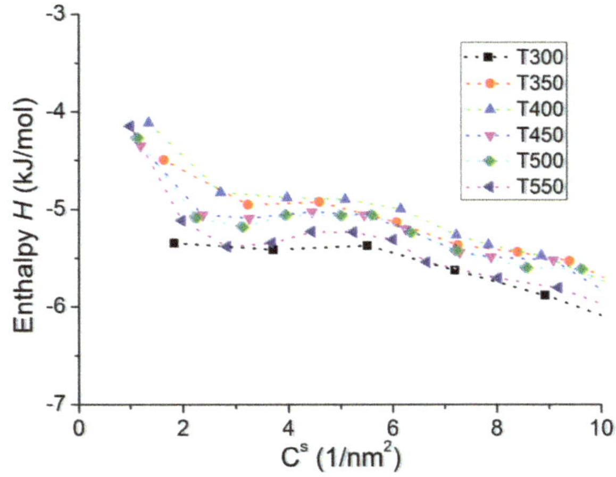

Figure 15.10. The enthalpy of adsorption of CO_2 on graphite as a function of the excess adsorption in layer 1 at temperatures between 300 and 550 K. Reproduced from [37] with permission from The Royal Society of Chemistry.

Table 15.2. Standard entropy and enthalpy of CO_2 adsorbed into two layers on a graphite surface

	Total layer	Layer 1	Layer 2
$\mu_i^{0,s}$, (kJ/mol)	-47 ± 1	-47 ± 1	-42 ± 1
$S_i^{0,s}$ (J/(K mol))	157 ± 3	145 ± 3	155 ± 3
$H_i^{0,s}$, (kJ/mol)	0 ± 1	-4 ± 1	3 ± 1

first layer. In the second layer, the positive enthalpy is compensated by the larger entropy, meaning that adsorption can still take place.

15.5 Data Reduction Procedure

We are now in position to write down the data reduction procedure used in this chapter:

1. After a characterization of the surface density, find $\Gamma^s(c^s; T)$ by SSM.
2. Find $K_H(T)$ for a suitable, possibly hypothetical, layer.
3. Find the surface standard state chemical potential for this choice using Eq. (15.8).

4. Compute $\gamma^{s}(c^{s};T)$ using Eq. (15.11).
5. Find the chemical potential using Eq. (15.12).
6. Find the entropy of the layer from minus the temperature derivative of the chemical potential.
7. Find the enthalpy from the chemical potential and entropy.

15.6 Concluding Remarks

We have seen how the SSM can play a central role in generating thermodynamic data for adsorbed states. The method is used to find the thermodynamic factor of one or two adsorbed layers, data which are difficult to obtain otherwise. The activity coefficient, the enthalpy and entropy of adsorption can then be calculated. Once data are obtained for the thermodynamic limit, one can make use of scaling laws for the thermodynamic factor, Eq. (15.4), to find corresponding data for small systems. This opens up the study of nanometric slit pores [120] and other pores. What is remarkable is that thermodynamic relations, properly defined, remain valid down to the molecular scale.

Chapter 16

Adsorption on Small, Curved Surfaces

Adsorption of gases on solid particles depends on the particle chemistry, but may also depend on their size and shape. The subdivision potential represents the energy needed to overcome the disadvantage of entropy reduction upon adsorption, a disadvantage that becomes larger when the system becomes smaller. We show how Hill's nanothermodynamics can be used to create a theory of adsorption, and how experimental and/or simulated data can be understood in terms of size-dependent properties.

Forest fires, volcanic eruptions, cars, trucks and industries may all produce particles in the atmosphere. These can next serve as adsorption- or nucleation sites [125]. Understanding adsorption on small surfaces is therefore central to understanding natural as well as man-made systems. In general, adsorption depends on particle type, but also size and shape. The dependence is relevant for gases that condense on particles in the atmosphere.

Research on size-effects in thermodynamics started in the 1930s. The literature on nucleation is most abundant [52, 126–131]. Gibbs' theory of heterogeneous systems [20] has been essential, while Hill's method [4] is little explored. In their description of curved surfaces, the methods of Hill and Gibbs are equivalent [36], see Chapter 12. The purpose of this chapter is to formulate adsorption equations for size- and shape-dependent adsorption, and to show how the equations can be used with simulated or experimental data to find size-dependent properties. The chapter builds on the work of Strøm *et al.* [38] on CO_2 adsorption on a spherical graphite-like adsorbent. This adsorption can be investigated using molecular dynamics simulations. System properties that are normally referred to as intensive start to depend on system size when the system becomes small. Differential thermodynamic properties for a small system are no longer Euler homogeneous functions of degree one, and cannot be integrated directly as before. This dependence can be investigated in a systematic manner with nanothermodynamics. We construct an ensemble of small systems and determine its properties. The system is completely characterized by the temperature, T, pressure, p, chemical potential, μ, and amount of a single adsorbent, N_A. It is thus an osmotic-like ensemble, cf. Section 3.5 and definitions that follow.

In what follows we derive the equations for this ensemble, introduce a reference state that is commonly used in adsorption thermodynamics, and apply the equilibrium condition to an adsorbed film and the gas above. A description follows on how simulation data can be obtained and analyzed. In the end we present the size-dependence of some central thermodynamic properties.

16.1 An Osmotic-like Ensemble

The idea of Hill [3, 4] was to introduce an ensemble of replicas of the small system. The ensemble of replicas has in this case a total internal energy U_t, a total entropy S_t, a total volume V_t, a total number of particles $N_{A,t}$ of the single adsorbent A (like graphite), a total number of particles N_t of a single adsorbate (like CO_2) that can adsorb on the adsorbent, and a number of replicas, $\mathcal{N}$. By construction, the ensemble variables U_t, S_t, V_t, $N_{A,t}$, N_t, $\mathcal{N}$ are Euler homogeneous functions of degree one. The total differential of U_t follows the Hill–Gibbs relation, cf. Eq. (1.3):

$$dU_t = TdS_t - pdV_t + \mu_A dN_{A,t} + \mu dN_t + \varepsilon d\mathcal{N} \tag{16.1}$$

The partial derivatives of U_t define the temperature, pressure and chemical potentials as follows:

$$\left(\frac{\partial U_t}{\partial S_t}\right)_{V_t,N_{A,t},N_t,\mathcal{N}} = T, \quad \left(\frac{\partial U_t}{\partial V_t}\right)_{S_t,N_{A,t},N_t,\mathcal{N}} = -p$$

$$\left(\frac{\partial U_t}{\partial N_{A,t}}\right)_{S_t,V_t,N_t,\mathcal{N}} = \mu_A, \quad \left(\frac{\partial U_t}{\partial N_t}\right)_{S_t,V_t,N_{A,t},\mathcal{N}} = \mu \tag{16.2}$$

The new thermodynamic derivative, ε, is defined by

$$\left(\frac{\partial U_t}{\partial \mathcal{N}}\right)_{S_t,V_t,N_{A,t},N_t} = \varepsilon \tag{16.3}$$

with the ensemble variables $S_t, V_t, N_{A,t}, N_t$. The *subdivision potential* ε was introduced by Hill, cf. Eq. (1.3). It measures the increase in the total internal energy as the number of replicas increases, keeping S_t, V_t, $N_{A,t}$, N_t constant. But S_t, V_t, $N_{A,t}$, N_t are subdivided over more replicas, which explains the name subdivision potential. The subdivision potential ε may be positive or negative, depending on whether reversible work is needed or gained when we introduce a new replica, thereby changing the ratio of the surface area to the volume of the ensemble.

We consider an ensemble of $\mathcal{N}$ independent replicas in which all replicas have the same amount of adsorbent N_A. This implies that $N_{A,t} = N_A\mathcal{N}$. We refer to Section 3.5 for a comparison. The Hill–Gibbs relation for the ensemble, written in terms of the set of independent variables S_t, V_t, N_t, N_A, and $\mathcal{N}$, becomes

$$dU_t = T\,dS_t - p\,dV_t + \mu\,dN_t + \mu_A\mathcal{N}\,dN_A + X\,d\mathcal{N} \tag{16.4}$$

where U_t is the total internal energy. The replica energy, X, is defined by

$$X \equiv \varepsilon + \mu_A N_A = \left(\frac{\partial U_t}{\partial \mathcal{N}}\right)_{S_t,V_t,N_t,N_A} \tag{16.5}$$

The replica energy X describes how a change $d\mathcal{N}$ in the number of replicas will change the internal energy of the entire ensemble when S_t, V_t, N_t and N_A are kept constant. The average entropy, $S \equiv S_t/\mathcal{N}$, average volume, $V \equiv V_t/\mathcal{N}$ and average amount of adsorbed component $N \equiv N_t/\mathcal{N}$ then decrease when $\mathcal{N}$ increases. In the process, the ensemble gains the amount $N_A \, d\mathcal{N}$ of adsorbent. In the thermodynamic limit, $X \to \mu_A N_A$. This motivates the definition of the integral chemical potential $\widehat{\mu}_A$ away from the limit by

$$\widehat{\mu}_A N_A \equiv X = \varepsilon + \mu_A N_A \tag{16.6}$$

In the thermodynamic limit, $\widehat{\mu}_A \to \mu_A$. We refer to μ_A as the differential chemical potential. Both $\widehat{\mu}_A$ and μ_A differ from the corresponding thermodynamic limit value at the same T and p. For the given choice of variables, the gas is not a part of the small system, which implies that N counts the number of absorbed particles only. When the adsorbent is spherical, N_A is related to the radius. In general, N_A is a measure of the size of the adsorbent.

The small system can therefore be said to differ from a large one, by the subdivision potential

$$\varepsilon = (\widehat{\mu}_A - \mu_A) N_A \tag{16.7}$$

The potential was viewed as a correction to classical thermodynamics in Chapter 3. It is called the subdivision potential because the process of subdividing the ensemble into a larger number of smaller systems, keeping S_t, V_t, N_t and $N_{A,t}$ constant, requires the energy $\varepsilon \, d\mathcal{N}$. It is characteristic of a size-dependent scaling law [40, 48]. The replica energy, the subdivision potential and the scaling laws are particular for the variables that control the ensemble.

The total variables U_t, S_t, V_t and N_t are Euler homogeneous functions of degree one in the number of replicas, so we can integrate Eq. (16.4) at constant T, p, μ, N_A. By dividing the resulting Euler equation by $\mathcal{N}$, we obtain the mean internal energy of a single replica

$$U \equiv U_t/\mathcal{N} = TS - pV + \mu N + \widehat{\mu}_A N_A \tag{16.8}$$

where $S \equiv S_t/\mathcal{N}$, $V \equiv V_t/\mathcal{N}$ and $N \equiv N_t/\mathcal{N}$. By introducing the definitions for U, S, V and N in Eq. (16.4) and eliminating $\widehat{\mu}_A N_A$

by Eq. (16.8), we obtain

$$dU = T\,dS - p\,dV + \mu\,dN + \mu_A\,dN_A \tag{16.9}$$

This differential change in internal energy of a single small system has the same form as the normal Gibbs relation for a macroscopic system, with the important distinction that the intensive properties are functions of the system size. The energy U is therefore not an Euler homogeneous function of S, V, N and N_A. And, we cannot obtain Eq. (16.8) by integration of Eq. (16.9). This can be seen if we rewrite Eq. (16.8) as

$$U = TS - pV + \mu N + \mu_A N_A + \varepsilon \tag{16.10}$$

where we have introduced the subdivision potential $\varepsilon = (\widehat{\mu}_A - \mu_A)N_A$. The equation is the same as we would expect for a large system except for the extra term ε.

This general property of small systems has several implications. Adsorption isotherms obtain one more variable (explained in Eq. (16.12)). It will also lead to a modification of Clausius–Clapeyron-type equations (cf. Eq. (16.40)). We shall examine these changes for single component adsorption, by computing the small system Gibbs adsorption isotherms, see Eq. (16.54), as well as the adsorption entropy and enthaply, see Eq. (16.40).

By differentiating Eq. (16.8) and subtracting Eq. (16.9), we obtain the Hill–Gibbs–Duhem equation

$$d(\widehat{\mu}_A N_A) = -S\,dT + V\,dp - N\,d\mu + \mu_A\,dN_A \tag{16.11}$$

This is the function characteristic of a single system in the osmotic ensemble. The form of the equation is practical, because it contains the environmental (controllable) variables as independent variables. By subtracting $d(\mu_A N_A)$ from Eq. (16.11), we obtain

$$d\varepsilon = -S\,dT + V\,dp - N\,d\mu - N_A\,d\mu_A \tag{16.12}$$

This equation shows that a small system has one additional independent variable compared to a macroscopic system. In the thermodynamic limit $\varepsilon = 0$, and Eq. (16.12) becomes the normal Gibbs–Duhem equation. This relation between the intensive variables

T, p, μ, μ_A means that only three of them are independent in the macroscopic limit. From Eqs. (16.11) and (16.12), we obtain

$$d\widehat{\mu}_A = -\frac{S}{N_A}\,dT + \frac{V}{N_A}\,dp - \frac{N}{N_A}\,d\mu - \frac{\varepsilon}{N_A^2}\,dN_A \tag{16.13}$$

The change in the adsorbent chemical potential, $d\mu_A$, is obtained by considering that μ_A is a function of T, p, μ and N_A, and by writing the general expression for the differential.

$$\begin{aligned} d\mu_A = &-\left(\frac{\partial S}{\partial N_A}\right)_{T,p,\mu} dT + \left(\frac{\partial V}{\partial N_A}\right)_{T,p,\mu} dp \\ &-\left(\frac{\partial N}{\partial N_A}\right)_{T,p,\mu} d\mu - \frac{1}{N_A}\left(\frac{\partial \varepsilon}{\partial N_A}\right)_{T,p,\mu} dN_A \end{aligned} \tag{16.14}$$

The first three differential coefficients were obtained from Maxwell relations that follow from Eq. (16.11). The average Gibbs energy and enthalpy of a single replica are

$$\begin{aligned} G &\equiv U - TS + pV = \mu N + \widehat{\mu}_A N_A \\ &= \mu N + \mu_A N_A + \varepsilon \equiv \widehat{G} + \varepsilon \end{aligned} \tag{16.15}$$

$$\begin{aligned} H &\equiv U + pV = TS + \mu N + \widehat{\mu}_A N_A \\ &= TS + \mu N + \mu_A N_A + \varepsilon \equiv \widehat{H} + \varepsilon \end{aligned} \tag{16.16}$$

These equations provide the basis needed to describe adsorption on a small adsorbent.

16.2 The Reference State

A reference state needs to be defined to make the equations above operational. It is common to choose as reference the state of pure adsorbent (amount N_A) with volume V_{0A} at external pressure, p, and temperature T. By pure adsorbent is meant the adsorbent with clean surface in the absence of adsorption, $N_0 = 0$.

The quantities T, p and N_A were defined above. The properties of the adsorbate can then be defined as the properties of the total system relative to those of the reference state. In practice, say for simulation purposes, we subtract properties of the pure adsorbent from the properties of the total system, while keeping the interaction energy with the film molecules. The equivalents of Eqs. (16.8), (16.9) and (16.11) for the reference state are

$$U_{0A} = TS_{0A} - pV_{0A} + \widehat{\mu}_{0A} N_A \tag{16.17}$$

$$dU_{0A} = T\, dS_{0A} - p\, dV_{0A} + \mu_{0A}\, dN_A \tag{16.18}$$

$$d(\widehat{\mu}_{0A} N_A) = -S_{0A}\, dT + V_{0A}\, dp + \mu_{0A}\, dN_A \tag{16.19}$$

By subtracting Eq. (16.17) from (16.8), Eq. (16.18) from (16.9), and Eq. (16.19) from (16.11), we obtain the following equations for the film:

$$U_s = TS_s - pV_s + \mu N_s - \widehat{\Phi} N_A \tag{16.20}$$

$$dU_s = T\, dS_s - p\, dV_s + \mu\, dN_s - \Phi\, dN_A \tag{16.21}$$

$$d(\widehat{\Phi} N_A) = S_s\, dT - V_s\, dp + N_s\, d\mu + \Phi\, dN_A \tag{16.22}$$

where Eq. (16.22) is the characteristic equation for the adsorbed film in the given environment. The film properties are defined by $S_s \equiv S - S_{0A}$, $V_s \equiv V - V_{0A}$, $N_s \equiv N - N_0 = N$, $\widehat{\Phi} \equiv \widehat{\mu}_{0A} - \widehat{\mu}_A$ and $\Phi \equiv \mu_{0A} - \mu_A$. The property N_s is defined for consistency in notation because N is already the amount of adsorbed component only. In the thermodynamic limit U_s, S_s, V_s, N_s and N_A are proportional to Ω.

16.3 Size-Dependent Thermodynamic Properties

From now on we assume that the adsorbent is unaffected by the adsorption. In general, the adsorbent volume is a function of T, p and N_A. Here we shall take this volume and the surface area as functions of N_A only. For a sphere of constant radius (shape), there is now only one independent variable among the adsorbent volume, surface area and N_A. We choose this variable to be the adsorbent surface area, Ω, since we want to describe the properties of the film. We therefore eliminate N_A in Eqs. (16.21) and (16.22), using $dN_A = (dN_A/d\Omega)\, d\Omega$, and obtain

$$dU_s = T\, dS_s - p\, dV_s + \mu\, dN_s - \varphi\, d\Omega \tag{16.23}$$

$$d(\widehat{\varphi}\Omega) = S_s\, dT - V_s\, dp + N_s\, d\mu + \varphi\, d\Omega \tag{16.24}$$

where we have re-defined the differential and integral chemical potential differences $\varphi \equiv \Phi(dN_A/d\Omega)$ and $\widehat{\varphi} \equiv \widehat{\Phi}(N_A/\Omega)$, respectively. The property φ is the spreading pressure in adsorption thermodynamics [132]. The properties $\widehat{\varphi}$ and φ are related by the subdivision potential ε_s by $\varepsilon_s \equiv \varepsilon - \varepsilon_0 = (\varphi - \widehat{\varphi})\Omega$. Equivalent forms of Eqs. (16.13) and (16.14) are obtained as follows:

$$d\widehat{\varphi} = \frac{S_s}{\Omega}\,dT - \frac{V_s}{\Omega}\,dp + \frac{N_s}{\Omega}\,d\mu + \frac{\varepsilon_s}{\Omega^2}\,d\Omega \tag{16.25}$$

$$\begin{aligned} d\varphi = {} & \left(\frac{\partial S_s}{\partial \Omega}\right)_{T,p,\mu} dT - \left(\frac{\partial V_s}{\partial \Omega}\right)_{T,p,\mu} dp \\ & + \left(\frac{\partial N_s}{\partial \Omega}\right)_{T,p,\mu} d\mu + \frac{1}{\Omega}\left(\frac{\partial \varepsilon_s}{\partial \Omega}\right)_{T,p,\mu} d\Omega \end{aligned} \tag{16.26}$$

The integral form of the coefficients in Eq. (16.25) and the differential form of the corresponding coefficients in Eq. (16.26), motivate the use of the names "integral spreading pressure" for $\widehat{\varphi}$ and "differential spreading pressure" for φ. From Eqs. (16.15) and (16.16), we see that

$$G - G_{0A} = \mu N_s - \widehat{\varphi}\Omega = \mu N_s - \varphi\Omega + \varepsilon_s \tag{16.27}$$

$$H - H_{0A} = TS_s + \mu N_s - \widehat{\varphi}\Omega = TS_s + \mu N_s - \varphi\Omega + \varepsilon_s \tag{16.28}$$

By comparing to the common formulation of adsorption thermodynamics, we recognize the term $\varphi\Omega$ as the analogue of pV. We therefore define the film functions G_s and H_s by

$$G_s \equiv G - G_{0A} + \varphi\Omega = \mu N_s + \varepsilon_s \tag{16.29}$$

$$H_s \equiv G_s + TS_s = TS_s + \mu N_s + \varepsilon_s \tag{16.30}$$

The choice will eventually lead to the condition for equilibrium in terms of H_s and S_s (see Eq. (16.42)) which we are seeking. The condition Eq. (16.42) reduces to the classical formula when the system is macroscopic and $\varepsilon_s = 0$.

By using Eqs. (16.25), (16.26) and the relation $\varepsilon_s = (\varphi - \widehat{\varphi})\Omega$, we can write $d\varepsilon_s$ in terms of the environmental variables T, p, μ, Ω

as follows:

$$d\varepsilon_s = \Omega^2 \left(\frac{\partial S_s/\Omega}{\partial \Omega}\right)_{T,p,\mu} dT - \Omega^2 \left(\frac{\partial V_s/\Omega}{\partial \Omega}\right)_{T,p,\mu} dp$$
$$+\Omega^2 \left(\frac{\partial N_s/\Omega}{\partial \Omega}\right)_{T,p,\mu} d\mu + \Omega \left(\frac{\partial \varphi}{\partial \Omega}\right)_{T,p,\mu} d\Omega \qquad (16.31)$$

The differential coefficients of type $\partial(Y/\Omega)/\partial\Omega$ in Eq. (16.31), where $Y = S_s, V_s, N_s$, can be expanded to

$$\left(\frac{\partial Y/\Omega}{\partial \Omega}\right)_{T,p,\mu} = \frac{1}{\Omega}\left[\left(\frac{\partial Y}{\partial \Omega}\right)_{T,p,\mu} - \frac{Y}{\Omega}\right] \qquad (16.32)$$

This emphasizes the distinction between differential and integral quantities for small systems, one of the key characteristics of nanothermodynamics. In the thermodynamic limit, the Euler homogeneous functions of the type $\partial Y/\partial\Omega = Y/\Omega$ are again true and the square bracket term vanishes. It follows directly from Eq. (16.31) that the effect of size on the variables S_s/Ω, V_s/Ω, N_s/Ω, and φ can be related to ε_s and changes in ε_s in

$$\left(\frac{\partial S_s/\Omega}{\partial \Omega}\right)_{T,p,\mu} = \frac{1}{\Omega^2}\left(\frac{\partial \varepsilon_s}{\partial T}\right)_{p,\mu,\Omega} \qquad (16.33)$$

$$\left(\frac{\partial V_s/\Omega}{\partial \Omega}\right)_{T,p,\mu} = -\frac{1}{\Omega^2}\left(\frac{\partial \varepsilon_s}{\partial p}\right)_{T,\mu,\Omega} \qquad (16.34)$$

$$\left(\frac{\partial N_s/\Omega}{\partial \Omega}\right)_{T,p,\mu} = \frac{1}{\Omega^2}\left(\frac{\partial \varepsilon_s}{\partial \mu}\right)_{T,p,\Omega} \qquad (16.35)$$

$$\left(\frac{\partial \varphi}{\partial \Omega}\right)_{T,p,\mu} = \frac{1}{\Omega}\left(\frac{\partial \varepsilon_s}{\partial \Omega}\right)_{T,p,\mu} \qquad (16.36)$$

From Eq. (16.25), we have

$$\left(\frac{\partial \widehat{\varphi}}{\partial \Omega}\right)_{T,p,\mu} = \frac{\varepsilon_s}{\Omega^2} \qquad (16.37)$$

These relations are all derived from Hill's formalism. They can be actively used in descriptions of film properties. The subdivision potential is a central variable. It can be regarded as a direct measure of the system smallness, here and in the equations to follow. The relations are not directly obtainable in Gibbs' treatment of adsorption. From Eq. (16.30), it follows, using Eqs. (16.33) and (16.35), that

$$\left(\frac{\partial H_s/\Omega}{\partial \Omega}\right)_{T,p,\mu} = \frac{T}{\Omega^2}\left(\frac{\partial \varepsilon_s}{\partial T}\right)_{p,\mu,\Omega} + \frac{\mu}{\Omega^2}\left(\frac{\partial \varepsilon_s}{\partial \mu}\right)_{T,p,\Omega} + \left(\frac{\partial \varepsilon_s/\Omega}{\partial \Omega}\right)_{T,p,\mu} \tag{16.38}$$

16.4 Single-Component Adsorption

The special case of an adsorbed film in equilibrium with a macroscopic gas at T, p will be used to illustrate the equations that were derived in the previous sections.

The chemical potential μ is controlled by the gas chemical potential μ_G. It follows that μ is completely determined by T and p and that a change $d\mu$ is given by $d\mu = -s_G\, dT + v_G\, dp$, where s_G and v_G are the entropy and volume per gas molecule. Equation (16.25) then gives the integral spreading pressure as

$$d\widehat{\varphi} = (s_s - s_G)\Gamma_s\, dT - (v_s - v_G)\Gamma_s\, dp + \varepsilon_s/\Omega^2\, d\Omega \tag{16.39}$$

where $s_s \equiv S_s/N_s$, $v_s \equiv V_s/N_s$ and $\Gamma_s \equiv N_s/\Omega$. The entropy per film molecule relative to the entropy per gas molecule (the adsorption entropy) at the same conditions is then given by the Clausius–Clapeyron-type equation:

$$s_s - s_G = (v_s - v_G)\left(\frac{\partial p}{\partial T}\right)_{\widehat{\varphi},\Omega} \tag{16.40}$$

The two equations together with Eq. (16.50) that follows will be applied to calculate the integral and differential spreading pressures (Fig. 16.1), the subdivision potential and the adsorption entropy (Fig. 16.2).

From the condition $\mu = \mu_G$, it follows that

$$h_s - Ts_s - \varepsilon_s/N_s = h_G - Ts_G \tag{16.41}$$

or

$$h_s - h_G = T(s_s - s_G) + \varepsilon_s/N_s \tag{16.42}$$

These equations will be used in what follows to calculate the adsorption enthalpy; i.e. the enthalpy per film molecule relative to the gas.

The chemical potential of the film follows from Eq. (16.30). It is equal to

$$\mu = h_s - Ts_s - \varepsilon_s/N_s \tag{16.43}$$

where $h_s = H_s/N_s$. We see how the subdivision potential alters the standard equations, as it shifts adsorption equilibria.

Size-dependence of variables

For completeness we give also equations that relate the effect of size on intensive variables in a simpler form. Equation (16.31) becomes

$$d\varepsilon_s = \Omega^2 \left[\frac{\partial(s_s - s_G)\Gamma_s}{\partial\Omega}\right]_{T,p} dT - \Omega^2 \left[\frac{\partial(v_s - v_G)\Gamma_s}{\partial\Omega}\right]_{T,p} dp + \Omega\left(\frac{\partial\varphi}{\partial\Omega}\right)_{T,p} d\Omega \tag{16.44}$$

The operational equivalents of Eqs. (16.33)–(16.36), now follow directly from Eq. (16.44)

$$\left[\frac{\partial(s_s - s_G)\Gamma_s}{\partial\Omega}\right]_{T,p} = \frac{1}{\Omega^2}\left(\frac{\partial\varepsilon_s}{\partial T}\right)_{p,\Omega} \tag{16.45}$$

$$\left[\frac{\partial(v_s - v_G)\Gamma_s}{\partial\Omega}\right]_{T,p} = -\frac{1}{\Omega^2}\left(\frac{\partial\varepsilon_s}{\partial p}\right)_{T,\Omega} \tag{16.46}$$

$$\left(\frac{\partial\varphi}{\partial\Omega}\right)_{T,p} = \frac{1}{\Omega}\left(\frac{\partial\varepsilon_s}{\partial\Omega}\right)_{T,p} \tag{16.47}$$

From Eq. (16.39), we have the operational equivalent of Eq. (16.37)

$$\left(\frac{\partial\widehat{\varphi}}{\partial\Omega}\right)_{T,p} = \frac{\varepsilon_s}{\Omega^2} \tag{16.48}$$

From Eq. (16.42), we have the operational equivalent of Eq. (16.38) by using Eq. (16.45):

$$\left[\frac{\partial(h_s - h_G)\Gamma_s}{\partial\Omega}\right]_{T,p} = \frac{T}{\Omega^2}\left(\frac{\partial\varepsilon_s}{\partial T}\right)_{p,\Omega} + \left(\frac{\partial\varepsilon_s/\Omega}{\partial\Omega}\right)_{T,p} \tag{16.49}$$

Hill–Gibbs' adsorption isotherm

The adsorption isotherm is central in the analysis of experimental and numerical data. The first step in any procedure is thus to relate the particular surface excess absorbate (Γ_s) to properties of the fluid (p, T) in equilibrium with the surface. With this information we can integrate Eq. (16.39) at constant T and Ω and obtain $\widehat{\varphi}$ for a given adsorbent size. This is the Hill–Gibbs isotherm for the integral spreading pressure:

$$\widehat{\varphi} = \int_0^p \Gamma_s(v_G - v_s)\,dp \quad (T,\ \Omega \text{ const.}) \tag{16.50}$$

The subdivision potential is not visible in Eq. (16.50), but it is there, cf. Eq. (16.46).

The choice of reference state, used to define the film properties, can be motivated by this equation. As the gas pressure decreases toward zero, the system state approaches that of the reference state. Therefore, $\widehat{\varphi}$ vanishes at the lower integration limit. When the pure adsorbent is used as a reference state, and the criterion for the distinction between a free and an adsorbed adsorbate is such that $V_s = 0$, the small system adsorption isotherm is given by

$$d\varphi = \Gamma_s\,d\mu \quad (T \text{ const.}) \tag{16.51}$$

With equilibrium between ideal gas and film, $d\mu = d\mu_G = v_G\,dp \approx kT\,d\ln p$, and the integrated form of Eq. (16.51) gives for the spreading pressure

$$\varphi = k_{\mathrm{B}}T\int_0^p \Gamma_s\,d\ln p, \quad (T \text{ const.}) \tag{16.52}$$

The same expression for the spreading pressure was derived by Myers [133] and Hill [134].

The procedure leading to Eq. (16.52) can, however, *not* be used to obtain the small system equation. The variables of the small system are not Euler homogeneous functions, so direct integration of Eq. (16.23) is not possible. Instead, we use the ensemble procedure of Hill and obtain the Hill–Gibbs–Duhem equation

$$d\varepsilon_s = d[(\varphi - \widehat{\varphi})\Omega] = -S_s\, dT + V_s\, dp - N_s\, d\mu + \Omega d\varphi \qquad (16.53)$$

By integrating at constant T and Ω, we obtain Eq. (16.50). Assuming that the gas is ideal, and that the criterion adopted for the free and adsorbed adsorbates is such that $V_s = 0$, we find the analogue of Eq. (16.52) for a small system from the more general Eq. (16.50). The result is an expression for the integral spreading pressure:

$$\widehat{\varphi} = k_{\mathrm{B}}T \int_0^p \Gamma_s\, d\ln p \quad (T,\, \Omega \text{ const.}) \qquad (16.54)$$

Equations (16.52) and (16.54) become identical in the macroscopic limit. Equation (16.53) shows directly the role of the subdivision potential.

These equations can be used to decompose results from experiments or molecular simulations. We present an example from the following simulations.

16.5 Simulating Adsorption on a Small Sphere

Molecular simulations have been performed with the open source software LAMMPS (version 20 August 2019) [66] using the grand canonical Monte Carlo simulation technique. For details, see [38].

All simulations were bounded by a cubic simulation box with periodic boundary conditions. The size of the box was set for each state. Gas properties were sampled sufficiently far away from the adsorbent surface to ensure that the gas obtained bulk properties. The interaction between gas molecules was described by the Mie potential [135].

$$u_G(r) = B\epsilon_G\left[\left(\frac{\sigma_G}{r}\right)^{\gamma_r} - \left(\frac{\sigma_G}{r}\right)^{\gamma_a}\right], \quad B = \left(\frac{\gamma_r}{\gamma_r - \gamma_a}\right)\left(\frac{\gamma_r}{\gamma_a}\right)^{\left(\frac{\gamma_a}{\gamma_r - \gamma_a}\right)} \qquad (16.55)$$

where ϵ_G is the energy parameter of the interaction, σ_G is the length parameter of the interaction, γ_r is the repulsive exponent, and γ_a is the attractive exponent. Parameters from [136] were used to represent single site coarse-grained CO_2 molecules.

Reduced units were used. For energy, we used as unit the strength of the Mie interaction ($361.69k_B$) J, where k_B is the Boltzmann constant. For the length, we used as unit the diameter of the CO_2 segment, 3.741 Å. For mass, we used as unit the molecular mass of CO_2. This gives by the definition of reduced units $\sigma_G = 1$, $\epsilon_G = 1$ and $\gamma_r = 23.0$, $\gamma_a = 6.66$, and the reduced potential cut-off distance $r_{c,\text{mie}} = 4.0$.

The interaction between the external field, functioning as the adsorbent, and a gas molecule was represented by a spherical colloid potential located at the box center. The expression for this potential followed by integration of the pairwise interactions between a gas molecule and the adsorbent constituent atoms over the adsorbent volume.

The interaction between a gas molecule and the adsorbent constituent atom was given by the standard Lennard-Jones potential

$$u(r') = 4\epsilon \left[\left(\frac{\sigma}{r'}\right)^{12} - \left(\frac{\sigma}{r'}\right)^{6} \right] \tag{16.56}$$

Here, $u(r')$ is the interaction energy between adsorbent and gas, r' is the center-to-center distance between them, ϵ is the energy parameter of the interaction and σ is the length parameter of the interaction. By integrating Eq. (16.56) over the spherical adsorbent, we find

$$\mathcal{U}(r) = \frac{16\pi\epsilon c_A \sigma^3}{3} \left[\frac{(15R^3r^6 + 63R^5r^4 + 45R^7r^2 + 5R^9)\sigma^9}{15(r^2 - R^2)^9} - \frac{R^3\sigma^3}{(r^2 - R^2)^3} \right] \quad r > R \tag{16.57}$$

where $\mathcal{U}(r)$ is the interaction energy between the adsorbent and gas, r is the center-to-center distance between them, c_A is the adsorbent number density and R is the adsorbent radius. The cut-off distance r_c for the potential was defined by the relation $\mathcal{U}(r_c)/\min(\mathcal{U}) = 5.5 \times 10^{-3}$, where $\min(\mathcal{U})$ is the potential minimum. The value of r_c, satisfying the relation, in reduced units, was approximated by $r_c = 1.23R + 3.0$ for each adsorbent radius. The reduced units are those introduced for the gas.

We used the conventional definition of the Hamaker constant $\mathcal{A}_{12} \equiv \pi^2 C_{12} c_1 c_2$ for two arbitrary components, where C_{12} is the coefficient of the dispersion energy $u_{disp}(r') = -C_{12}/(r')^6$. For a potential of the type in Eq. (16.56), it follows that $C_{12} = 4\epsilon_{12}\sigma_{12}^6$ and

$$\mathcal{A}_{12} = 4\pi^2 \epsilon_{12} c_1 c_2 \sigma_{12}^6 \tag{16.58}$$

From Eq. (16.58) and the mixing rules $\sigma = (\sigma_G + \sigma_A)/2$ and $\epsilon = \sqrt{\epsilon_G \epsilon_A}$, we calculated $\epsilon c_A \sigma^3$ in terms of σ_G, σ_A, ϵ_G and $\mathcal{A}_A$ for use in Eq. (16.57). Here σ_A and ϵ_A are the length and energy parameters of the dispersion energy of two interacting adsorbent atoms, and $\mathcal{A}_A$ is the Hamaker constant for the interaction between two graphite bodies.

We took $\sigma_A = 3.4$ Å for the carbon atom–atom interaction, and $\mathcal{A}_A = 4.7 \times 10^{-19}$ J for the Hamaker constant [137]. It follows that $\epsilon c_A \sigma^3 = [\sqrt{\epsilon_G \mathcal{A}_A}/(16\pi)][(\sigma_G + \sigma_A)/\sigma_A]^3 \approx 1.79$ in reduced units for the graphite adsorbent system. In all simulations, we used $p = 0.011$ and $T = 1.08$ in reduced units, or 1.05 MPa and 250 K, respectively. Simulations were run, varying the radius of the adsorbent from 6 to $25 \cdot 10^{-10}$ m. In this manner only Ω was varied.

In each simulation the system was first equilibrated for a number of cycles that depended on the state we were simulating. The properties of interest were monitored to make sure they fluctuated around a steady value. This was followed by a number of cycles where samples for calculation of ensemble averages and errors were collected. The number of cycles varied with the state simulated, and were chosen in each case such that the estimated error was less than 1.5% of the mean with 95% confidence for all sampled properties as well as the calculated property Γ_s. The strongest correlation between samples was found in Γ_s. This property therefore determined the number of cycles to use. In estimating the standard error, we accounted for correlation in the data by block analysis.

16.6 Procedure for Data Reduction

There is presently not much literature, besides that given by Hill, to guide reduction of simulated data in terms of nanothermodynamics. In what follows we show how to acquire and reduce simulation data for a range of state variables. The control variables are T, μ, Ω in all simulations and we sample the gas pressure and density from the region of macroscopic gas properties. The total amount of the adsorbate component is also sampled. The rest of the properties follow from the thermodynamic relations of the preceding sections. Data can be evaluated using the scientific computing library SciPy [138].

The following procedure was adopted to find the properties from numerical data. We refer to Strøm *et al.* [38] for more details, e.g., on the accuracy of the calculation procedure.

1. **Establish the adsorption isotherm** $\Gamma_s(p;T,\Omega)$ by sampling the system in equilibrium. Define surface excess properties [20] in order to distinguish between adsorbed and free molecules. The spherical mathematical dividing surface was located at the radial distance, a, from the box center by the criterion $\mathcal{U}(a) = 0$. The adsorbent volume is then $V_A \equiv (4/3)\pi a^3$, and the adsorbant area $\Omega = 4\pi a^2$. The number of adsorbed molecules is $N \equiv N_B - c_G(V_B - V_A)$, where N_B is the number of adsorbate molecules in the box, c_G is the gas density and V_B is the box volume.
2. **Find the integral spreading pressure** from Eq. (16.50). We interpolated adsorption isotherms $\Gamma_s(p;T,\Omega)$ using univariate splines, constraining the splines to be linear for the lowest pressures according to Henry's law. We then extrapolated the linear region to zero pressure to allow the integration in Eq. (16.50). We evaluated the integral analytically for the lowest pressures, and numerically for the remaining pressures. We integrated the product of $\Gamma_s(p;T,\Omega)$ and $(v_G - v_s)$ over the pressure variation. The resulting function, $\widehat{\varphi}(p;T,\Omega)$, applies to one area Ω. We repeated the procedure for a range of sphere sizes to obtain the functional relation shown in Fig. 16.1.

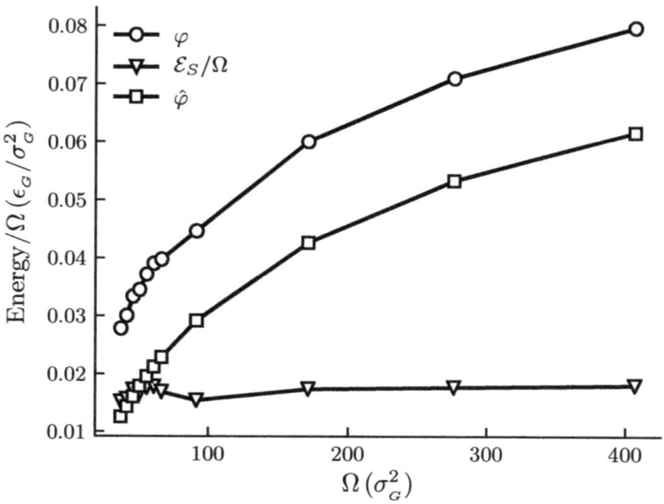

Figure 16.1. Film spreading pressures, φ and $\hat{\varphi}$, and specific subdivision potential ε_s/Ω for graphite adsorbent, as functions of adsorbent area Ω (all in reduced units) at constant temperature T and pressure p. The points indicate the simulated adsorbent sizes. The lines are added as visual aid. Reproduced from [38] with permission from MDPI Publishers.

3. **Obtain the differential spreading pressure**, φ. This spreading pressure was obtained from the function $\widehat{\varphi}(p;T,\Omega)$ for each area, using $\varphi = \widehat{\varphi} + \Omega(\partial\widehat{\varphi}/\partial\Omega)_{T,p}$. This follows from Eq. (16.39) at constant T, p and the definition $\varepsilon_s \equiv (\varphi - \widehat{\varphi})/\Omega$. The derivative $\Omega(\partial\widehat{\varphi}/\partial\Omega)_{T,p}$ was approximated by $\Omega(\Delta\widehat{\varphi}/\Delta\Omega)_{T,p}$, which can be calculated from the functions $\widehat{\varphi}(p;T,\Omega_1)$ and $\widehat{\varphi}(p;T,\Omega_2)$ for values of Ω_1 and Ω_2, which are not too far apart.
4. **Obtain the subdivision potential.** The curve for ε_s/Ω can finally be obtained from the difference $\varphi - \widehat{\varphi}$.
5. **Obtain the adsorption entropy and enthalpy**. The entropy and enthalpy of adsorption, shown in Fig. 16.2, can finally be obtained from Eqs. (16.40) and (16.42), respectively. The derivative term is with $(\partial p/\partial T)_{\widehat{\varphi},\Omega} \approx (\Delta p/\Delta T)$, which was calculated from two functional values of $p(\widehat{\varphi};T_1,\Omega)$ and $p(\widehat{\varphi};T_2,\Omega)$ from values of T_1 and T_2, which are not too far apart. The pressures were obtained by inversion of $\widehat{\varphi}(p;T_1,\Omega)$ and $\widehat{\varphi}(p;T_2,\Omega)$.

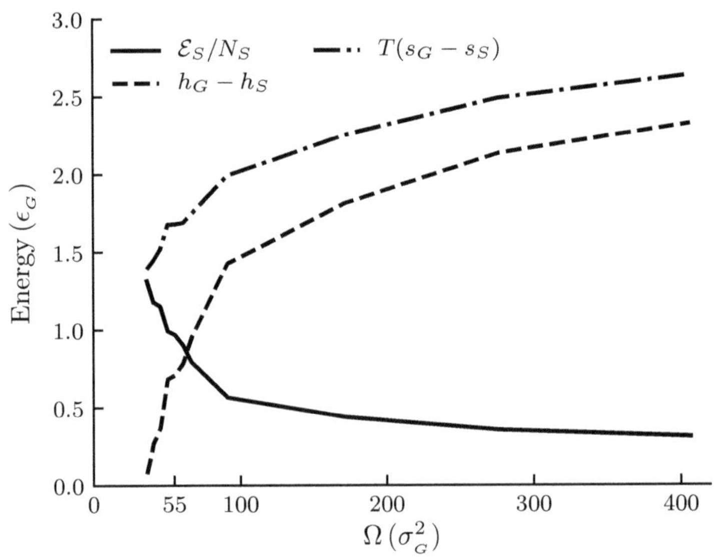

Figure 16.2. Film enthalpy, entropy and subdivision potential per film molecule as function of the adsorbent area Ω at constant temperature $T = 1.080$ and pressure $p = 0.011$ (reduced units). Enthalpy and entropy for a film molecule are given relative to the gas. Reproduced from [38] with permission from MDPI Publishers.

16.7 Films on Small Surfaces

Results for the integral and differential spreading pressures, and the entropy and enthalpy of adsorption are given for a graphite-like adsorbent of varying surface areas in Figs. 16.1 and 16.2. The adsorption subdivision potential is illustrated in both figures.

We see how the integral spreading pressure computed from the isotherms varies with adsorbent surface area. Equation (16.50) has the same form as the one usually encountered in adsorption thermodynamics, cf. [132, 134], with the important distinction that $\widehat{\varphi}$ and φ are not the same. The spreading pressures, φ and $\widehat{\varphi}$, as well as the subdivision potential per unit of area depend clearly on the surface area. The last area value corresponds to a surface area of 400 σ_G^2. After an initial steep increase in φ or $\widehat{\varphi}$, the rate of increase slows down. For larger areas, there is in this system still a significant difference between φ and $\widehat{\varphi}$. The subdivision potential per unit of area, ε_s/Ω, appears to be fairly constant in the figure, but in fact it is on decline for larger areas (not shown). These results are therefore consistent with the theoretical prediction that $\varphi = \widehat{\varphi}$ or ε_s approaches zero in the thermodynamic limit.

The second figure shows the enthalpy and entropy of adsorption, along with the subdivision potential per particle, ε_s/N_s. Also this figure shows that all properties depend strongly on the absorbent area. Equation (16.40) is the same as the one usually encountered in adsorption thermodynamics, Eq. (21) in [134], except for the fact that the function that is kept constant, when we take the derivative, is $\widehat{\varphi}$ instead of φ. This entropy is the entropy typically discussed in relation to the degree of order of adsorbed molecules. Equation (16.42) differs from the usual equation by the term ε_s/N_s. We observe that for small systems, $h_s - h_G \neq T(s_s - s_G)$. This is because $\varepsilon_s \neq 0$.

The expressions referred to above become the usual expressions in the thermodynamic limit where $\varepsilon_s = 0$ and $\widehat{\varphi} = \varphi$. We place the results first in the context of this limit. Consider therefore an infinitely large, flat film adsorbed on a flat adsorbent. The intensive thermodynamic properties of the macroscopic film do not depend on the surface area. There is then no difference between the integral spreading pressure $\widehat{\varphi}$ and the differential spreading pressure φ, and no difference between $T(s_G - s_s)$ and $h_G - h_s$, because $\varepsilon_s = 0$, cf. Eq. (16.42). Both figures document that the systems are far away from the thermodynamic limit, since $\varepsilon_s \neq 0$. The systems that we have considered here are therefore small in Hill's sense. The area available for adsorption must therefore be regarded as rather small. On the scale used, the thermodynamic limit has not yet been reached for an area of $400\sigma_G^2$.

We are thus far from the thermodynamic limit, but the results appear as expected. We see from Fig. 16.2 that the rate at which the plotted properties change with increasing area goes down. The difference ε_s/N_s, or the deviation from the corresponding macroscopic system, decreases as the adsorbent area becomes larger. This is the expected behavior.

The quantity ε_s/N_s is a direct measure of deviation from the thermodynamic limit. We can use it to investigate why the deviation occurs. The quantities in Fig. 16.2 are plotted as $T(s_G - s_s)$ and $h_G - h_s$ for convenience, so that all quantities are positive. We see that the surface entropy becomes less negative, the smaller the system becomes. This fits with the understanding that less states become available for adsorption as the system becomes smaller. *The subdivision potential can then be seen as a measure of the energy needed to overcome the disadvantage of entropy reductions that arise from the relatively small number of available states in small systems.* This "energy of disadvantage" has to be supplied by a combination of interactions that change with changes in structure, shape and size.

It is practical to have one property as a direct measure of the deviation from the thermodynamic limit, and we conclude that the quantity ε_s/N_s serves well in this context, as we can readily investigate why the deviation occurs. The subdivision potential, the enthalpy and entropy, the spreading pressures, will of course all depend on the chemistry involved and the interaction potentials used. A change in the interaction potential will shift all curves [38]. Other changes are also likely.

16.8 Concluding Remarks

In this chapter, we have seen how properties of films adsorbed to small absorbents deviate significantly from the same adsorption properties of large surfaces. Using Hill's method for an osmotic-like ensemble, we have shown how to obtain in practice a direct measure of smallness through the subdivision potential, ε_s. The equilibrium condition between gas and adsorbed film $\mu = \mu_G$ resulted in the expression

$$h_s - h_G = T(s_s - s_G) + \varepsilon_s/N_s \tag{16.59}$$

We have presented a data reduction procedure, which can be used with simulated or experimental results to examine this condition.

The procedure shows how to derive thermodynamic properties from simulations, using CO_2 adsorption on a small graphite sphere as an example. These properties help us understand the factors that stabilize the film.

The procedure may be applied to studies of nucleation or adsorption on particles in the atmosphere.

Chapter 17

A Representative Elementary Volume of a Porous Medium

Porous media consist of a set of phases with contact areas and contact lines between them. A thermodynamic description of porous materials must therefore deal with the size- and shape-dependencies of variables. Here we define a representative elementary volume (REV) of a porous medium by means of variables which are additive, i.e., we compute each variable as a sum of bulk-, surface- and line-contributions. This allows us to write the Gibbs equation for the REV using these coarse-grained variables. We distinguish between variables that only depend on the material density and those that also depend on the material configuration.

Porous media represent a vast group of materials. Their pores can be nano- and micrometer-sized in the same material, and surface energies can be relatively large. Porous materials vary largely and can be found everywhere. The heterogeneous porous catalysts filled with reactants and products is one example. The human body with porous bones and all kinds of tissues is another. Membranes where a liquid–vapor phase transition takes place inside pores constitute a group. Central is that capillary effects are non-negligible. The problem is to find a continuous description of the flow through the material on the scale of its measurement. This flow is often of a viscous nature interwoven with diffusion. Heat flow and chemical reactions can also take place.

The variables that are used to describe flow on the macroscopic level need to be constructed to reflect the underlying events. A coarse-graining procedure is needed to reduce the number of pore scale variables. The art is to find *a representative elementary volume* (REV). Several approaches have been used over the years to define the REV and its coarse-grained variables [42, 139–146]. The attempt to find variables that also obey the Gibbs equation is recent [17]. Once there is a set of variables which obey the Gibbs equation, we can identify the entropy production and the relevant flux–force relations, see Chapter 19. The aim of this chapter is therefore to define a set of variables that allow us to use the Gibbs equation in porous media, enabling us to find proper flux–force relations.

Whenever a system calls for inclusion of size and shape as variables, Hill's ensemble construction [4, 14] becomes relevant. We shall use his idea of a small system under control of the environment to define the coarse-grained variables of the REV. The aim is to present a new systematic procedure for construction of a REV and its variables, extending some first attempts [17, 144].

In order to illustrate the procedure, consider an example which is typical in geology; namely two immiscible fluids in a porous rock. The case is illustrated in Fig. 17.1. The porous system consists of three components: one more wetting fluid, w, and one less wetting, n, both contained in a solid matrix, r. The red square in the figure gives the boundaries of a part of the porous medium which is representative of the whole, the REV. As shown in the figure, the REV must contain all phases that are present in the medium, their surface areas and three-phase contact lines. In short, all typical micro-states of the medium are present in the REV. It can thus be regarded as representative in a thermodynamic sense.

The REV has a given volume and is open to the environment. The most relevant environment has constant temperature and chemical potentials. This means that the REV is controlled by the grand canonical (GC) ensemble, and the Gibbs equation is

$$dU^{\mathrm{GC,REV}} = TdS^{\mathrm{GC,REV}} - p^{\mathrm{GC,REV}}dV^{\mathrm{REV}} + \mu^n dM^{n,\mathrm{REV}} + \mu^w dM^{w,\mathrm{REV}} + \mu^r dM^{r,\mathrm{REV}} \quad (17.1)$$

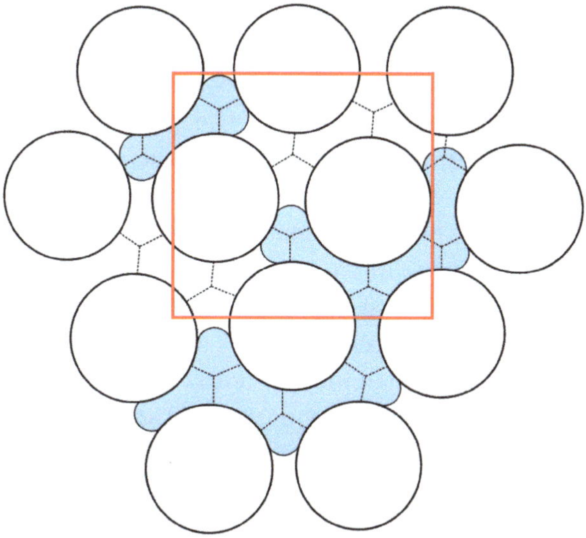

Figure 17.1. Illustration of an open representative elementary volume (REV) with two immiscible components (white and blue) in a matrix (of white spheres). The REV is a part of the total system. It contains all possible microstates of the total system. Its boundaries are here given in red.

At the outset, Gibbs equation applies to homogeneous solutions only. The variables of this equation do not refer to a homogeneous solution, however; we write the equation for the heterogenous REV of Fig. 17.1. The variables that obey this equation must be constructed in a special way. This is what we discuss in the sections that follow. Doing this, we propose a procedure for descriptions of porous media on the overall scale, reflecting underlying capillary effects [17, 42, 144].

We shall find that the variables of the REV which satisfy Eq. (17.1) are additive of nature, and have contributions from each material as well as from the configuration of the material contributions. Volume- and mass variables are obviously material contributions only. But the entropy also contains configurational contributions. In order to explain what we mean by this, consider the entropy of the REV in Fig. 17.1. The entropy has material contributions from all bulk phases, as well as excess surface and line contributions. But the entropy also has a configurational contribution. Consider the various volume contributions to the volume of the REV. There are several ways to arrange these. The entropy of the (say) blue volumes per se, i.e., the entropy of the sum of the phases, does not capture the entropy of the phase *arrangements*. The freedom in the arrangement, or the arrangement probability distribution, represents an additional number of states, or additional entropy. This type of additive contribution is of a configurational type, and will be called just that, i.e., configurational. In this sense we will be dealing with material content as well as material structure inside the REV, and we shall include both contributions in our search for a foundation for the Gibbs equation of the REV.

17.1 Additivity of Variables: Densities

Consider the REV-part of the large system, as pictured in Fig. 17.1. We want to determine its thermodynamic properties. Like for any small system, the environment plays a role in controlling the variables of the REV. In the continuous system that we are seeking to describe, the control variables are the temperature and chemical potentials (pressures). These may vary with time and position in a real process. For now, we are only concerned with their value when there is global equilibrium in a large system. When the REV is embedded in a large system which controls its temperature and chemical potentials, these intensive variables in the REV are constant. The relevant potential for the open REV is the grand potential. We shall therefore use the grand potential to find the set of variables needed to describe the REV here. The rock matrix has a constant porosity. But the system can be said to be homogeneous in the sense that the typical porosity and pore surface area are the same everywhere.

Porosity, saturation, surface area and contact line length

It is common in porous media science [139, 143] to use the porosity and the degree of saturation (popularly called just saturation) as variables. Contact area and contact line lengths and curvature have also been introduced as geometric variables [146].

The REV here is made up of three immiscible one-component phases. A chemical constituent is therefore synonymous with a phase. The state of the REV is characterized, among other variables, by the volumes of the fluid phases $V^{n,\mathrm{REV}}$, $V^{w,\mathrm{REV}}$ and of the solid $V^{r,\mathrm{REV}}$. The volume of the pores is

$$V^{p,\mathrm{REV}} \equiv V^{n,\mathrm{REV}} + V^{w,\mathrm{REV}} \tag{17.2}$$

while the volume of the REV is

$$V^{\mathrm{REV}} \equiv V^{r,\mathrm{REV}} + V^{p,\mathrm{REV}} + V^{nwr,\mathrm{REV}} \tag{17.3}$$

Superscript REV is used to indicate a property of the REV. The excess volumes of the surfaces are zero by definition. Excess properties of surfaces are defined following Gibbs, see [25] for further explanations. The last term in Eq. (17.3) is the excess volume of the three-phase contact lines [25]. This is not zero, as the dividing surfaces may cross each other at lines which have slightly different locations. The corresponding excess volume is often very small, meaning that the term can be neglected. The volume of the REV is a *material variable.* As Eq. (17.3) shows, the volume is an additive variable.

The porosity, ϕ, of the system, and the *saturation* variables of the fluids, $\tilde{S}^n$ and $\tilde{S}^w$, are defined by

$$\phi \equiv \frac{V^{p,\mathrm{REV}}}{V^{\mathrm{REV}}}, \quad \tilde{S}^n \equiv \frac{V^{n,\mathrm{REV}}}{V^{p,\mathrm{REV}}}, \quad \tilde{S}^w \equiv \frac{V^{w,\mathrm{REV}}}{V^{p,\mathrm{REV}}} \tag{17.4}$$

The saturations carry the tilde-symbol, to distinguish them from the entropy (in the following). The porosity and the saturations are intensive variables. They do not depend on the size of the REV. They have therefore no superscript. It follows from these definitions that

$$\tilde{S}^n + \tilde{S}^w = 1 \tag{17.5}$$

In addition to the volume variables, there are interfacial contact areas between the phases in the REV: $\Omega^{nw,\mathrm{REV}}$, $\Omega^{nr,\mathrm{REV}}$, $\Omega^{wr,\mathrm{REV}}$.

When the fluid-rock surface areas inside the REV are added, we obtain the total surface area of the pores:

$$\Omega^{p,\mathrm{REV}} = \Omega^{nr,\mathrm{REV}} + \Omega^{wr,\mathrm{REV}} \tag{17.6}$$

The total length of the three-phase contact lines is $\Lambda^{nwr,\mathrm{REV}}$. The interface areas and the contact line lengths are also additive. A REV of the double size of a particular REV has the double volumes, surface areas of various types and double contact line lengths.

Additive coarse-grained thermodynamic variables

So far, we have defined variables that depend on the geometry of the REV. We proceed to define the coarse-grained variables of the volume V^{REV}. The variables are constructed with the purpose of producing the Gibbs equation. A proposal to use Minkowski functionals is similar [145, 146], but assumes that Hadwiger's theorem holds, cf. Chapters 5 and 6 for a discussion.

We shall take the REV to be large in Hill's sense. The REV variables, which are additive, will then double by doubling the REV size. We indicate these variables by superscript REV. The densities, however, do not depend on the size of the REV, so we do not give them the superscript REV. Additivity of variables *inside* the REV is explained in what follows.

Mass variables are additive material variables. The mass of each component, n, w and r, in the REV is the sum of bulk, excess surface and excess contact line masses, respectively:

$$\begin{aligned}
M_n^{\mathrm{REV}} &= M_n^{n,\mathrm{REV}} + M_n^{nr,\mathrm{REV}} + M_n^{nw,\mathrm{REV}} + M_n^{nwr,\mathrm{REV}} \\
M_w^{\mathrm{REV}} &= M_w^{w,\mathrm{REV}} + M_w^{wr,\mathrm{REV}} + M_w^{nwr,\mathrm{REV}} \\
M_r^{\mathrm{REV}} &= M_r^{r,\mathrm{REV}}
\end{aligned} \tag{17.7}$$

There is some freedom in the allocation of mass to variables. Here, the equimass surface position of the rock phase is used to position the fluid–rock dividing surfaces, and the equimass surface of the wetting fluid is applied to the fluid–fluid dividing surface [25], see comment following Eq. (17.3). The contact lines are positioned such that $M_r^{nwr,\mathrm{REV}} = 0$. The total mass of each component in the REV is *independent* of the choice of location of the dividing surfaces or contact lines.

When we introduce the various mass densities, Eqs. (17.3)–(17.7) become

$$\begin{aligned} \rho_n V^{\mathrm{REV}} &= \rho_n^n V^{n,\mathrm{REV}} + \rho_n^{nr} \Omega^{nr,\mathrm{REV}} + \rho_n^{nw} \Omega^{nw,\mathrm{REV}} + \rho_n^{nwr} \Lambda^{nwr,\mathrm{REV}} \\ \rho_w V^{\mathrm{REV}} &= \rho_w^w V^{w,\mathrm{REV}} + \rho_w^{wr} \Omega^{wr,\mathrm{REV}} + \rho_w^{nwr} \Lambda^{nwr,\mathrm{REV}} \\ \rho_r V^{\mathrm{REV}} &= \rho_r^r V^{r,\mathrm{REV}} \end{aligned} \tag{17.8}$$

The mass densities have dimensions $\mathrm{kg\,m^{-3}}$ for the bulk phases, $\mathrm{kg\,m^{-2}}$ for the surfaces and $\mathrm{kg\,m^{-1}}$ for the contact line. The masses of the components are, like the volume, material variables.

These densities are defined for the REV. If we increase the REV size, for instance by doubling it, all variables with superscript REV will double. But this is not the case for the densities. They remain the same, independent of the size of the REV. This is true also for the densities of the bulk phases, surfaces and contact lines. This explains why superscript REV is not used for the densities.

Contributions from material density and structure

The entropy of the REV is the sum of bulk, excess interfacial and excess line entropies. In addition, there may be *configurational contributions* to the entropy arising from the distribution of phases in the REV. To see this, consider, e.g., the various volume contributions to a REV volume. There are several ways to obtain particular values of the volume, surface area and line length, by arranging the volume-contributions differently. This freedom in the arrangement and its distribution represents a configurational entropy.

The integral of the logarithm of the probability distribution of possible arrangements, times the Boltzmann constant, k_B, gives the configurational contribution to the entropy [147] for molecular systems. We derive an expression as an example of such a contribution in

Section 17.3. For the moment, we are justified to write the following:

$$S^{\mathrm{REV}} = S^{\mathrm{mat}} + S^{\mathrm{conf}} = S^{n,\mathrm{REV}} + S^{w,\mathrm{REV}} + S^{r,\mathrm{REV}} + S^{nr,\mathrm{REV}} + S^{wr,\mathrm{REV}} + S^{nw,\mathrm{REV}} + S^{nwr,\mathrm{REV}} + S^{\mathrm{conf}} \tag{17.9}$$

where S^{conf} is the configurational contribution to the REV entropy. The bulk, excess interfacial and excess line entropies constitute the *material* contributions to the entropy, S^{mat}. By introducing entropy densities, we obtain

$$\begin{aligned} sV^{\mathrm{REV}} &= \left(s^{\mathrm{mat}} + s^{\mathrm{conf}}\right) V^{\mathrm{REV}} \\ &= s^{n} V^{n,\mathrm{REV}} + s^{w} V^{w,\mathrm{REV}} + s^{r} V^{r,\mathrm{REV}} + s^{nr} \Omega^{nr,\mathrm{REV}} \\ &\quad + s^{wr} \Omega^{wr,\mathrm{REV}} + s^{nw} \Omega^{nw,\mathrm{REV}} + s^{nwr} \Lambda^{nwr,\mathrm{REV}} + s^{\mathrm{conf}} V^{\mathrm{REV}} \end{aligned} \tag{17.10}$$

Their dimensions are $\mathrm{J\,K^{-1}\,m^{-3}}$ (bulk and configurational densities), $\mathrm{J\,K^{-1}\,m^{-2}}$ (excess interfacial densities) and $\mathrm{J\,K^{-1}\,m^{-1}}$ (excess line entropy densities). In Eq. (17.10), we have assumed that there are no excess surface and line contributions to the configurational entropy. Relations similar to Eqs. (17.9) and (17.10) apply to all variables. But not all REV variables have a configurational contribution. We shall see which variables have this property in what follows.

The internal energy of the REV contains sums of bulk, excess interface, and excess line internal energies. The internal energy has a configurational contribution because there may be attractive or repulsive interactions between the volume elements. Already the names of the fluid phases, the wetting and non-wetting phases, indicate that there are preferences for one type of neighbor volume over another, and that the configuration is not completely random. This motivates us to write

$$U^{\mathrm{REV}} = U^{n,\mathrm{REV}} + U^{w,\mathrm{REV}} + U^{r,\mathrm{REV}} + U^{nr,\mathrm{REV}} + U^{wr,\mathrm{REV}} + U^{nw,\mathrm{REV}} + U^{nwr,\mathrm{REV}} + U^{\mathrm{conf}} \tag{17.11}$$

By introducing the internal energy densities, we obtain

$$\begin{aligned} uV^{\mathrm{REV}} &= \left(u^{\mathrm{mat}} + u^{\mathrm{conf}}\right) V^{\mathrm{REV}} \\ &= u^{n} V^{n,\mathrm{REV}} + u^{w} V^{w,\mathrm{REV}} + u^{r} V^{r,\mathrm{REV}} + u^{nr} \Omega^{nr,\mathrm{REV}} \\ &\quad + u^{wr} \Omega^{wr,\mathrm{REV}} + u^{nw} \Omega^{nw,\mathrm{REV}} + u^{nwr} \Lambda^{nwr,\mathrm{REV}} + u^{\mathrm{conf}} V^{\mathrm{REV}} \end{aligned} \tag{17.12}$$

The densities have dimensions $\mathrm{J\,m^{-3}}$ (bulk and configurational densities), $\mathrm{J\,m^{-2}}$ (excess interface densities) and $\mathrm{J\,m^{-1}}$ (excess line internal energy density).

The property of special interest in porous media science is the grand potential. A more colloquial name for this property is the compressional energy. The grand potential describes systems for which the temperature, chemical potentials and volume are controlled; a common situation of a REV in a reservoir. The GC ensemble is then central. The grand potential of the grand canonical ensemble of REVs is equal to Hill's replica energy, X, see Chapter 3. The grand potential of the REV is equal to minus the logarithm of the grand canonical partition function times $k_{\mathrm{B}}T$. The partition function can be used to understand the material contributions to the grand potential. In the next section, we show that the grand potential also has a configurational contribution. We have, similar to the expression for the entropy, Eq. (17.9);

$$\begin{aligned} X^{\mathrm{GC,REV}} &= X^{\mathrm{mat}} + X^{\mathrm{conf}} = X^{n,\mathrm{GC,REV}} + X^{w,\mathrm{GC,REV}} \\ &\quad + X^{r,\mathrm{GC,REV}} + X^{nr,\mathrm{GC,REV}} + X^{wr,\mathrm{GC,REV}} \\ &\quad + X^{nw,\mathrm{GC,REV}} + X^{nwr,\mathrm{GC,REV}} + X^{\mathrm{conf}} \end{aligned} \tag{17.13}$$

By introducing the densities of the grand potential, this equation becomes

$$\begin{aligned} x^{\mathrm{GC}} V^{\mathrm{REV}} &= \left(x^{\mathrm{mat}} + x^{\mathrm{conf}}\right) V^{\mathrm{REV}} = x^{n,\mathrm{GC}} V^{n,\mathrm{REV}} + x^{w,\mathrm{GC}} V^{w,\mathrm{REV}} \\ &\quad + x^{r,\mathrm{GC}} V^{r,\mathrm{REV}} + x^{nr,\mathrm{GC}} \Omega^{nr,\mathrm{REV}} + x^{wr,\mathrm{GC}} \Omega^{wr,\mathrm{REV}} \\ &\quad + x^{nw,\mathrm{GC}} \Omega^{nw,\mathrm{REV}} + x^{nwr,\mathrm{GC}} \Lambda^{nwr,\mathrm{REV}} + x^{\mathrm{conf}} V^{\mathrm{REV}} \end{aligned} \tag{17.14}$$

The densities have dimension $\mathrm{J\,m^{-3}}$ (bulk and configurational density), $\mathrm{J\,m^{-2}}$ (excess interfacial density) and $\mathrm{J\,m^{-1}}$ (excess line density). The density of the grand potential does not depend on the size of the REV, so we drop its superscript REV. Following Hill [4, 14] and Chapter 3, we define the integral pressures, surface and line tensions as follows:

$$
\begin{aligned}
&\widehat{p} = -x^{\mathrm{GC}}, \quad \widehat{p}^{\mathrm{mat}} = -x^{\mathrm{mat}}, \quad \widehat{p}^{\mathrm{conf}} = -x^{\mathrm{conf}} \\
&\widehat{p}^{n} = -x^{n,\mathrm{GC}}, \quad \widehat{p}^{w} = -x^{w,\mathrm{GC}}, \quad \widehat{p}^{r} = -x^{r,\mathrm{GC}} \\
&\widehat{\gamma}^{nr} = x^{nr,\mathrm{GC}}, \quad \widehat{\gamma}^{wr} = x^{wr,\mathrm{GC}}, \quad \widehat{\gamma}^{nw} = x^{nw,\mathrm{GC}}; \quad \widehat{\gamma}^{nwr} = x^{nwr,\mathrm{GC}}
\end{aligned}
\tag{17.15}
$$

These quantities depend typically on the size of the pore and are therefore unequal to the corresponding variable without a hat. By introducing Eq. (17.15) into Eq. (17.14), we obtain the REV grand potential

$$
\begin{aligned}
\widehat{p}V^{\mathrm{REV}} = \left(\widehat{p}^{\mathrm{mat}} + \widehat{p}^{\mathrm{conf}}\right)V^{\mathrm{REV}} &= \widehat{p}^{n}V^{n,\mathrm{REV}} + \widehat{p}^{w}V^{w,\mathrm{REV}} \\
&+ \widehat{p}^{r}V^{r,\mathrm{REV}} - \widehat{\gamma}^{nr}\Omega^{nr,\mathrm{REV}} - \widehat{\gamma}^{wr}\Omega^{wr,\mathrm{REV}} - \widehat{\gamma}^{nw}\Omega^{nw,\mathrm{REV}} \\
&- \widehat{\gamma}^{nwr}\Lambda^{nwr,\mathrm{REV}} + \widehat{p}^{\mathrm{conf}}V^{\mathrm{REV}}
\end{aligned}
\tag{17.16}
$$

In the analysis we have chosen the volume of the REV so large that $\widehat{p}$, $\widehat{p}^{\mathrm{mat}}$ and $\widehat{p}^{\mathrm{conf}}$ do not depend on V^{REV}. The material and the configurational pressures obey then

$$
\widehat{p} = p, \quad \widehat{p}^{\mathrm{mat}} = p^{\mathrm{mat}}, \quad \widehat{p}^{\mathrm{conf}} = p^{\mathrm{conf}} \tag{17.17}
$$

The other pressures and tensions may still depend on the pore size. Equation (17.16) for the REV grand potential reduces to

$$
\begin{aligned}
pV^{\mathrm{REV}} = \left(p^{\mathrm{mat}} + p^{\mathrm{conf}}\right)V^{\mathrm{REV}} &= \widehat{p}^{n}V^{n,\mathrm{REV}} + \widehat{p}^{w}V^{w,\mathrm{REV}} \\
&+ \widehat{p}^{r}V^{r,\mathrm{REV}} - \widehat{\gamma}^{nr}\Omega^{nr,\mathrm{REV}} - \widehat{\gamma}^{wr}\Omega^{wr,\mathrm{REV}} - \widehat{\gamma}^{nw}\Omega^{nw,\mathrm{REV}} \\
&- \widehat{\gamma}^{nwr}\Lambda^{nwr,\mathrm{REV}} + p^{\mathrm{conf}}V^{\mathrm{REV}}
\end{aligned}
\tag{17.18}
$$

In what follows we show why the grand potential $X^{\text{GC,REV}} = -\widehat{p}V^{\text{REV}} = -pV^{\text{REV}}$ and, as a consequence, $\widehat{p} = p$ have configurational contributions, when $T, V^{\text{REV}}, \mu^n, \mu^w, \mu^r$ are controlled.

Relative sizes

Equations like Eq. (17.12)–(17.16) can be used to find all REV densities. The various bulk, surface and line contributions can be computed, and their relative size can be found. By dividing Eq. (17.11) by V^{REV}, we obtain

$$u = u^n \frac{V^{n,\text{REV}}}{V^{\text{REV}}} + u^w \frac{V^{w,\text{REV}}}{V^{\text{REV}}} + u^r \frac{V^{r,\text{REV}}}{V^{\text{REV}}} + u^{nr} \frac{\Omega^{nr,\text{REV}}}{V^{\text{REV}}} + u^{wr} \frac{\Omega^{wr,\text{REV}}}{V^{\text{REV}}} + u^{nw} \frac{\Omega^{nw,\text{REV}}}{V^{\text{REV}}} + u^{nwr} \frac{\Lambda^{nwr,\text{REV}}}{V^{\text{REV}}} + u^{\text{conf}} \tag{17.19}$$

The bulk contributions are proportional to $V^{n,\text{REV}}/V^{\text{REV}}$, $V^{w,\text{REV}}/V^{\text{REV}}$, $V^{r,\text{REV}}/V^{\text{REV}}$, respectively, and are therefore generally of the same order of magnitude as u. The configurational contribution can be of the same order of magnitude as u. The surface contributions are proportional to $\Omega^{nr,\text{REV}}/V^{\text{REV}}$, $\Omega^{wr,\text{REV}}/V^{\text{REV}}$, $\Omega^{nw,\text{REV}}/V^{\text{REV}}$, respectively, and are therefore in general smaller than u. The line contribution is proportional to $\Lambda^{nwr,\text{REV}}/V^{\text{REV}}$, and is therefore in general small compared to the surface contributions, and much smaller than bulk contributions.

These statements on the relative size contributions depend on the magnitude of the excess densities. The excess mass of a surface-active material can be large. Similarly, the surface tension times the area can give a sizable contribution to the internal energy, leading to a difference in pressure across the phase boundary.

In the practical situation, a REV be must be chosen large enough to provide representative average values. The REV volume must therefore be larger than the single pore volume. In a study of a two-dimensional network [148], a REV with 20×20 links was large enough to document validity of the ergodic hypothesis. For a two-dimensional honeycomb lattice [149], we found that the REV required 60 × 40 links. The more regular the system the smaller, the REV. When the porous material consisted of a face-centered array of large solid spheres, we found that a unit cell was large enough to provide a REV [41]. For argon in a zeolite structure, Kjelstrup *et al.* [150] similarly found that a unit cell was necessary. The last findings are reasonable, as a unit cell includes all possible system microstates.

In a system with lattice symmetry, the REV may thus be rather small. In general, it may consist of nano-sized pores, but micro-sized pores can also be treated as described above. The definitions of the variables apply when the excess densities depend on the surface and three-phase contact line curvatures [146].

17.2 Estimate of the Configurational Contribution to the Entropy

The configurational contribution to the entropy is a contribution special to porous media. We can estimate the magnitude of this contribution using mean field theory and obtain some numerical insights. To do so, consider again the example in Fig. 17.1 of a porous rock with one wetting and one non-wetting fluid distributed throughout a porous rock.

We estimate first the number of possibilities to distribute the wetting and the non-wetting fluids in the pores. As variables, we use the diameter d of one pore, the volume of pores $V^{p,\mathrm{REV}} = \phi V^{\mathrm{REV}}$ and the saturation variable $\tilde{S}^n$, where $\tilde{S}^w = 1 - \tilde{S}^n$. The fluid elements in one pore have a volume of the order of magnitude $\approx d^3$.

Any particular system state is obtained by filling pores with wetting or non-wetting fluids. There are numerous ways to obey a given saturation. The total number of fluid elements in the REV is $V^{p,\mathrm{REV}}/d^3 = \phi V^{\mathrm{REV}}/d^3$. The number of volume elements with the wetting and the non-wetting fluids in the REV are $\tilde{S}^w V^{p,\mathrm{REV}}/d^3 = \phi \tilde{S}^w V^{\mathrm{REV}}/d^3$ and $\tilde{S}^n V^{p,\mathrm{REV}}/d^3 = \phi \tilde{S}^n V^{\mathrm{REV}}/d^3$.

Consider next that all filling states are equally probable. This means that they have the same energy. The total number of ways to fill the rock is

$$\frac{(\phi V^{\mathrm{REV}}/d^3)!}{(\phi \tilde{S}^w V^{\mathrm{REV}}/d^3)!(\phi \tilde{S}^n V^{\mathrm{REV}}/d^3)!} = (\tilde{S}^w)^{-\phi \tilde{S}^w V^{\mathrm{REV}}/d^3} (\tilde{S}^n)^{-\phi \tilde{S}^n V^{\mathrm{REV}}/d^3} \quad (17.20)$$

where we used Stirling's approximation to obtain the right-hand side. The configurational contribution to the entropy is accordingly

$$S^{\mathrm{conf}} = -k_{\mathrm{B}} \phi \frac{V^{\mathrm{REV}}}{d^3} [\tilde{S}^w \ln \tilde{S}^w + \tilde{S}^n \ln \tilde{S}^n] \quad (17.21)$$

This expression resembles the well-known formula for the entropy of an ideal mixture. Here, the entropy value depends on the pore diameter, the porosity and the saturation. The expression can serve as a first-order estimate of the value. In the *absence* of configurational internal energy, the configurational pressure can be defined as

$$p^{\mathrm{conf}} = T s^{\mathrm{conf}} = -k_{\mathrm{B}} T \frac{\phi}{d^3} [\tilde{S}^w \ln \tilde{S}^w + \tilde{S}^n \ln \tilde{S}^n] \quad (17.22)$$

When all distributions of wetting and non-wetting fluid elements are equally probable, S^{conf} does not depend on the temperature.

There may, however, be preferred positions of volume elements. The REV energy may be lowered, for instance, if we put the same wetting phase element next to another (as in ganglia formation). This will change the above expression for the pressure. Assume that E^w and E^n represent the interaction energy of neighboring volume elements w and n. A mean field approximation for the configurational pressure then becomes

$$p^{\mathrm{conf}} = -k_{\mathrm{B}} T \frac{\phi}{d^3} [\exp(-\beta E^w) \tilde{S}^w \ln \tilde{S}^w + \exp(-\beta E^n) \tilde{S}^n \ln \tilde{S}^n] \quad (17.23)$$

where $\beta = 1/(k_{\mathrm{B}}T)$. The corresponding configurational contribution to the entropy is given by

$$s^{\mathrm{conf}} = \frac{\partial p^{\mathrm{conf}}}{\partial T} = -k_{\mathrm{B}} \frac{\phi}{d^3} \times [(1 + \beta E^w) \exp(-\beta E^w) \tilde{S}^w \ln \tilde{S}^w + (1 + \beta E^n) \exp(-\beta E^n) \tilde{S}^n \ln \tilde{S}^n] \quad (17.24)$$

A criterion for ganglia formation is obtained from $|\beta E^w|$ or $|\beta E^n| > 1$. Equations (17.24) and (17.23) reduce to Eqs. (17.21) and (17.22) when these factors are negligible compared to unity. Ganglia are then not likely to form.

The mean field approximation for the configurational contribution to the internal energy is

$$\begin{aligned} u^{\text{conf}} &= Ts^{\text{conf}} - p^{\text{conf}} \\ &= -k_{\text{B}}\frac{\phi}{d^3}[\,\beta E^w \exp(-\beta E^w)\,\tilde{S}^w \ln \tilde{S}^w + \beta E^n \exp(-\beta E^n)\,\tilde{S}^n \ln \tilde{S}^n] \end{aligned} \tag{17.25}$$

The configurational contribution to the internal energy goes to zero for temperatures that are large compared to $|E^w|/k_{\text{B}}$ and $|E^n|/k_{\text{B}}$.

This simple model for configurational contributions has terms which are inversely proportional to the third power of the pore diameter. The configurational contributions will thus be comparatively more important for smaller pores.

17.3 Additivity of Variables: Statistical Mechanical Basis

We have constructed above a basis set of variables that are additive. This property can be motivated by the statistical properties of the grand potential, with as control variables, the temperature, T, the chemical potentials, μ^n, μ^w, μ^r, and the volume of the REV, V^{REV}. The REV replicas do not interact, and we use the grand canonical ensemble to calculate the grand canonical partition function, $\Xi(T, V^{\text{REV}}, \mu^n, \mu^w, \mu^r)$ of the REV, as we explained in Section 4.5. We can do this for small or large REVs. The grand potential is given in terms of this partition function by

$$X^{\text{GC,REV}}(T, V^{\text{REV}}, \mu^n, \mu^w, \mu^r) = -k_{\text{B}}T \ln \Xi(T, V^{\text{REV}}, \mu^n, \mu^w, \mu^r) \tag{17.26}$$

The statistical mechanical treatment of a subsystem within the REV follows the standard procedure of the larger system. A normalization constant is included, obtained from the full distribution in phase space.

Subsystems of the REV can be said to be *weakly coupled* when their interaction energy is negligible compared to the energies of the separate subsystems. The REV probability distribution in phase space is then equal to the product of the probability distributions of the subsystems as follows:

$$\begin{aligned}\exp(-\mathcal{H}/k_{\rm B}T) &= \exp(-\mathcal{H}^n/k_{\rm B}T) \times \exp(-\mathcal{H}^w/k_{\rm B}T)\\ &\quad\times \exp(-\mathcal{H}^r/k_{\rm B}T) \times \exp(-\mathcal{H}^{nr}/k_{\rm B}T)\\ &\quad\times \exp(-\mathcal{H}^{wr}/k_{\rm B}T) \times \exp(-\mathcal{H}^{nw}/k_{\rm B}T)\\ &\quad\times \exp(-\mathcal{H}^{nwr}/k_{\rm B}T) \qquad (17.27)\end{aligned}$$

For a given distribution of bulk fluids, the state of the REV is described solely by the contributing volumes, surface areas or contact lines. It follows that the grand canonical partition function of the REV is a product of the partition functions of all contributions. This is then *compatible with the assumption of additivity of variables.* We may say that material contributions arise in the REV grand potential because the wetting and non-wetting fluids, the rock, the surfaces and the contact line are *weakly coupled.*

The origin of the configurational contributions, on the other hand, can be traced to a probability distribution in phase space. This distribution contains also the distribution of bulk fluid components over all possible choices of subvolumes of the REV. By integrating over phase space, we must integrate over these distributions as well, resulting in a separate partition function. All of this taken together gives

$$\begin{aligned}&\Xi(T, V^{\rm REV}, \mu^n, \mu^w, \mu^r)\\ &\quad= \Xi^n(T, V^{n,\rm REV}, \mu^n) \times \Xi^w(T, V^{w,\rm REV}, \mu^w)\\ &\quad\quad\times \Xi^r(T, V^{r,\rm REV}, \mu^r) \times \Xi^{nr}(T, \Omega^{nr,\rm REV}, \mu^n, \mu^r)\\ &\quad\quad\times \Xi^{wr}(T, \Omega^{wr,\rm REV}, \mu^w, \mu^r) \times \Xi^{nw}(T, \Omega^{nw,\rm REV}, \mu^n, \mu^w)\\ &\quad\quad\times \Xi^{nwr}(T, \Lambda^{nwr,\rm REV}, \mu^n, \mu^w, \mu^r) \times \Xi^{\rm conf}(T, V^{\rm REV}) \qquad (17.28)\end{aligned}$$

The mass of a component can, as a variable, have no configurational contribution. It follows that Ξ^{conf} does not depend on the chemical potentials. When we take the logarithm of Eq. (17.28), multiply with $-k_{\text{B}}T$, and identify the contributions of the subsystems with the corresponding contributions to the grand potential, we obtain the following from Eq. (17.13):

$$
\begin{aligned}
&X^{\text{GC,REV}}(T, V^{\text{REV}}, \mu^n, \mu^w, \mu^r) \\
&\quad = X^{n,\text{GC,REV}}(T, V^{n,\text{REV}}, \mu^n) + X^{w,\text{GC,REV}}(T, V^{w,\text{REV}}, \mu^w) \\
&\qquad + X^{r,\text{GC,REV}}(T, V^{r,\text{REV}}, \mu^r) + X^{nr,\text{GC,REV}}(T, \Omega^{nr,\text{REV}}, \mu^n, \mu^r) \\
&\qquad + X^{wr,\text{GC,REV}}(T, \Omega^{wr,\text{REV}}, \mu^w, \mu^r) \\
&\qquad + X^{nw,\text{GC,REV}}(T, \Omega^{nw,\text{REV}}, \mu^n, \mu^w) \\
&\qquad + X^{nwr,\text{GC,REV}}(T, \Lambda^{nwr,\text{REV}}, \mu^n, \mu^w, \mu^r) + X^{\text{conf}}(T, V^{\text{REV}})
\end{aligned}
\tag{17.29}
$$

The parentheses are used to indicate the variables that the contributions depend on. In the present example, we have used immiscible bulk fluids. When the fluids are miscible, the material contributions will depend on all chemical potentials. Equations (17.14)–(17.18) are direct consequences of Eq. (17.29).

The autonomous nature of the surfaces and the three-phase contact lines can be understood as a direct consequence of the fact that each subsystem is weakly coupled to all other subsystems. Examples of weakly coupled subsystems can be found everywhere. The catalytic surface with its own temperature, and the liquid–vapor interface, are two examples.

The volume and the mass of a component has only material-type contributions. This is not a consequence of weak coupling, however. The analysis in the next section will show that the Gibbs energy also has only material contributions. All other thermodynamic energies, like the internal energy, the Helmholtz energy and the enthalpy, have in addition configurational contributions.

17.4 Gibbs' Equation

To summarize so far, we have proposed a way to construct a basis set of coarse-grained variables, and provided a statistical mechanical scenario for when the set may be applied. Coarse-grained variables are needed for a description of transport on the macro-scale, and the present choice of variables includes the effects of surfaces and contact lines, in addition to bulk-type contributions. The following steps were taken to find a basis set of extensive (additive) coarse-grained variables:

1. The set of coarse-grained variables was chosen, composed of contributions from all relevant phases, surfaces and three-phase contact lines. Variables are additive and have material and possibly also configurational contributions.
2. A statistical basis for the additivity could be formulated; reasoning that the interaction energies of the various phases, surfaces and contact lines are small so that they are weakly coupled.
3. The densities so constructed served to describe a REV on the continuum level if the REV is large enough to represent all the microstates of the porous medium.
4. The REV can be composed of several systems that are small in the sense of Hill. Each small system is subject to the same set of control variables (T, μ^n, μ^w, μ^r) as the REV is.

With these premises, we have described a REV in equilibrium with the surroundings. The surroundings control the REV properties $T, V^{\text{REV}}, \mu^n, \mu^w$ and μ^r. The small system contributions to the REV are in the GC ensemble. Any REV will then satisfy the Gibbs relation, as expressed in the introduction to this chapter:

$$dU^{\text{GC,REV}} = TdS^{\text{GC,REV}} - pdV^{\text{REV}} + \mu^n dM^{n,\text{GC,REV}} + \mu^w dM^{w,\text{GC,REV}} + \mu^r dM^{r,\text{GC,REV}} \tag{17.30}$$

Each additive variable in this equation is constructed as described in the preceding sections. For large enough REVs, $U^{\text{GC,REV}}$ is an Euler homogeneous function of degree one in the size variable. In that case, we can use the densities and Eq. (17.30) becomes

$$du^{\mathrm{GC}} = Tds^{\mathrm{GC}} + \mu^{n} d\rho^{n,\mathrm{GC}} + \mu^{w} d\rho^{w,\mathrm{GC}} + \mu^{r} d\rho^{r,\mathrm{GC}} \tag{17.31}$$

We have argued that V^{REV}, $M^{n,\mathrm{GC,REV}}$, $M^{w,\mathrm{GC,REV}}$, $M^{r,\mathrm{GC,REV}}$ have material contributions, while $X^{\mathrm{GC,REV}}$, p and $S^{\mathrm{GC,REV}}$ also have configurational contributions.

For small REVs, $U^{\mathrm{GC,REV}}$ is *not* an Euler homogeneous function of degree one in the size variable. With the chosen type of variables, the Hill–Gibbs–Duhem's equation then applies (see Section 3.4):

$$\begin{aligned} dX^{\mathrm{GC,REV}} &= -d(\widehat{p}V^{\mathrm{REV}}) = -S^{\mathrm{GC,REV}} dT - pdV^{\mathrm{REV}} \\ &\quad - M^{n,\mathrm{GC,REV}} d\mu^{n} - M^{w,\mathrm{GC,REV}} d\mu^{w} - M^{r,\mathrm{GC,REV}} d\mu^{r} \end{aligned} \tag{17.32}$$

It follows that

$$\begin{aligned} S^{\mathrm{GC,REV}} &= -\frac{\partial X^{\mathrm{GC,REV}}}{\partial T}, \\ p &= -\frac{\partial X^{\mathrm{GC,REV}}}{\partial V^{\mathrm{REV}}}, \quad M^{n,\mathrm{GC,REV}} = -\frac{\partial X^{\mathrm{GC,REV}}}{\partial \mu^{n}} \\ M^{w,\mathrm{GC,REV}} &= -\frac{\partial X^{\mathrm{GC,REV}}}{\partial \mu^{w}}, \quad M^{r,\mathrm{GC,REV}} = -\frac{\partial X^{\mathrm{GC,REV}}}{\partial \mu^{r}} \end{aligned} \tag{17.33}$$

All equations in the rest of the chapter apply also for REVs that are small in Hill's sense.

We know that $M^{n,\mathrm{GC,REV}}$, $M^{w,\mathrm{GC,REV}}$, $M^{r,\mathrm{GC,REV}}$ have no configurational contribution. The configurational contribution X^{conf} to $X^{\mathrm{GC,REV}}$ must therefore be independent of $\mu^{n}, \mu^{w}, \mu^{r}$. Given that $S^{\mathrm{GC,REV}}$ has the configurational contribution S^{conf}, it follows that the contribution X^{conf} depends on T. The derivative gives the configurational contribution to S^{conf}

$$S^{\mathrm{conf}} = -\frac{\partial X^{\mathrm{conf}}}{\partial T} = V^{\mathrm{REV}} \frac{\partial \widehat{p}^{\mathrm{conf}}}{\partial T} \tag{17.34}$$

Furthermore, it follows that

$$
\begin{aligned}
p &= -\frac{\partial X^{\mathrm{GC,REV}}}{\partial V^{\mathrm{REV}}} = \frac{\partial(\widehat{p}V^{\mathrm{REV}})}{\partial V^{\mathrm{REV}}} = \widehat{p} + V^{\mathrm{REV}}\frac{\partial \widehat{p}}{\partial V^{\mathrm{REV}}} \\
p^{\mathrm{conf}} &= -\frac{\partial X^{\mathrm{conf}}}{\partial V^{\mathrm{REV}}} = \frac{\partial(\widehat{p}^{\mathrm{conf}}V^{\mathrm{REV}})}{\partial V^{\mathrm{REV}}} = \widehat{p}^{\mathrm{conf}} + V^{\mathrm{REV}}\frac{\partial \widehat{p}^{\mathrm{conf}}}{\partial V^{\mathrm{REV}}}
\end{aligned} \tag{17.35}
$$

When $\widehat{p}$ and $\widehat{p}^{\mathrm{conf}}$ depend on V^{REV}, $p \neq \widehat{p}$ and $p^{\mathrm{conf}} \neq \widehat{p}^{\mathrm{conf}}$. When V^{REV} is large enough, so that $\widehat{p}$ and $\widehat{p}^{\mathrm{conf}}$ do not depend on V^{REV}, $p = \widehat{p}$ and $p^{\mathrm{conf}} = \widehat{p}^{\mathrm{conf}}$. In this situation, x^{conf} and s^{conf} depend only on T. We observe that V^{REV}, $M^{n,\mathrm{GC,REV}}$, $M^{w,\mathrm{GC,REV}}$, $M^{r,\mathrm{GC,REV}}$ have no configurational contribution, while $X^{\mathrm{GC,REV}}$, p and $S^{\mathrm{GC,REV}}$ have such contributions.

The internal energy $U^{\mathrm{GC,REV}}$, the Helmholtz energy $F^{\mathrm{GC,REV}}$, the enthalpy $H^{\mathrm{GC,REV}}$ and the Gibbs energy $G^{\mathrm{GC,REV}}$ are given by

$$
\begin{aligned}
U^{\mathrm{GC,REV}} &= TS^{\mathrm{GC,REV}} - \widehat{p}V^{\mathrm{REV}} + \mu^n M^{n,\mathrm{GC,REV}} + \mu^w M^{w,\mathrm{GC,REV}} \\
&\quad + \mu^r M^{r,\mathrm{GC,REV}} \\
F^{\mathrm{GC,REV}} &= -\widehat{p}V^{\mathrm{REV}} + \mu^n M^{n,\mathrm{GC,REV}} + \mu^w M^{w,\mathrm{GC,REV}} \\
&\quad + \mu^r M^{r,\mathrm{GC,REV}} \\
H^{\mathrm{GC,REV}} &= TS^{\mathrm{GC,REV}} + \mu^n M^{n,\mathrm{GC,REV}} + \mu^w M^{w,\mathrm{GC,REV}} \\
&\quad + \mu^r M^{r,\mathrm{GC,REV}} \\
G^{\mathrm{GC,REV}} &= \mu^n M^{n,\mathrm{GC,REV}} + \mu^w M^{w,\mathrm{GC,REV}} + \mu^r M^{r,\mathrm{GC,REV}}
\end{aligned} \tag{17.36}
$$

It follows that also $G^{\mathrm{GC,REV}}$ has no configurational contribution. For the configurational contributions to $u^{\mathrm{GC,REV}}$, $f^{\mathrm{GC,REV}}$, $h^{\mathrm{GC,REV}}$ and $g^{\mathrm{GC,REV}}$, we obtain

$$
\begin{aligned}
u^{\mathrm{conf}} &= Ts^{\mathrm{conf}} - \widehat{p}^{\mathrm{conf}} = T\frac{\partial \widehat{p}^{\mathrm{conf}}}{\partial T} - \widehat{p}^{\mathrm{conf}} \\
f^{\mathrm{conf}} &= -\widehat{p}^{\mathrm{conf}} \\
h^{\mathrm{conf}} &= Ts^{\mathrm{conf}} = T\frac{\partial \widehat{p}^{\mathrm{conf}}}{\partial T} \\
g^{\mathrm{conf}} &= 0
\end{aligned} \tag{17.37}
$$

where we used Eq. (17.34). For large REVs, $\widehat{p}^{\mathrm{conf}}$ is equal to p^{conf}.

A REV in local equilibrium

So far in this chapter we have discussed a REV in global equilibrium with its surroundings. Away from global equilibrium, the neighboring REVs are not necessarily in equilibrium with each other. The assumption of local equilibrium covers this situation. It implies thermodynamic equilibrium within the REV, but not between neighboring REVs.

With the assumption of local equilibrium and the Gibbs equation in place, we can proceed as normal in nonequilibrium thermodynamics theory. For the time-variation in the coarse-grained variables, we shall use Eq. (17.31) plus the balance equations for mass and energy to find the entropy production of the REV and the entropy balance. The constitutive equations, the flux–force relations will then follow, see Chapter 19.

17.5 Concluding Remarks

We have proposed a way to describe porous media on the nanoscale, extending the method of Bedeaux, Kjelstrup *et al.* [17, 42, 144]. The control variables T, μ^n, μ^w, μ^r are well defined by the environment (a reservoir). The REV of volume V^{REV} is then defined by the grand canonical ensemble. It is described by a basis set of coarse-grained thermodynamic variables. The set is expected to obey the Gibbs equation, as they are additive. The Gibbs equation, so constructed, will serve as a basis for the derivation of the entropy production and constitutive equations in Chapter 19.

The thermodynamic properties on the REV scale have a combination of material and configurational contributions. The material ones are a sum of contributions from the solid and fluid phases and of excesses on their interfaces and contact line. The configurational ones are due to the different ways that the fluid phases can be configured in the pores. The configurational contributions were estimated using mean field theory. The coarse-grained variables will be continuous on the REV scale and may be used to characterize processes in porous media.

Chapter 18

Integral and Differential Pressures in a Slit-Pore

We examine a slit pore of molecular dimensions which contains a two-phase fluid. We find that the integral pressure along the pore is the same in both phases. This is compatible with the laws of Young and Young–Laplace. The integral and differential pressures are connected by a geometric scaling law.

A variation in the dimensions of a nanopore, which leads to changes in molecular interactions at the pore wall, can promote changes in surface adsorption, vapor pressure and phase stability in the pore [151–155]. Given the importance of porous media for applications, say in fuel cells [156], batteries [157] and membranes for drinking water production [158–160], it is important to understand how properties change upon confinement, e.g., when the pore walls come closer and closer to each other. In particular, it is of interest to predict how thermodynamic properties of the fluid in a pore change with pore size. This type of scaling was presented in Chapters 5, 6 and 15, see also [2, 37, 48]. In this chapter, we derive a scaling law for the subdivision potential that applies to a two-phase fluid in a thin slit pore.

The literature offers several definitions of the pressure inside a nano-porous material [161, 162]. In order to obtain a well-defined pressure, we first need to identify the system we assign it to. The pressure is an intensive variable in thermodynamics, normally assigned to a control volume in a bulk phase. For a heterogeneous system, with significant surface energies, different specifications are needed. We shall refer to a *representative elementary volume* (REV), which captures all states representative of the system, see previous chapter, and define the differential and integral pressures of the REV [24]. We shall explain possible REVs for a slit pore that contains a liquid in equilibrium with its vapor, building on the work of [40]. The relations are applied in Chapter 20 [41].

The system under study is controlled in the grand canonical ensemble. The integral pressure, $\widehat{p}$, and the differential pressure, p, are then related by the subdivision potential (see Section 3.4), and this relation provides the scaling law

$$p - \widehat{p} = \varepsilon / V \tag{18.1}$$

The subdivision potential is the signature of the small system. In the thermodynamic limit, it is zero. The integral pressure can differ significantly from the differential pressure in confined systems [41, 49].

The difference in the differential and integral pressures depends on the system shape or geometry. Here, we investigate a slit pore of nanoscale dimensions. The wall distances vary between 3 and 5 nm. The pore is filled with a vapor in equilibrium with its liquid phase inside the pore. The pore is illustrated in Fig. 18.1. The figure shows the liquid droplet and the cartesian coordinates of the pore. The simple geometry has motivated the choice of system, because this geometry makes it possible to find and test the scaling law that is particular to slit pores. Some results will also be given for the cylindrical pore.

A scaling law relates properties of one system size to another size with the same shape. Such knowledge is helpful in a situation where measurements are lacking. It follows from the law that the thermodynamic properties of the two-phase equilibrium changes as the walls become closer. We shall see that a new condition can be formulated for the phase equilibrium, compatible with existing descriptions in terms of Young's and Young–Laplace's laws.

The tool to investigate the scaling law is again molecular dynamics simulations [49, 163]. The pore *shape* is kept constant in the

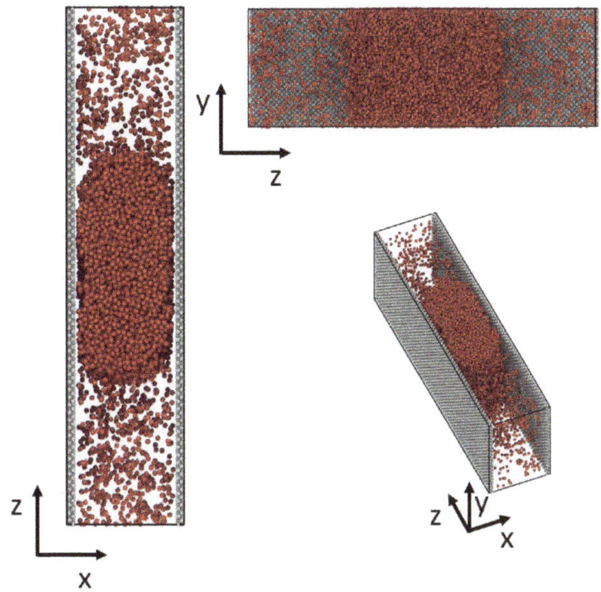

Figure 18.1. Illustration of a slit pore. Walls are at a distance $2a$ apart. The droplet in the middle of the pore is surrounded by its vapor at equilibrium on both sides. Reproduced from [40] with permission from Nanomaterials.

derivations. We recapitulate first the equations leading to Eq. (18.1). The relation is next tested using molecular dynamics simulations.

18.1 Thermodynamic Relations for Fluids in Small Pores

The ensemble formulation of Hill [4] deals with the size and shape-dependency of small systems. Consider therefore an ensemble of slit pores. We have taken slit pores as an example, but the equations apply equally well to cylindrical pores.

The number of identical and independent replicas of the slit pore is $\mathcal{N}$. The extended version of the Gibbs equation for the REV, here called Hill–Gibbs equation, is

$$dU_t = TdS_t - p\mathcal{N}dV + \sum_j \mu_j dN_{j,t} - \widehat{p}Vd\mathcal{N} \tag{18.2}$$

The total energy, entropy and fluid particle number are U_t, S_t and N_t, respectively. Furthermore, T is the temperature, p is the differential pressure, μ_j is the chemical potential of component j and $\widehat{p}$ is the integral pressure. The term $-\widehat{p}Vd\mathcal{N}$ is the reversible work needed to change the total volume, by changing the number of small systems, keeping S_t, V and $N_{t,j}$ constant. The equation motivates the pressure names; p is relevant when the volume changes by a differential amount dV, $\widehat{p}$ is relevant when the volume changes by a multiple of V. As explained in Chapter 3, all changes of the volume are isomorphic.

We shall elucidate the difference between p and $\widehat{p}$, the differential and integral pressures of a fluid confined to a pore. The grand canonical ensemble applies. The average values in this ensemble of a single small system are

$$U = \frac{U_t}{\mathcal{N}}, \quad N_j = \frac{N_{j,t}}{\mathcal{N}}, \quad S = \frac{S_t}{\mathcal{N}} \tag{18.3}$$

By introducing these equations into Eq. (18.2), following the procedure in Section 3.4, we obtain the Hill–Gibbs–Duhem equation for the REV

$$d(\widehat{p}V) = SdT + \sum_j N_j d\mu_j + pdV \tag{18.4}$$

The equation reduces to the usual Gibbs–Duhem equation when $\widehat{p} = p$. Equation (18.4) gives the following relation between the integral and differential pressure [4, 14]:

$$p = \left(\frac{\partial \widehat{p}V}{\partial V}\right)_{T,\mu_j} = \widehat{p} + V\left(\frac{\partial \widehat{p}}{\partial V}\right)_{T,\mu_j} \tag{18.5}$$

The volume change is isomorphic. The expression shows how p and $\widehat{p}$ are related for small systems, where $\widehat{p}$ depends on the volume. In a macroscopic system, $\widehat{p}$ is independent of the volume, so that p and $\widehat{p}$ are the same. The variation in the integral pressure with the volume is therefore a property of the small system, and provides us with the geometric scaling law.

The above equations will be applied to a two-phase system contained in a slit pore and a cylindrical pore. Inside the pore, the differential pressure of the liquid differs from that of the vapor phase. For the time being, we introduce the notation p^v and p^l for the differential pressure of the vapor and the liquid, respectively, and p^f to indicate either of the two. The difference in the differential and the integral pressure times the volume is equal to the subdivision potential [4, 14], ε^f. From Eq. (18.5) we obtain

$$p^f - \widehat{p} = \frac{\varepsilon^f(T, V, \mu_j)}{V} = V\left(\frac{\partial \widehat{p}}{\partial V}\right)_{T,\mu_j} \tag{18.6}$$

where $p = p^f$. The subdivision potential is an additive thermodynamic property [4, 14]. When such a property is divided by the small system volume, a scaling law appears, of a similar type as we have found earlier [2, 48]. The ratio ε^f/V is proportional to the inverse characteristic length, $\Omega/V = 1/L_\Omega$ of the system, where Ω is the surface area between the fluid and the solid, see Eq. (5.2) in Chapter 5. In general terms,

$$\begin{aligned} p^f - \widehat{p} = \frac{\varepsilon^f}{V} &\equiv \zeta^f = \zeta^{f,\infty} + \zeta^{f,s}\left(\frac{\Omega}{V}\right) + \zeta^{f,se}\left(\frac{\Omega}{V}\right)^2 + \cdots \\ &= \zeta^{f,s}\left(\frac{1}{L_\Omega}\right) + \zeta^{f,se}\left(\frac{1}{L_\Omega}\right)^2 + \cdots \end{aligned} \tag{18.7}$$

The coefficients $\zeta^{f,\infty}, \zeta^{f,s}$ and $\zeta^{f,se}$ do not depend on $L_\Omega \equiv V/\Omega$, but are functions of T and μ_j. As p^f and $\widehat{p}$ are equal in the thermodynamic limit, it follows that the coefficient $\zeta^{f,\infty} = 0$. The expansion applies to slit as well as cylindrical pores.

18.2 Representative Elementary Volumes

In a bulk system, we refer to control volumes when we define system average properties, also away from equilibrium. In a porous medium, we refer to a chosen REV for the same reason [144]. Consider a two-phase fluid in a slit pore. A liquid droplet is in equilibrium with its vapor inside a slit pore, see Fig. 18.2 for an illustration. The pore is so narrow that the surface energy is comparable to the energy of the bulk fluid.

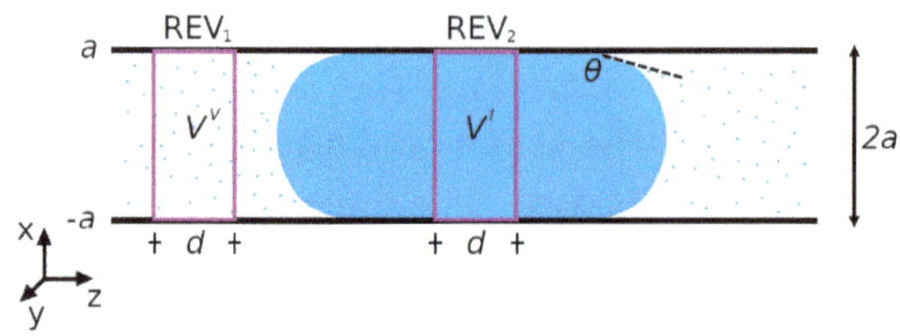

Figure 18.2. Slit pore of width $2a$. The liquid droplet in the middle is in equilibrium with a vapor phase on both sides. Examples of representative volume elements are REV_1 (vapor of volume V^v) and REV_2 (liquid of volume V^l), both with length d. The z-axis is located at the center of the pore. Reproduced from [40] with permission from Nanomaterials.

Any REV needs to cover the whole cross-section of the pore, as in Fig. 18.2. At global equilibrium, one possibility is the REV_1 that contains the vapor phase. Another is REV_2 that contains the liquid phase. Both REVs contain the surface of the wall. A third possibility is to include all phases into one REV. Let us proceed with the simplest possibilities, REV_1 and REV_2, both positioned away from the surface between liquid and vapor, see Fig. 18.2. When we integrate along a slit pore, we will encounter these REVs.

We now assume that the surface energy depends only weakly on the bulk energy. Weak coupling was discussed in the previous chapter. The grand canonical partition function for each REV, Ξ, contains therefore, the product of the partition functions for each of the subsystems, of the surface (Ξ_1) and of the bulk (Ξ_2), respectively.

$$\Upsilon \equiv -k_{\text{B}}T\ln(\Xi_1\Xi_2) = -k_{\text{B}}T\ln\Xi \tag{18.8}$$

The k_{B} is as usual the Boltzmann constant. The grand potential Υ provides the statistical mechanical link to particle energies. It provides the tool for pressure computations, because the compressional energy, $\widehat{p}V$, is defined by the grand potential, Υ [4, 14].

$$\widehat{p}V \equiv -\Upsilon \tag{18.9}$$

This relation defines the integral pressure of an isothermal REV with constant chemical potential. The compressional energy is additive when Eq. (18.8) applies. Similar to $\widehat{p}V$, ε is additive. It is in general a sum of contributions from all phases and interfaces present within the REV. This property shall next be applied.

18.3 The Integral Pressure

In the example, $\widehat{p}V$ is a sum of contributions from all phases and interfaces present within the REV. The compressional energy for REV_n, with $n = 1, 2$, is then

$$\widehat{p}_{\text{REV}_n} V = \widehat{p}^f V^f - \widehat{\gamma}^{fs} \Omega^{fs} \tag{18.10}$$

where V is the volume of the REV, V^f is the volume of the fluid in the REV, Ω^{fs} is the surface area between the fluid and the wall. Furthermore $\widehat{p}^f$ is the integral pressure in the fluid and $\widehat{\gamma}^{fs}$ is the integral surface tension at the wall. In the present case, the two REVs are completely filled with one fluid; a vapor (REV_1) or a liquid (REV_2). The REV volume in both cases obeys $V = V^f = V^v = V^l = 2adw$, where $2a$ is the slit pore wall distance, d, the REV length and w, the width of the REV. Superscripts v and l refer again to the vapor and the liquid. The surface area between fluid and solid is $\Omega^{fs} = 2dw$. This implies that $L_\Omega = a$, i.e., half the distance between the walls in the pore. The surface tensions between fluid and pore surface are $\widehat{\gamma}^{ls}$ and $\widehat{\gamma}^{vs}$, respectively, where the hat indicates that the system can be small, meaning that the surface tension is a function of the contact area and system geometry.

For sufficiently large values of $L_\Omega = a$, the fluid in the center of the pore has bulk properties. In that case, the integral pressure in the vapor $\widehat{p}^v$ and the liquid $\widehat{p}^l$ are independent of V^v or V^l, respectively, and

$$\widehat{p}^v = p^v, \quad \widehat{p}^l = p^l \tag{18.11}$$

Here, p^l and p^v are bulk pressures of the liquid and vapor, respectively. When the distance between the walls ($L_\Omega = a$) is sufficiently large, the fluid–solid surface tension also shows macroscopic behavior, and

$$\widehat{\gamma}^{vs} = \gamma^{vs}, \quad \widehat{\gamma}^{ls} = \gamma^{ls} \tag{18.12}$$

By introducing Eqs. (18.11) and (18.12), the volume and the fluid–solid surface area, into Eq. (18.10), we obtain the integral pressure of the liquid and vapor REVs as follows:

$$\widehat{p}_{\mathrm{REV}_2} = p^l - \frac{\gamma^{ls}}{a}, \quad \widehat{p}_{\mathrm{REV}_1} = p^v - \frac{\gamma^{vs}}{a} \tag{18.13}$$

We now make the hypothesis that the integral pressure is constant at equilibrium along the pore. This means that

$$\widehat{p}_{\mathrm{REV}_2} = \widehat{p}_{\mathrm{REV}_1} \tag{18.14}$$

or, that

$$p^l = p^v + \frac{\gamma^{ls} - \gamma^{vs}}{a} \tag{18.15}$$

Young's law derives from a force balance at phase equilibrium. It takes the form

$$\gamma^{ls} = \gamma^{vs} + \gamma^{lv}\cos\theta \tag{18.16}$$

where γ^{lv} is the liquid–vapor surface tension, and θ is the contact angle at the pore wall, see Fig. 18.2. We introduce this law into Eq. (18.15) and obtain

$$p^l = p^v + \frac{\gamma^{lv}}{a}\cos\theta = p^v + \frac{\gamma^{lv}}{R} \tag{18.17}$$

This is the Young–Laplace equation. The radius of curvature of the liquid–vapor surface, $R = a/\cos\theta$, was introduced in the last equality.

We see that the hypothesis in Eq. (18.14) is true if the laws of Young and Young–Laplace apply. We expect the last to hold true, and by implication, Eq. (18.14) is also true. We have thus derived a new way to express phase equilibrium in nanopores.

Consider again a slit pore with a two-phase fluid at equilibrium. The integral pressures of the fluids are given by Eq. (18.13). For each fluid, the difference in the differential and integral pressures follow from Eq. (18.5) as follows:

$$p^f - \widehat{p} = \frac{\varepsilon^f}{V} = \zeta^f = a^3\left(\frac{\partial \widehat{p}}{\partial a^3}\right)_{T,\mu} = \frac{\gamma^{fs}(\alpha)}{3a} \tag{18.18}$$

This is the geometric scaling law for the pressure difference, similar to the one proposed from the Small System Method, see, e.g., Chapters 5, 6 [2, 48] and Eq. (18.7). For large systems, when $a \to \infty$, the difference between p^f and $\widehat{p}$ disappears, leading to $p = \widehat{p} = p^f$ in the REV. For small systems, the difference between p^f and $\widehat{p}$ is described by the ratio between fluid-solid surface tension and the half channel width (γ^{fs}/a) divided by 3.

The surface tension may depend on a parameter α which characterizes the molecular interactions of the fluid and the wall. The shape is a variable in small system thermodynamics, and must remain constant when we differentiate. To keep the shape constant in the derivation means to take the isomorphic derivative of the integral pressure as in Eq. (18.18).

We are now in a position to find the scaling law for cylindrical pores with the same fluid conditions. The proportionality factor in the scaling law in Eq. (18.18) will change. The ratio of the area over the volume in the REV of a cylindrical pore is double the value of a slit pore. The ratio of the slope in a cylindrical pore over that of a slit pore is therefore equal to 2. The scaling law for a cylindrical pore becomes

$$p^f - \widehat{p} = a^3 \left(\frac{\partial \widehat{p}}{\partial a^3}\right)_{T,\mu} = \frac{2\gamma^{fs}(\alpha)}{3a} \tag{18.19}$$

Using the thermodynamic scheme of Hill may seem more complicated than using Young's and Young–Laplace's laws. Hill's theory shows, however, how the two Eqs. (18.18) and (18.19) are linked to one another and to new variables (e.g., ε^f). This is useful. Equation (18.18) can be tested by molecular dynamics simulations [40]. Results will be shown in what follows, confirming that the equilibrium condition $\widehat{p}_{\mathrm{REV}_1} = \widehat{p}_{\mathrm{REV}_2}$ holds.

18.4 Pressure Components

We have seen above that the integral pressure of two fluid phases in equilibrium is constant if Young's and Young–Laplace's laws apply. The integral and differential pressures must accordingly be related to the mechanical pressure tensor. This relation is now of interest. The mechanical pressure tensor and its components will from now on be denoted with upper case P, while the thermodynamic pressures are denoted as before, with lower case p.

Consider again the liquid droplet in a slit pore, as shown in Fig. 18.2. The slit pore wall is parallel to the yz-plane. The pressure tensor component acting perpendicular to the wall is the xx-component, $P_\perp = P_{xx}$. There are two parallel components of the pressure tensor, the yy-component and the zz-component, which are equal in equilibrium. The parallel component is given as the average of the two, $P_\parallel = \frac{1}{2}(P_{yy} + P_{zz})$.

The surface tension is defined as minus the excess of the parallel component [164],

$$\gamma^{fs} = \frac{1}{2}\int_{-a}^{a} (P_\perp(x) - P_\parallel(x))dx \tag{18.20}$$

The factor of one half accounts for the fact that the integral covers two fluid–solid surfaces. By introducing the fluid–solid surface tension into the equation for the integral pressure of a single fluid phase, cf. Equation (18.13), we obtain

$$\widehat{p} = p^f - \frac{1}{2a}\int_{-a}^{a} (P_\perp(x) - P_\parallel(x))dx \tag{18.21}$$

where superscript f again refers to either the liquid or the vapor. The perpendicular component of the mechanical pressure tensor is constant at flat surfaces at mechanical equilibrium. The integral of the perpendicular component divided by $2a$ is therefore equal to $P_\perp$. We assume that the perpendicular component is equal to the differential pressure of the bulk fluid in the center of the pore, $p^f = P_\perp$. In what follows, the assumption is shown to hold. The differential pressure of the bulk fluid and the integral over the perpendicular component of the pressure cancel each other, and the integral pressure becomes equal to the average parallel component,

$$\widehat{p} = \frac{1}{2a}\int_{-a}^{a} P_\parallel(x)dx = \langle P_T\rangle \tag{18.22}$$

The equation shows that the integral pressure of a bulk fluid in a slit pore at mechanical equilibrium can be understood as the average parallel pressure component.

The differential pressure can be related to the mechanical pressure tensor. The difference between the differential and integral pressures is equal to the fluid–solid surface tension divided by $3a$. By introducing Eq. (18.20) for the fluid–solid surface tension, we obtain

$$p - \widehat{p} = \frac{1}{6a}\int_{-a}^{a}(P_{\perp}(x) - P_{\parallel}(x))dx \tag{18.23}$$

By re-organizing and introducing the average of the parallel pressure for the integral pressure, we obtain

$$p = \frac{1}{3}\left(\langle P_{\perp}\rangle + 2\langle P_{\parallel}\rangle\right) = \frac{1}{3}\mathrm{Tr}\langle P\rangle \tag{18.24}$$

The differential pressure is equal to the average of the trace of the mechanical pressure tensor (trace divided by 3). The thermodynamic pressures can thus all be understood in mechanical terms.

18.5 Molecular Simulations

System

Molecular dynamic simulations were carried out for a droplet in a slit pore. The droplet was created in the middle of the pore, surrounded by its vapor phase on both sides (see Fig. 18.1). The simulation box, with side lengths $L_x = L_y \neq L_z$ had periodic boundary conditions in all directions.

The simulation box was divided into n rectangular layers along the z-axis with a layer thickness of $\Delta z = L_z/n$. Both the pore-wall (gray) and the fluid (red) consisted of Lennard-Jones/spline particles [165], see Appendix D. The particles of the wall were immobilized, such that the particles were able to interact with the fluid particles, but not move. Due to periodic boundary conditions, the wall thickness was $t = 3.5\sigma_0$ and had a reduced density of $\rho^*_{\mathrm{wall}} = 1.05$. The fluid inside the pore was initialized in a way that the overall reduced density inside the pore was $\rho^*_{\mathrm{fluid}} = 0.27$, thus in the two-phase regime [165]. The reduced average density of the bulk liquid was $\rho^*_{\text{liquid}} = 0.75 \pm 0.0015$ and the one of the vapor, $\rho^*_{\text{vapor}} = 0.03 \pm 0.0025$. We used Lennard-Jones reduced units, which are indicated by superscript * [18], see Appendix D.

The simulations were carried out using LAMMPS [166] in the canonical (NVT) ensemble at a reduced temperature of $T^* = 0.70$. The Nosé–Hoover thermostat [167] was used to keep the temperature constant. The system was simulated for four slit pore widths, $a = 8.7\sigma_0$, $10.45\sigma_0$, $12.2\sigma_0$ and $13.7\sigma_0$. The half channel widths were chosen so that both, the pressure of the fluid in the center, p^f, as well as the fluid-solid surface tension, γ^{fs}, are independent of the distance a, i.e., they show macroscopic bulk properties. Furthermore, the slit widths are large enough to avoid the effect of the disjoining pressure, see [152].

Particle interaction potential

The interaction between particles of type i and j was defined by the Lennard-Jones/spline potential [18, 165], $\phi_{\text{LJ/s}}(r_{ij})$, where r_{ij} is the distance between the particles, see Appendix D. The interaction parameters, ε_{ij} and σ_{ij}, in reduced units were set to 1 for all particle pairs (Lennard-Jones units). The interaction parameter α_{ij} is used to control the attractive interaction between the wall and the fluid, where we used $\alpha_{\text{sf}} = 0.05$, 0.15 and 0.25. For wall–wall and fluid–fluid interactions, the interaction parameter was set to $\alpha_{\text{ss}} = \alpha_{\text{ff}} = 1$. For all α_{sf} values, the interaction between wall and fluid is strongly repulsive, thus of a non-wetting type [165].

Computing pressures and wall–fluid surface tension

The mechanical pressure components in any layer l or combination of layers, or in a REV (core plus all layers included), were computed following Kirkwood [164]:

$$P_{\alpha\beta} = \frac{1}{3V}\sum_{i\in l} m_i(v_{i,\alpha} - v_{m,\alpha})(v_{i,\beta} - v_{m,\beta}) - \frac{1}{6V}\sum_{i\in l}\sum_{j=1}^{N}(r_{ij,\alpha} f_{ij,\beta}) \tag{18.25}$$

The subscripts α and β denote the Cartesian coordinates, x, y and z, while V is the volume of the layer, m_i is the mass of particle i, $v_{i,\alpha}$ is the velocity of fluid particle i in the α-direction and $v_{m,\alpha}$ is the average velocity in the α-direction. This velocity is zero in our equilibrium simulations. The properties $r_{ij,\alpha}$ and $f_{ij,\alpha}$ are, respectively, the distance and the force between particle i and j in the α-direction. The first term is the kinetic energy contribution from the particles, and the second term arises from pairwise interactions. Because of isotropy, the tensor is diagonal.

Half the value of the pairwise energy contribution was assigned to the layer that contains particle i, while the other half was assigned to the layer that contains particle j. A more precise distribution of the contributions is available [168–171], but the present method has sufficiently small errors for the volumes used.

The pressure of a fluid volume element was calculated, starting with the core of the pore (lightest color in Fig. 18.3), which has

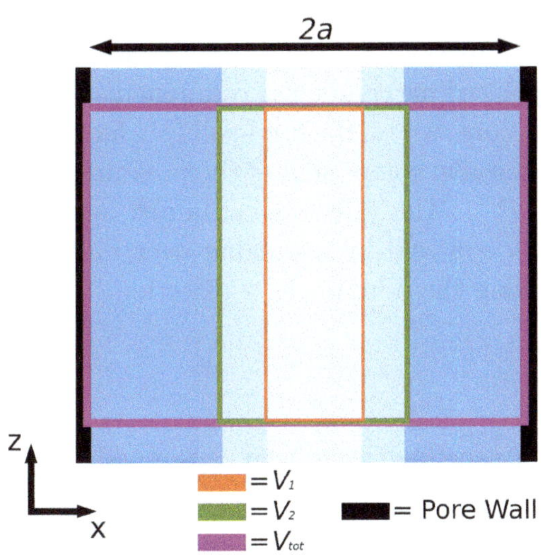

Figure 18.3. The core layer of the slit pore (lightest color) and subsequent layers (progressively darker color) used in the calculation of the pressure tensor of the fluid. Symbols V_1 and V_2 denote the volumes of the core and the core plus one layer, respectively. The total volume, covering all interactions with the pore wall, is V_{tot}. Reproduced from [40] with permission from MDPI Publishers.

volume V_1. Volume was gradually added to the core, including in the end also the wall (black color in Fig. 18.3). In this manner, the volume of the core was growing, layer by layer, until in the end $s-1$ layers were added to the central core. The pressure was computed for the core (V_1), for the core plus one subsequent layer (V_2) and for the total volume (V_{tot}). The total volume is equal to the volume V of the REV (see Fig. 18.2). Thus, the pressure was computed in growing pore core-volumes until the volume was large enough to cover the whole width ($2a$) of the pore. This *growing-core technique* [40] enabled an investigation of the pressure in the center of the pore (the core) and the averaged pressure of the whole cross-section of the pore, including all interactions with the pore wall. When all layers plus the wall inside the pore were included (V_{tot}), we obtained $\widehat{p}$ of the REV.

As long as there was no impact of the wall (the core plus layers is sufficiently far away from the wall), we observed the bulk fluid pressure, p^f. This procedure was applied to REV_1 and REV_2.

The fluid–solid surface tension was computed as the excess of the parallel pressure tensor component [164] from Eq. (18.20).

The extrapolated perpendicular component $P_\perp$ was interpreted as the pressure in core, where the fluid-solid surface has no effect on the pressure. The parallel component of the pressure was calculated from $P_\| = \frac{1}{2}(P_{yy} + P_{zz})$. The value was obtained using a low resolution-value in x-direction and by averaging over multiple chunks of the liquid phase along the pore in the z-direction.

Case studies

Three types of repelling walls were examined, for $\alpha_{\text{sf}} = 0.05, 0.15$ and 0.25. The attractive interaction increases with increasing value of α_{sf}. The corresponding reduced liquid–solid surface tensions are given in Table 18.1. Pore widths varied between $17.4\sigma_0$ and $27.4\sigma_0$. These widths ensure that there is always bulk fluid present in the core of the pore. The bulk pressure varied between 6.0 and 0.02 in reduced units. Within this range of pressures, all systems (as given by parameters α_{sf} and a) are in the two-phase region.

Table 18.1. Reduced Surface Tensions for the liquid/solid and vapor/solid interfaces.

α_{sf}	γ^{ls}	γ^{vs}
0.05	0.36 ± 0.02	0.0067 ± 0.0026
0.15	0.33 ± 0.01	0.005 ± 0.0035
0.25	0.31 ± 0.01	0.004 ± 0.0045

Notes: Values are presented for three interaction parameters α_{sf} in the Lennard-Jones potential used in the computations.

In the derivation of Eq. (18.22), the pressure tensor perpendicular to the slit pore wall, $P_{\perp}$, is assumed to be equal to the bulk pressure p^f. This assumption was tested using Gibbs ensemble Monte Carlo simulations, and found to be true for two values of p^f (0.2 and 6.0 reduced units) [40]. The method was taken from [49].

18.6 Pressure Variations Perpendicular and Parallel to the Pore Wall

The computed pressure components perpendicular and parallel to the pore wall are shown in Figs. 18.4 and 18.5. The first figure shows the pressure variation from the center of the slit to the wall, in the direction perpendicular to the wall (Fig. 18.4). The pressure components parallel to the wall are given in Fig. 18.5.

The variation in the pressure component perpendicular to the walls in Fig. 18.4 is shown for the liquid (top) and the vapor phases (bottom) for pore widths $2a = 17.4\sigma_0$ (left) and $2a = 27.4\sigma_0$ (right). In all cases, $\alpha_{\mathrm{sf}} = 0.15$ (repelling interactions).

The horizontal straight line in all sub-figures is the value of the xx-component extrapolated to the wall. We know that there is bulk fluid in the core, as the pressures in V_1 and V_2 are the same, so the component has the value of the bulk pressure. The other curves in the figure, the parallel components of the pressures (the yy- and zz-components), fall on top of each other, as expected. The distances $2a$, b and c between the four dashed lines and the pore wall (gray) in the figure represent three fluid volumes, of increasing size (V_1, V_2

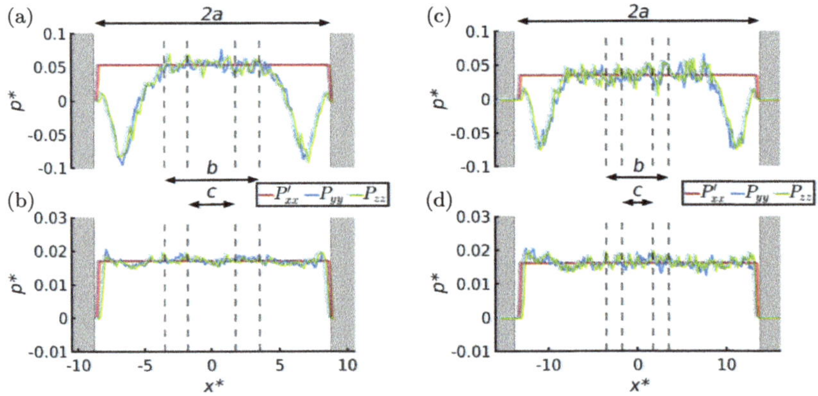

Figure 18.4. Pressure variation of the three pressure components of the liquid (a and c) and of the vapor phase (b and d). The distance between the walls was $17.4\sigma_0$ (a or b) and $27.4\sigma_0$ (c and d). In all cases, $\alpha_{sf} = 0.15$. The perpendicular component of the pressure tensor, P'_{xx}, is the value obtained by extrapolation from the center of the slit pore. Reproduced from [40] with permission from MDPI Publishers.

and V_{tot}). These are the core volume (width c, volume V_1), the core plus layer 1 (width b, volume V_2) and the whole volume (width $2a$, volume V_{tot}). The distance $2a$ is the width of the channel. In the liquid phase (top), the parallel pressure component became negative close to the surface. The variation was used to compute the surface tension between solid and liquid from Eq. (18.20), knowing P'_{xx}. The surface tension was found, within the accuracy of the calculation, to be independent of the distance between the two surface walls, see Table 18.1.

For the vapor phase (bottom), there was almost no negative dip in the parallel pressure component. In the vapor, not only the extrapolated perpendicular pressure component, P'_{xx}, but also in reasonable approximation the two parallel components, P_{yy} and P_{zz}, are constant and independent of the distance between the two surfaces. The resulting surface tension between the vapor and the wall is almost two orders of magnitude smaller than between the liquid and the wall. Given the numerical error, it could be neglected.

The pressure in the pore core depends on the distance between the walls, i.e., it is larger in the pore with a small wall distance (left)

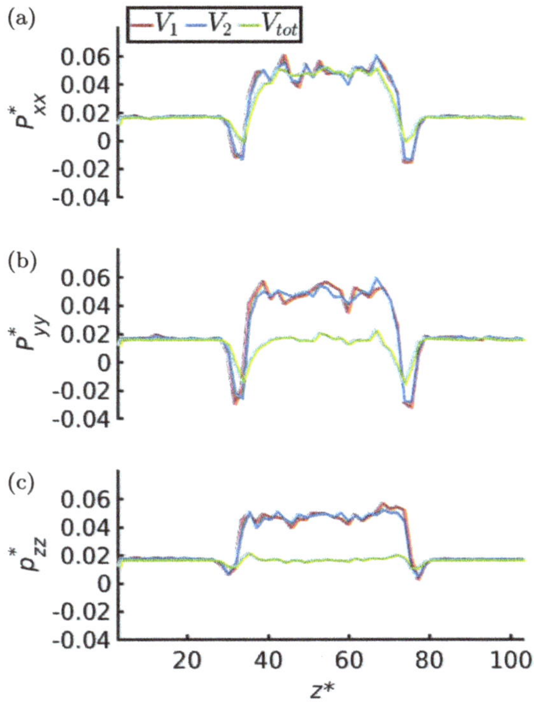

Figure 18.5. Pressure components (a) P_{xx}, (b) P_{yy} and (c) P_{zz}, along the z-direction, computed for three regions defined using the growing-core technique. The liquid droplet is positioned in the center of the box, from $z = 30\sigma_0$ to $z = 75\sigma_0$. On each side of the liquid droplet is a vapor phase in equilibrium with the droplet. Reproduced from [40] with permission from Nanomaterials.

than in the one with a larger wall distance (right). The bulk fluid pressures are related to the curvature of the liquid-vapor meniscus, the inverse of the slit pore width $2a$, by Young–Laplace's law.

Pressure tensor components were computed along the pore channel for three volumes of increasing size (V_1, V_2 and V_{tot}) using the growing core technique. Here V_1 and V_2 are subvolumes in the center of the REV and V_{tot} is the volume covering the whole REV, i.e., $V_{\text{tot}} = V^l = V^v$, see Figs. 18.2 and 18.3. The pressure components are plotted as a function of the z-direction in Fig. 18.5. The z-direction is parallel to the pore wall and passes through the vapor and the liquid droplet.

We see first that the vapor pressures are all the same, near 0.2 in reduced units. All pressure components are also constant along the liquid droplet. The yy- and zz-components are the same along the liquid droplet, as they should be. For V_{tot}, they are as expected also equal to the pressure in the vapor. The xx-component in the liquid droplet is not equal to the vapor pressure.

The volumes V_1 and V_2 contain bulk fluid, since the pressure components in these volumes of the box are the same. This was also documented by an independent calculation [40]. This gives the bulk pressures of the fluids.

The volume V_{tot} includes the fluid and its interface with the pore wall. The pressure component perpendicular to the fluid wall is still equal to the fluid pressure, $P_{xx} = p^f$. The average pressure components parallel to the fluid wall, in the bulk liquid and bulk vapor, for V_{tot}, are equal to the integral pressure, $\widehat{p} = \langle P_{\parallel} \rangle$. Finally, we see that the integral pressure is the same in the liquid and vapor phases according to Eq. (18.22). This confirms the formulation of the equilibrium condition in terms of the integral pressure, $\widehat{p} =$ constant.

18.7 Pressure Differences Across the Liquid–Vapor Surface

Figure 18.6(a) shows the difference between the perpendicular components of the bulk pressure in the liquid and vapor. The pressure difference is shown as a function of $1/a$, for three values of interaction parameter α_{sf}. Figure 18.6(b) shows the difference in the parallel components of the pressures between the liquid and vapor, i.e., the integral pressure $\widehat{p}$ (dashed lines) and the bulk pressure p^f (solid lines). Results are shown as a function of the inverse half-channel width between the two pore walls, $1/a$, and three fluid–solid interactions parameters α. These plots allow us to test Eq. (18.14).

We first observe that the difference in integral pressure for the two fluids is zero, meaning that the integral pressure is constant along the pore, see Fig. 18.6(b). The results support Eq. (18.14). In the bulk phases, the parallel and perpendicular components are the same. This also confirms that our description is consistent.

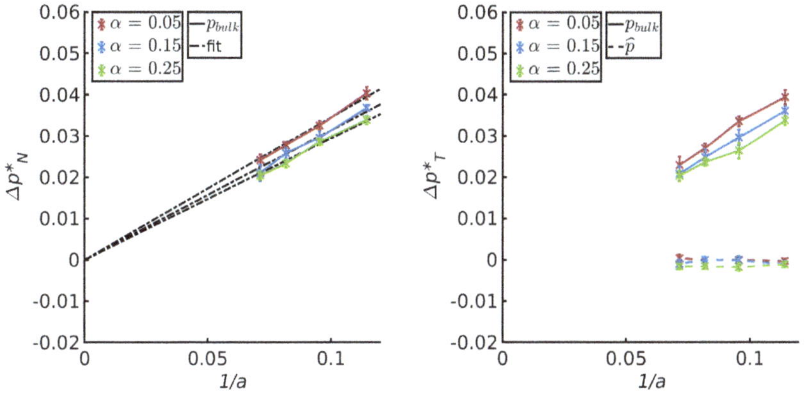

(a) (b)

Figure 18.6. Difference between (a) the perpendicular components of the pressure of the bulk liquid and vapor and (b) the difference in the parallel component of the pressure of the bulk liquid and vapor, for three different solid–fluid interactions (α) plotted as a function of the inverse half-pore width. The bulk pressure, p^f and the integral pressure are represented by solid and dashed lines, respectively. Reproduced from [40] with permission from MDPI Publisher.

The figure shows furthermore that the difference in bulk pressure components depends on the distance between the two pore walls and the interaction parameter α. This is probably a consequence of a decrease of the liquid–vapor curvature. Uncertainties in the results did not allow us to determine θ. The liquid–solid and vapor–solid surface tensions, computed with Eq. (18.20), were given in Table 18.1 for the three $\alpha_{\rm sf}$-values. The surface tension changes only slightly with $\alpha_{\rm sf}$.

The slope of the three data sets was estimated. Lines to guide the eye were fitted through zero in the left-hand side figure. The values of the slopes were 0.35/0.31/0.29, for $\alpha_{\rm sf} = 0.05/0.15/0.25$, respectively. The slope is equal to $\gamma^{lv}\cos\theta$ which is approximately equal to γ^{ls} since $\gamma^{vs} \approx 0$ (see Eq. (18.16) and Table 18.1).

The slopes depended on the interaction parameter $\alpha_{\rm sf}$. The value increased for decreasing values of $\alpha_{\rm sf}$. This is expected: the more repelling the surface, the higher the surface tension. The slopes are in agreement with the liquid–solid surface tensions given in Table 18.1.

18.8 A Scaling Law for Pressures

The difference between the differential pressure, p, and the integral pressure, $\widehat{p}$, was given by Eq. (18.18), or $\varepsilon/V = \gamma^{fs}/(3a)$. The relation is an example of a small system scaling law, cf. Eq. (18.7).

To investigate this law further, the differential pressure, p, of the liquid REV minus the integral pressure $\widehat{p}$ of the same was plotted in Fig. 18.7 as a function of the inverse half-channel width for $\alpha_{\rm sf} = 0.15$. The difference between p and $\widehat{p}$ was proportional to $1/a$, and the slope was as predicted by the liquid/solid surface tension from Table 18.1 divided by the half channel width, a, and a factor 3. The line goes through zero, when $1/a$ becomes zero. This is the thermodynamic limit value. All is as expected thus.

Erdős *et al.* [49] reported slopes for cylindrical pores in addition to slit pores. The ratio of slopes from slit and cylindrical pores was 0.52–0.58. From Eqs. (18.18) and (18.19), the ratio is 0.50, when the surface tension is constant. The surface tension is, however, a function of the interaction parameter as we have seen. The surface tension may also depend on curvature. The results from the two pores

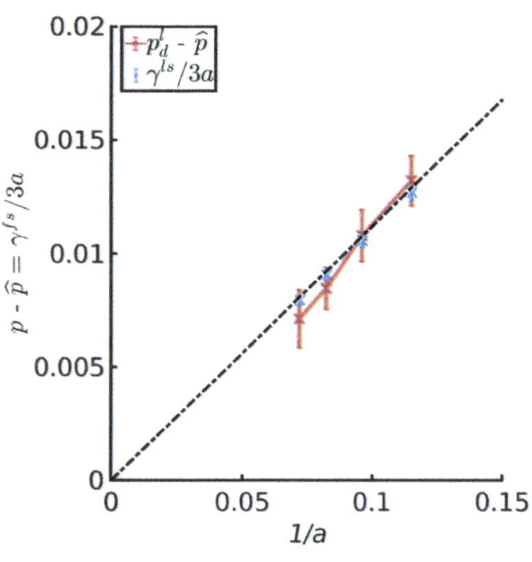

Figure 18.7. Difference between p and $\widehat{p}$ for the liquid REV_2, as a function of the inverse half-channel width for $\alpha = 0.15$. Reproduced from [40] with permission from MDPI Publishers.

with different geometries, but same α_{sf} (surface tension), fall on top of each other.

The scaling law that relates the integral and differential pressures of an REV offers a way to compute the surface tension. The well-defined dependence on the inverse characteristic length may help distinguish between other types of dependencies (i.e., curvature).

18.9 Concluding Remarks

In this chapter, we have shown, by theory and simulations, that the integral pressure of a nanoconfined single-component two-phase fluid is constant along a pore in equilibrium.

$$\widehat{p} = \text{constant} \tag{18.26}$$

This theoretical result was obtained by demanding that Young's and Young–Laplace's equations apply.

We have shown that the integral pressure in mechanical equilibrium can be understood as the average parallel component of the pressure tensor. It is this component that is constant along a pore at equilibrium. The differential pressure at mechanical equilibrium can be understood as the average of the trace of the mechanical pressure tensor (trace divided by 3).

From the subdivision potential, we have developed a small system scaling law for nanopores. Equation (18.7) applies to a slit pore as well as a cylindrical pore, but the coefficients are specific to the system. For a slit pore of half width a, the scaling law reads

$$p^f - \widehat{p} = \frac{\varepsilon^f}{V} = \zeta^f = a^3 \left(\frac{\partial \widehat{p}}{\partial a^3}\right)_{T,\mu} = \frac{\gamma^{fs}(\alpha)}{3a} \tag{18.27}$$

One may imagine using the law for more complex geometries, where higher order terms are relevant (line tension contributions). This opens up a new way to study confined systems.

Chapter 19

Entropy Production in a Porous Medium

We derive the entropy production for transport of immiscible fluids in a non-deformable porous medium exposed to differences in pressure, temperature and chemical potentials. The representative elementary volume and its variables are defined by a coarse-graining procedure, which involves a basis set of additive variables, and intensive variables controlled by the environment. The constitutive equations follow from the entropy production. Fluctuation–dissipation theorems are formulated. We show how they can be used to find transport coefficients.

A macroscale description of porous media transport must encounter several simultaneous processes. The macroscale in geological sciences is often called the Darcy scale. On this scale, there may be a pressure variation in the gravitational field. A gradient in composition has also been observed. In the fractured carbonaceous Ekofisk oil field, the mole fraction of methane varied by 0.05 units per 100 m depth. This field was also exposed to a geothermal gradient [172]. In general, for most porous media, the interplay of all driving forces is a question of interest. How can we derive a thermodynamic theory of transport when two or more driving forces interact on immiscible fluids in a porous medium?

At the core of the development of a porous media transport theory is the representative elementary volume (REV) and its variables. The construction of these coarse-grained variables was discussed in Chapter 17. The variables were defined by a coarse-graining procedure that involves adding bulk, surface and line properties plus intensive variables controlled by the environment. In this chapter we will not consider the so-called configurational contributions to the variables.

In a transport problem, we want to integrate over variables on the Darcy scale. This means to integrate over a series of REVs, each in local equilibrium. For the single REV, we have been able to formulate a Gibbs equation which reflects the underlying surface and line energies characteristic of the medium. This was the topic of Chapter 17. In this chapter, we proceed to find the entropy production of a REV in local equilibrium. The present approach is special in the way that balance equations are used on the REV level.

In the first development of nonequilibrium thermodynamics for multi-phase flow, the balance equations were formulated, not for the REV, but for each phase in the REV. Therefore, no expression was developed for a Gibbs equation on the REV level. The validity of this equation on the REV level was questioned [140, 141], and its validity is still by no means obvious.

In order to reduce the number of variables of the REV, we introduced a set of coarse-grained variables of the REV in Chapter 17. We next argued, using the additive nature of the chosen variables, and the possibility of environmental control of intensive variables, that the Gibbs equation may apply for these variables of the REV, cf. Eq. (17.1). This sets the stage for the next step, to introduce the balance equations for the REV into the Gibbs equation. Doing that, we can obtain the entropy production the normal way [17, 23], and the independent constitutive equations follow. The fluctuation-dissipation theorems are a consequence of this. They provide a new way to compute Onsager conductances and permeabilities of porous media, without resorting to numerous variables.

This chapter presents such equations for a multi-component non-isothermal transport problem. Subsequently, we give the example of two immiscible isothermal fluids. The overall aim is to describe transport due to thermal, chemical, mechanical and gravitational forces in a systematic manner, using coarse-grained variables that produce transport equations sufficiently simple for practical use.

19.1 The Entropy Production

Several phenomena take place when immiscible liquids are pushed into a porous medium. Some are fast, some are slow. In order for various phenomena to interact, they need to relax on (approximately) the same time scale. This interaction is synonymous with coupling in the language of nonequilibrium thermodynamics. The rules of coupling, or the constitutive equations, can be derived from the entropy production.

In Chapter 17, we defined the basis set of REV variables. We shall now assume that such a set of coarse-grained additive variables exists, and that they fluctuate and relax on one time scale. This assumption is by no means obvious, and must be tested by independent means or through application of the results. The derivation of the entropy production follows, however, the standard procedure [25]: Conservation equations for mass and energy (and possibly momentum) are introduced into the Gibbs equation on density form, cf. Eq. (17.31).

Fluxes into and out of the REV produce changes in the variables inside a REV. The changes are not so large that they disturb the local equilibrium of each REV. There is sufficient time for equilibrium to be re-established during the transport. The conservation equations for the masses of the components and the internal energy of a REV are

$$\begin{aligned}\frac{\partial \rho_i}{\partial t} &= -\frac{\partial}{\partial x} J_i \\ \frac{\partial u}{\partial t} &= -\frac{\partial}{\partial x} J_u = -\frac{\partial}{\partial x}\left[J_q^{'} + \sum_{i=1}^{n} J_i H_i\right]\end{aligned} \tag{19.1}$$

We consider here transport in the x-direction only. The mass fluxes, J_i, and the flux of internal energy, J_u, are all REV-scale fluxes. The internal energy flux is by definition the sum of the measurable (or sensible) heat flux, $J_q^{'}$, and the partial specific enthalpies (latent heat), H_i (in J kg^{-1}) times the component fluxes, J_i, see [23, 25, 141] for further explanations. As a consequence, also $J_q^{'}$ is a REV-scale flux. Component r (the porous medium) is not moving and is the convenient frame of reference for the fluxes. The balance equations for the masses and the internal energy have their usual form, but the variables are the coarse-grained ones, as defined in Chapter 17 for the REV.

Entropy is not conserved, and the balance equation for the REV is

$$\frac{\partial s}{\partial t} = -\frac{\partial}{\partial x} J_s + \sigma \tag{19.2}$$

Here, J_s is the entropy flux, and σ is the entropy production. The entropy production is positive definite for the REV, $\sigma \geq 0$ (the second law of thermodynamics).

Gibbs' equation is written in the local form (using densities per unit of volume), Eq. (17.31). During the time interval dt, the entropy density change is

$$\frac{\partial s}{\partial t} = \frac{1}{T}\frac{\partial u}{\partial t} - \frac{1}{T}\sum_{i=1}^{n} \mu_i \frac{\partial \rho_i}{\partial t} \tag{19.3}$$

The conservation equations are introduced into this equation, see [25] for details. By comparing the result with Eq. (19.3), we identify first the entropy flux

$$J_s = \frac{J_u - \sum_{i=1}^{n} \mu_i J_i}{T} = \frac{1}{T} J_q' + \sum_{i=1}^{n} J_i S_i \tag{19.4}$$

and next the entropy production, σ. The entropy production can be expressed using the energy flux, J_u, or the measurable heat flux, J_q'. Here, $S_i = H_i - \mu_i/T$ is the partial specific entropy of component i. The choice is normally motivated by practical wishes; what is measurable or computable. We obtain

$$\sigma = J_u \frac{\partial}{\partial x}\left(\frac{1}{T}\right) - \sum_{i=1}^{n} J_i \frac{\partial}{\partial x}\left(\frac{\mu_i}{T}\right) = J_q' \frac{\partial}{\partial x}\left(\frac{1}{T}\right) - \frac{1}{T}\sum_{i=1}^{n} J_i \frac{\partial}{\partial x}\mu_{i,T} \tag{19.5}$$

The entropy production is invariant to the choice of fluxes. It is clear from Eq. (19.5) that different options are possible. When we choose J_u as flux with the conjugate force $\partial(1/T)/\partial x$, the mass fluxes J_i are driven by minus the gradient in the Planck potentials μ_i/T. When we choose $J_q' \equiv J_u - \sum_{i=1}^{n} J_i H_i$ as flux with the conjugate force $\partial(1/T)/\partial x$, the mass fluxes J_i are driven by minus the gradient in the chemical potentials at constant temperature, $d\mu_{i,T} \equiv d\mu_i + S_i dT$.

Equation (19.5) gives a thermodynamic basis for the transport laws that follow.

A comment on the chemical driving force

In the presence of compositional changes, we need explicit expressions for the differentials of the chemical potentials. The differential of the chemical potential of a component in a bulk system is given by [23]

$$d\mu_i = -S_i dT + V_i dp + \sum_{j=1}^{n} \left(\frac{\partial \mu_i}{\partial M_j} \right)_{p,T,M_i} dM_j \tag{19.6}$$

The partial specific entropy and volume are

$$S_i = -\left(\frac{\partial \mu_i}{\partial T} \right)_{p,M_j}, \quad V_i = \left(\frac{\partial \mu_i}{\partial p} \right)_{T,M_j} \tag{19.7}$$

The derivatives $(\partial \mu_i / \partial M_j)_{p,T,M_i}$ are also partial specific quantities. The change in the chemical potentials at constant temperature and pressure in Eq. (19.6) is denoted by

$$d\mu_i^c = \sum_{j=1}^{n} \left(\frac{\partial \mu_i}{\partial M_j} \right)_{p,T,M_i} dM_j \tag{19.8}$$

The change in the chemical potential at constant temperature, which enters the entropy production, is equal to the differential of the full chemical potential plus the entropic contribution

$$d\mu_{i,T} \equiv d\mu_i + S_i dT \tag{19.9}$$

The differential of the chemical potential at constant temperature is therefore

$$\frac{d\mu_{i,T}}{dx} = \frac{d\mu_i^c}{dx} + V_i \frac{dp}{dx} \tag{19.10}$$

In a porous medium, it may be relevant to include the action of the gravitational field, and of a thermal gradient. When geological time frames are considered, there is often equilibrium in the gravitational field. The pressure gradient then becomes $dp/dx = -\rho g$, where ρ is the total mass density and g is the acceleration of free fall [173].

The separation of components in the gravitational field is then obtained by setting $d\mu_{i,T} = 0$, as follows:

$$\frac{d\mu_i^c}{dx} = \frac{RT}{W_i}\frac{d\ln(\widetilde{S}_i y_i)}{dx} = V_i \rho g \tag{19.11}$$

Here, W_i is the molar mass (in $\text{kg}\,\text{mol}^{-1}$), $\widetilde{S}_i$, the saturation and y_i, the activity coefficient of component i. The gas constant, R, has dimension $\text{J}\,\text{K}^{-1}\,\text{mol}^{-1}$. We understand that components may separate in the geothermal field. Much work has been done to clarify this [172].

It follows from Euler homogeneity that the coarse-grained chemical potentials of a large REV are related by $0 = SdT - Vdp + \sum_{j=1}^{n} \rho_j d\mu_j$ (Gibbs–Duhem's equation). By introducing Eq. (19.9) into this equation, we obtain the equivalent expression

$$0 = \sum_{j=1}^{n} \rho_j d\mu_j^c \tag{19.12}$$

19.2 Constitutive Equations

The entropy production in Eq. (19.5) in terms of J_u dictates the following flux–force relations:

$$\begin{aligned} J_u &= \ell_{uu}\frac{\partial}{\partial x}\left(\frac{1}{T}\right) - \sum_{i=1}^{n} \ell_{ui}\frac{\partial}{\partial x}\left(\frac{\mu_i}{T}\right) \\ J_j &= \ell_{ju}\frac{\partial}{\partial x}\left(\frac{1}{T}\right) - \sum_{i=1}^{n} \ell_{ji}\frac{\partial}{\partial x}\left(\frac{\mu_i}{T}\right) \end{aligned} \tag{19.13}$$

Following Onsager, the conductivity matrix is symmetric

$$\ell_{ju} = \ell_{uj}, \quad \ell_{ji} = \ell_{ij} \tag{19.14}$$

When we use J'_q as a variable, we obtain

$$\begin{aligned} J'_q &= \widetilde{L}_{qq}\frac{\partial}{\partial x}\left(\frac{1}{T}\right) - \frac{1}{T}\sum_{i=1}^{n} \widetilde{L}_{qi}\frac{\partial}{\partial x}\mu_{i,T} \\ J_j &= \widetilde{L}_{jq}\frac{\partial}{\partial x}\left(\frac{1}{T}\right) - \frac{1}{T}\sum_{i=1}^{n} \widetilde{L}_{ji}\frac{\partial}{\partial x}\mu_{i,T} \end{aligned} \tag{19.15}$$

Following Onsager, this conductivity matrix is also symmetric

$$\widetilde{L}_{jq} = \widetilde{L}_{qj}, \quad \widetilde{L}_{ji} = \widetilde{L}_{ij} \tag{19.16}$$

By using also the relation $J'_q \equiv J_u - \sum_{i=1}^{n} J_i H_i$, we can express the ℓ-matrix in the $\widetilde{L}$-matrix. The laws of Fourier, Fick and Darcy are contained in these equations.

19.3 Fluctuation–Dissipation Theorems

We have so far discussed how fluid flow through a porous medium is governed by the entropy production. All fluxes, densities, temperatures, pressures and chemical potentials have been defined for the REV (and not on the pore scale). The description can next be extended to incorporate fluctuations. We do this by adding fluctuating contributions to the fluxes. Fluctuating contributions are indicated by subscript R. These fluctuating contributions are crucial for the formulation of fluctuation–dissipation theorems. They represent the underlying, quickly changing molecular contributions to the fluxes, superimposed on a slowly changing collective movement. Their temporal and spacial correlations are on a molecular scale. By coarse-graining, we obtain δ-functions in space and time on the macroscopic space- and time-scales. The strength of the fluctuations is given by $2k_B$ times the Onsager conductivity matrix, as we shall see.

We write each total flux (with second subscript *tot*) as a combination of a collective contribution (no second subscript) and a random contribution (second subscript R). The set of the internal energy flux and the mass fluxes are

$$\begin{aligned} J_{u,tot} &= J_u + J_{u,R} \\ J_{j,tot} &= J_j + J_{j,R} \end{aligned} \tag{19.17}$$

The set of the measurable heat flux and the mass fluxes are

$$\begin{aligned} J'_{q,tot} &= J'_q + J'_{q,R} \\ J_{j,tot} &= J_j + J_{j,R} \end{aligned} \tag{19.18}$$

All fluxes in these equations are *local-valued* fluxes, and not fluxes integrated over the cross-section of the porous medium normal to the direction of flow. They apply or refer to a REV. The averages of the *random* contributions to the fluxes are always equal to zero, as follows:

$$\langle J_{u,R}\rangle = \left\langle J'_{q,R}\right\rangle = \langle J_{j,R}\rangle = 0 \tag{19.19}$$

This implies that

$$\begin{aligned}\langle J_{u,tot}\rangle &= \langle J_u\rangle\\ \left\langle J'_{q,tot}\right\rangle &= \left\langle J'_q\right\rangle\\ \langle J_{j,tot}\rangle &= \langle J_j\rangle\end{aligned} \tag{19.20}$$

The average of the total flux is the average of the collective effort to move in one direction. The equations apply for steady as well as unsteady states.

On the REV time-scale, the random contributions will appear as Gaussian white noise. Their spacial correlations are short-range, when compared to macroscale phenomena. Their second moments follow from the statistical mechanical description and satisfy the fluctuation–dissipation theorem [174–176]. The conditions for Onsager symmetry apply, giving

$$\begin{aligned}\left\langle J_{u,R}(\mathbf{r},t)J_{u,R}(\mathbf{r}',t')\right\rangle &= 2k_{\mathrm{B}}\ell_{uu}\delta(\mathbf{r}-\mathbf{r}')\delta(t-t')\\ \left\langle J_{u,R}(\mathbf{r},t)J_{j,R}(\mathbf{r}',t')\right\rangle &= \left\langle J_{j,R}(\mathbf{r},t)J_{u,R}(\mathbf{r}',t')\right\rangle\\ &= 2k_{\mathrm{B}}\ell_{uj}\delta(\mathbf{r}-\mathbf{r}')\delta(t-t')\\ \left\langle J_{i,R}(\mathbf{r},t)J_{j,R}(\mathbf{r}',t')\right\rangle &= \left\langle J_{j,R}(\mathbf{r},t)J_{i,R}(\mathbf{r}',t')\right\rangle\\ &= 2k_{\mathrm{B}}l_{ij}\delta(\mathbf{r}-\mathbf{r}')\delta(t-t')\end{aligned} \tag{19.21}$$

or

$$\begin{aligned}\langle J'_{q,R}(\mathbf{r},t)J'_{q,R}(\mathbf{r}',t')\rangle &= 2k_{\mathrm{B}}\widetilde{L}_{qq}\delta(\mathbf{r}-\mathbf{r}')\delta(t-t')\\ \langle J'_{q,R}(\mathbf{r},t)J_{j,R}(\mathbf{r}',t')\rangle &= \langle J_{j,R}(\mathbf{r},t)J'_{q,R}(\mathbf{r}',t')\rangle\\ &= 2k_{\mathrm{B}}\widetilde{L}_{qj}\delta(\mathbf{r}-\mathbf{r}')\delta(t-t')\\ \langle J_{i,R}(\mathbf{r},t)J_{j,R}(\mathbf{r}',t')\rangle &= \langle J_{j,R}(\mathbf{r},t)J_{i,R}(\mathbf{r}',t')\rangle\\ &= 2k_{\mathrm{B}}\widetilde{L}_{ij}\delta(\mathbf{r}-\mathbf{r}')\delta(t-t') \qquad (19.22)\end{aligned}$$

The fluctuations, and therefore the correlation of a flux pair, are characteristic of the pair. The correlation of fluctuation in the mass flux $J_{r,R}$ with the internal energy flux differs from the correlation with the measurable heat flux and the mass flux. Self-correlations give information on diagonal coefficients in the Onsager matrix of coefficients, while cross-correlations give coupling-coefficients. The fluctuating flux–flux correlation functions decay fast on the macroscopic time scale.

The fluctuation–dissipation theorem is general, and does not refer to the nature of the flux-force relationships. The theorems apply not only in equilibrium but also away from equilibrium [177]. At equilibrium, they are the well-known Green–Kubo relations. One set of fluctuation–dissipation relations can be derived from the other, and vice versa. They do not depend on steady state conditions.

19.4 Non-isothermal Transport of Two Immiscible Fluids

In order to illustrate more aspects of the equations, consider again non-isothermal transport of two immiscible fluid phases in a large REV. According to Eq. (19.5), the entropy production becomes

$$\sigma = J'_q\frac{\partial}{\partial x}\left(\frac{1}{T}\right) - J_w\frac{1}{T}\frac{\partial}{\partial x}\mu_{w,T} - J_n\frac{1}{T}\frac{\partial}{\partial x}\mu_{n,T} \qquad (19.23)$$

The flow of two immiscible phases is characterized by viscous dissipation or diffusion. Droplets can form at high flow rates, while ganglia may occur at low rates. We have seen in Chapter 17 that there is a minimum size of the REV for which the Gibbs equation can be written. When we assume that the Gibbs equation applies, we implicitly assume that there exists a uniquely defined state of the REV, also during transport. The existence of such a state was first postulated by Hansen and Ramstad [178]. Experimental evidence for the assumption was found by Erpelding [179], while Savani *et al.* [148] presented numerical evidence.

Application to porous media

We convert the equations above to a terminology more common in porous media science, see Hansen *et al.* [180].

The volume flows through the REV are denoted $Q_n \equiv AJ_nV_n$ and $Q_w \equiv AJ_wV_w$, where the cross-sectional area of the apparatus is $A \equiv V/\ell$, and ℓ is the length of the REV in the flow direction. Hansen *et al.* [180] assumed that the total volume flow was an Euler homogeneous function of degree one in the fractional areas, $A^n \equiv V_n/\ell = (V_n/V_p)V_p/\ell = \widetilde{S}_n A^p$ and $A^w \equiv V_w/\ell = (V_w/V_p)V_p/\ell = \widetilde{S}_w A^p$. Here $\widetilde{S}_n = V_n/V_p = A_n/A_p$ and $\widetilde{S}_w = V_w/V_p = A_w/A_p$ represent the saturation of the non-wetting and wetting fluid, respectively. The porosity becomes $\phi = V_p/V = A_p/A$. We express the total measurable heat flow $Q_q \equiv AJ'_q$. The REV is homogeneous in the y, z-directions, so the flows are integrals over the cross-sectional area of the REV normal to the flow direction. We introduce the total flows in the expression for the total entropy production, Eq. (19.23), and obtain

$$\Sigma \equiv A\sigma = Q_q \frac{\partial}{\partial x}\left(\frac{1}{T}\right) - Q_w \frac{1}{TV_w}\frac{\partial}{\partial x}\mu_{w,T} - Q_n \frac{1}{TV_n}\frac{\partial}{\partial x}\mu_{n,T} \tag{19.24}$$

The total fluxes as well as the corresponding thermodynamic forces depend only on x and t, and not on y and z.

Constitutive equations

The entropy production given by Eqs. (19.23) and (19.24) offer two equivalent choices for constitutive equations. For the sake of completeness, we give both of them.

Equation (19.23) dictates the following flux–force relations:

$$J_q' = \widetilde{L}_{qq}\frac{\partial}{\partial x}\left(\frac{1}{T}\right) - \frac{1}{T}\widetilde{L}_{qw}\frac{\partial}{\partial x}\mu_{w,T} - \frac{1}{T}\widetilde{L}_{qn}\frac{\partial}{\partial x}\mu_{n,T}$$

$$J_w = \widetilde{L}_{wq}\frac{\partial}{\partial x}\left(\frac{1}{T}\right) - \frac{1}{T}\widetilde{L}_{ww}\frac{\partial}{\partial x}\mu_{w,T} - \frac{1}{T}\widetilde{L}_{wn}\frac{\partial}{\partial x}\mu_{n,T} \qquad (19.25)$$

$$J_n = \widetilde{L}_{nq}\frac{\partial}{\partial x}\left(\frac{1}{T}\right) - \frac{1}{T}\widetilde{L}_{nw}\frac{\partial}{\partial x}\mu_{w,T} - \frac{1}{T}\widetilde{L}_{nn}\frac{\partial}{\partial x}\mu_{n,T}$$

These are the equations given in Eq. (19.15) for the special case of two fluids. According to Onsager [84, 85], the conductivity matrix is symmetric.

Equation (19.24) dictates relations for the total flows;

$$Q_q = L^P_{qq}\frac{\partial}{\partial x}\left(\frac{1}{T}\right) - \frac{L^P_{qw}}{TV_w}\frac{\partial}{\partial x}\mu_{w,T} - \frac{L^P_{qn}}{TV_n}\frac{\partial}{\partial x}\mu_{n,T}$$

$$Q_w = L^P_{wq}\frac{\partial}{\partial x}\left(\frac{1}{T}\right) - \frac{L^P_{ww}}{TV_w}\frac{\partial}{\partial x}\mu_{w,T} - \frac{L^P_{wn}}{TV_n}\frac{\partial}{\partial x}\mu_{n,T} \qquad (19.26)$$

$$Q_n = L^P_{nq}\frac{\partial}{\partial x}\left(\frac{1}{T}\right) - \frac{L^P_{nw}}{TV_w}\frac{\partial}{\partial x}\mu_{w,T} - \frac{L^P_{nn}}{TV_n}\frac{\partial}{\partial x}\mu_{n,T}$$

Also, this conductivity matrix is symmetric. From the relations between the fluxes, we find relations between the conductivity matrices of Eqs. (19.25)–(19.26):

$$L^P_{qq} = A\widetilde{L}_{qq}, \quad L^P_{qw} = AV_w\widetilde{L}_{qw}, \quad L^P_{qn} = AV_n\widetilde{L}_{qn}$$
$$L^P_{ww} = AV_w^2\widetilde{L}_{ww}, \quad L^P_{wn} = AV_wV_n\widetilde{L}_{wn}, \quad L^P_{nn} = AV_n^2\widetilde{L}_{nn} \qquad (19.27)$$

Fluctuation–Dissipation Theorems

As we discussed above, there are fluctuations in the macroscale fluxes. The total fluxes $J'_{q,\mathrm{tot}}$, $J_{w,\mathrm{tot}}$, $J_{n,\mathrm{tot}}$ can be formulated in terms of their mean value plus a fluctuating, random contribution as follows:

$$\begin{aligned} J'_{q,\text{tot}} &= J'_q + J'_{q,R} \\ J_{w,\text{tot}} &= J_w + J_{w,R} \\ J_{n,\text{tot}} &= J_n + J_{n,R} \end{aligned} \tag{19.28}$$

Again, the expressions apply for steady or non-steady states, independent of the nature of the flux–force relationships. The theorems apply not only in equilibrium but also away from equilibrium [23]. On the REV time scale, the random contributions are typically Gaussian white noise. Their spacial correlations are short-range compared to the macroscale phenomena. The averages of the fluctuating contributions are zero, and their second moments satisfy the fluctuation–dissipation theorem.

$$\begin{aligned} \left\langle J'_{q,R}\left(\mathbf{r},t\right) J'_{q,R}\left(\mathbf{r}',t'\right)\right\rangle &= 2k_{\text{B}}\widetilde{L}_{qq}\delta(\mathbf{r}-\mathbf{r}')\delta\left(t-t'\right) \\ \left\langle J'_{q,R}(\mathbf{r},t) J_{j,R}\left(\mathbf{r}',t'\right)\right\rangle &= \left\langle J_{j,R}\left(\mathbf{r},t\right) J'_{q,R}\left(\mathbf{r}',t'\right)\right\rangle \\ &= 2k_{\text{B}}\widetilde{L}_{qj}\delta\left(\mathbf{r}-\mathbf{r}'\right)\delta\left(t-t'\right) \\ \left\langle J_{i,R}\left(\mathbf{r},t\right) J_{j,R}\left(\mathbf{r}',t'\right)\right\rangle &= \left\langle J_{j,R}\left(\mathbf{r},t\right) J_{i,R}\left(\mathbf{r}',t'\right)\right\rangle \\ &= 2k_{\text{B}}\widetilde{L}_{ij}\delta\left(\mathbf{r}-\mathbf{r}'\right)\delta\left(t-t'\right) \end{aligned} \tag{19.29}$$

where i, j are either w or n.

All fluxes in the expressions above are *local*, in the sense that they are given per unit of area. They are not fluxes integrated over the cross-sectional area perpendicular to the direction of transport. Therefore, it should be possible to determine the random fluctuations via local optical techniques in a Hele-Shaw cell.

For Q_q, Q_w, Q_n, we have

$$\begin{aligned} Q_{q,\text{tot}} &= Q_q + Q_{q,R} \\ Q_{w,\text{tot}} &= Q_w + Q_{w,R} \\ Q_{n,\text{tot}} &= Q_n + Q_{n,R} \end{aligned} \tag{19.30}$$

The averages of the random contributions are zero. Their second moments satisfy the fluctuation–dissipation theorem

$$\begin{aligned}\langle Q_{i,R}(x,t)Q_{j,R}(x',t')\rangle &= \langle Q_{j,R}(x,t)Q_{i,R}(x',t')\rangle \\ &= 2k_{\mathrm{B}}L_{ij}^{P}\delta(x-x')\delta(t-t') \end{aligned} \tag{19.31}$$

where i, j are either q, w or n. This set of fluctuation–dissipation relations can be derived from Eq. (19.29) by integrating over the cross-section.

Winkler *et al.* [149] determined the transport coefficients of a network, simulating two-phase flow in a honeycomb lattice. They found that the Onsager reciprocal relations were obeyed. The simulations yielded fluctuating flux–flux correlation functions that decayed fast compared to the macroscopic time scale.

19.5 Concluding Remarks

We have derived the entropy production for transport of non-isothermal multiphase flow in porous media. The derivations followed the standard non-equilibrium thermodynamic procedure [25], with one exception. The only, but essential, change is the use of a basis set of coarse-grained variables. These were additive variables defined for the REV. On the REV level, the system is a mixture of small-sized subsystems. The REV is large in Hill's sense. While the derivations rest on the hypothesis that the coarse-graining procedure is possible, one may say that "the proof of the pudding lies in its eating". Experimental results as well as simulations indicate so far that this may be a possible route to conductivity coefficients. Tests can be explicitly formulated, from the constitutive equations and from the fluctuation–dissipation theorems. One advantage of the present formulation is that it can be used to formulate the transport problem on the scale of a flow experiment in accordance with the second law of thermodynamics, with a small number of variables.

Chapter 20

Transport of a Single Fluid Through a Porous Medium

We show how to obtain the driving force for transport of a single fluid in a porous medium. We choose a representative elementary volume (REV), and define the coarse-grained pressure and the integral pressure, for a set of control variables. The integral pressure is found by exploiting its dependence on the volume and the fluid–solid contact area. A scaling law relates the integral- and the differential pressure of the solid. With the gradient in the integral pressure of the REV as a driving force, we determine the fluid permeability of the porous medium.

In search for a description of fluid transport through porous media, there are many problems to address, as we have seen in the preceding chapters. A major one is how to define the control volume of the medium, also called the representative elementary volume (REV). This is needed for definition of average properties that can enter a continuum description [139–141, 162, 181, 182]. Relevant variables of the REV for such a description were presented in Chapter 17 and applied to describe the slit pore in Chapter 18. They were used in Chapter 19 to formulate constitutive equations and fluctuation-dissipation theorems. Here, we shall see how the general theory of Chapter 19 can be used to describe transport of a single fluid in a porous medium.

As example, we take a fluid confined in a regular face-centered cubic (fcc) lattice of solid grains. This was described by Galteland *et al.* [41, 183]. In such a lattice, the REV can be as small as half the unit cell. This is sufficient, for a sampling of all available microstates in molecular dynamics simulations. For better statistics, we shall choose a whole unit cell, however. A REV is normally large compared to the typical pore size in order to produce a statistically representative volume element [24, 41, 148]. But the REV also needs to be relatively small compared to the size of the porous medium, to provide an accurate path of integration [139]. This was already discussed in Chapter 17. A two-dimensional REV was illustrated in Fig. 17.1. The REV was then indicated by a red square, a part of the larger network.

Here we construct the REV in a regular porous medium illustrated in Fig. 20.1. The figure shows fluid particles in an fcc lattice of solid grains. The solid matrix (denoted r) is shown as blue particles, while the fluid particles (denoted f) are colored in red. A rectangular slab of a porous material is shown, with sides L_x, L_y, L_z. We shall compute properties along the x-axis in layers of width Δx. A question is then; how many layers are needed to make up a REV? The layer division can be regarded as a computational tool used to answer this question. One layer may be too thin to serve as an REV.

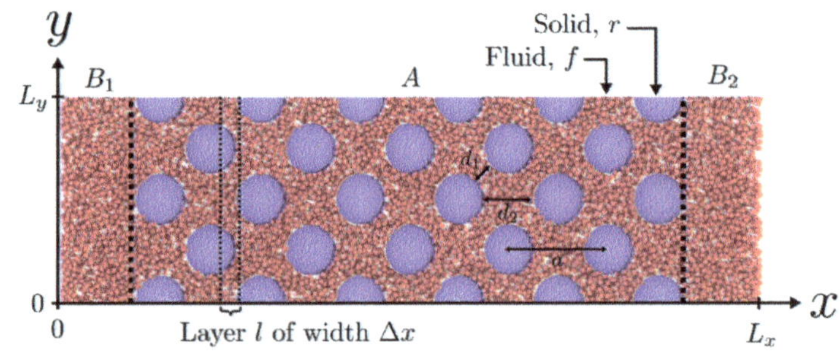

Figure 20.1. A porous medium fcc latice (blue grains) with a single fluid (red particles). The box has side lengths L_x, L_y, L_z. Properties are calculated along the x-axis in layers of width Δx. Reproduced from [41] with permission from Frontiers.

As stated above, it needs to contain a sufficient number of microstates to be representative, see [148].

Once we know the REV and its properties, we can construct the driving force between neighbouring REVs. In the present case, with a small system controlled in the grand canonical ensemble, the driving force is the gradient of the integral pressure. With knowledge of the integral pressure, we can formulate the equivalent to Darcy's law. The permeability of the single fluid in a medium of varying porosity will be examined.

20.1 The Variables of the Representative Elementary Volume

The basis set of variables of a porous medium are additive in the sense that they are sums of contributions of all bulk phases, interfaces and three-phase contact lines within the REV, see Chapter 17. The REV is controlled by an environment with constant temperature and chemical potentials, as described in Chapter 4. The REV belongs thus to the grand canonical ensemble.

The REV pressure

In the grand canonical ensemble, the grand potential Υ is

$$\Upsilon = -k_{\mathrm{B}}T \ln \Xi \tag{20.1}$$

The grand potential is also given by the integral pressure of the REV as

$$\Upsilon = -\widehat{p}V \tag{20.2}$$

In this expression, $\widehat{p}$ and V are the integral pressure and the volume of the REV. A more colloquial name of the grand potential is the compressional energy, and we shall use this. We are seeking an expression for $\widehat{p}$.

Consider the porous medium in Fig. 20.1 and the phases f and r. Between the phases is an interface. The partition function of the REV can be written as the product of the partition function of the separate phases and of the surfaces. The Υ of Eq. (20.1) is then the sum of contributions from all terms. For the present case, we obtain [24, 42]

$$\widehat{p}V = \widehat{p}^f V^f + \widehat{p}^r V^r - \widehat{\gamma}^{fr}\Omega^{fr} \tag{20.3}$$

where $\widehat{p}^f$ and V^f are the integral pressure and the subvolume of the fluid f, $\widehat{p}^r$ and V^r are the integral pressure and the subvolume of the rock r, and $\widehat{\gamma}^{fr}$ and Ω^{fr} are the integral surface tension and the surface area between the fluid and the rock, all for one REV. Presence of additional fluid phases would contribute by similar terms, as well as by contributions from three-phase contact lines. Equation (20.3) can be regarded as a definition of the integral pressure of the REV, $\widehat{p}$, in terms of its separate contributions. The hat on the symbol signifies that we are dealing with a small system variable.

The grand potential of a nanoporous medium is in general different from pV, cf. Eq. (20.4). In the large system limit, $\widehat{p}$ does not depend on V so that $\widehat{p} = p$. Equation (3.52) gives the general relation between the integral and differential pressures

$$p(V) = \left[\frac{\partial(\widehat{p}(V)V)}{\partial V}\right]_{T,\mu^f} = \widehat{p}(V) + \left[V\frac{\partial(\widehat{p}(V))}{\partial V}\right]_{T,\mu^f} \tag{20.4}$$

We shall find the two pressures of the system using knowledge of the geometric properties of the REV. In order to bring out principles, we study first a single spherical grain in a fluid, i.e., Case I. Case II is a more realistic model of a nanoporous medium, see Fig. 20.1 for an illustration. The REV of Case II will become evident.

Case I: A spherical grain in a confined fluid

Consider first the spherical grain r inside a box A surrounded by a fluid phase f. The small system, A, is illustrated in Fig. 20.2. It is in equilibrium with pure fluid outside A, in the box B. Phase f has volume V^f and phase r has volume V^r. The total volume is $V = V^f + V^r$. The surface area between phases f and r is Ω^{fr}. The system A is so large that

$$\widehat{p}^A = p^A \tag{20.5}$$

Remark 20.1. The meaning of Eq. (20.5) can be illustrated with an example. The pressure of a macroscopic system, say milk in a

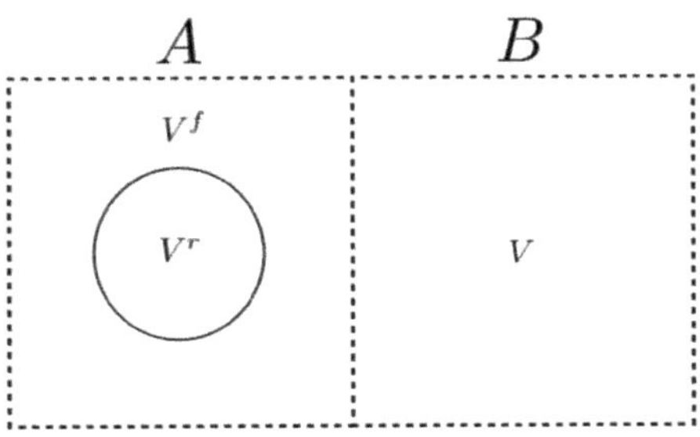

Figure 20.2. A fluid confined to the volume V_f in A in equilibrium with a bulk fluid of volume V in B. Reproduced from [41] with permission from Frontiers.

bottle, is constant on the REV level. The REV will contain many small grains (e.g., colloidal particles), but the REV is so large that the pressure is independent of its size.

The compressional energy of system A has contributions from all phases and surfaces, giving

$$p^A V = \widehat{p}^f V^f + \widehat{p}^r V^r - \widehat{\gamma}^{fr}\Omega^{fr} \tag{20.6}$$

The integral pressure of the fluid in A, $\widehat{p}^f$, is next assumed to be independent of V^f. It is therefore equal to p^f, giving $\widehat{p}^f = p^f$. We shall likewise assume that the surface tension does not depend on the contact area, Ω^{fr}, i.e., we neglect curvature dependencies. Only $\widehat{p}^r$ depends on the volume of the phase, V^r. The compressional energy becomes

$$p^A V = p^f V^f + \widehat{p}^r V^r - \gamma^{fr}\Omega^{fr} \tag{20.7}$$

System B, in contact with A, has the same volume V as A, and contains a pure fluid. Here, γ^{fr} is the differential surface tension. The volume is so large that $p^B = p^f$. The pressure is tuned so that it is in thermodynamic equilibrium with A, meaning that the fluid pressure p^f is the same in phases A and B, as follows:

$$p^B = p^f = p^A \tag{20.8}$$

We therefore obtain

$$p^f V = p^f V^f + \widehat{p}^r V^r - \gamma^{fr}\Omega^{fr} \tag{20.9}$$

By rearranging the terms, we obtain

$$\widehat{p}^r = p^f + \frac{\gamma^{fr}\Omega^{fr}}{V^r} = p^f + \frac{3\gamma^{fr}}{R} \tag{20.10}$$

where $V^r + V^f = V$ and $\Omega^{fr}/V^r = 3/R$ for the spherical grain.

Expression Eq. (20.4) applies to the r-phase. We use this to find p^r, keeping in mind that we are dealing with a grand canonical ensemble.

$$p^r = \frac{\partial(\widehat{p}^r V^r)}{\partial V^r} \tag{20.11}$$

When this is combined with Eq. (20.10), we find the familiar law of Young and Laplace

$$p^r - p^f = \frac{2\gamma^{fr}}{R} \tag{20.12}$$

In other words, validity of the law of Young–Laplace is equivalent to the condition in Eq. (20.8). It is the differential pressure that satisfies Young–Laplace's law; the integral pressure does not!

By subtracting Eq. (20.12) from Eq. (20.10), we obtain the scaling law for the r-phase of the form

$$\frac{\varepsilon^r}{V^r} \equiv \widehat{p}^r - p^r = \frac{\gamma^{fr}}{R} \tag{20.13}$$

The expression relates the difference of the integral and differential pressure of a spherical grain r with the radius R to the subdivision potential. The difference is equal to the subdivision potential divided by the volume of the present ensemble, ε^r/V^r, cf. Chapter 3. The expression relates properties of one small system to another.

Remark 20.2. The scaling law is of the same type as presented in Chapter 5, cf. Eq. (5.1). Both laws refer to the grand canonical ensemble.

We have established that the following equilibrium condition applies:

$$\widehat{p}^A = p^A = p^B = \widehat{p}^B \tag{20.14}$$

The integral pressure of system A is equal to the integral pressure of system B, cf. Eq. (20.8). The whole system A is a proper choice of the REV, when the system contains one solid grain. System B is a proper choice of the REV, when the system contains fluid only.

Case II: Lattice with confined fluids

The lattice of spherical grains with a confined fluid, as pictured in Fig. 20.1, can be seen as a very simple model of a porous medium. We neglect the possible dependence of the surface tension on the curvature [52, 53]. The pressure of an REV with n grains is then given by

$$p^A V = p^f V^f + \sum_{i=1}^{n} \widehat{p}_i^r V_i^r - \sum_{i=1}^{n} \gamma_i^{fr} \Omega_i^{fr} \tag{20.15}$$

The sum is taken over all grains in the volume. For each grain, we follow the same derivation of the integral and differential pressure as above. Analogous to Eq. (20.10), we obtain

$$\widehat{p}_i^r = p^f + \gamma_i^{fr} \frac{\Omega_i^{fr}}{V_i^r} = p^f + \frac{3\gamma_i^{fr}}{R_i} \tag{20.16}$$

where we have omitted summation signs. The last identity applies to spherical grains only. The differential pressure of the grains is given by a generalization of Eq. (20.12), as follows:

$$p_i^r = \left[\frac{\partial(\widehat{p}_i^r V_i^r)}{\partial V_i^r}\right]_{T,\mu^f} = p^f + \gamma_i^{fr} \frac{\partial \Omega_i^{fr}}{\partial V_i^r} = p^f + \frac{2\gamma_i^{fr}}{R_i} \tag{20.17}$$

where the last identity applies to spherical grains only. The differential pressures satisfy the Young–Laplace law.

By subtracting Eq. (20.17) from Eq. (20.16), we obtain the scaling law for the grains:

$$\frac{\varepsilon_i^r}{V_i^r} \equiv \widehat{p}_i^r - p_i^r = \frac{\gamma_i^{fr}}{R_i} \tag{20.18}$$

This expression relates the difference in the integral- and differential pressure of the spherical grains to the radius of the grains. This difference is equal to their subdivision potential over the volume, cf. Chapter 3. Equations (20.12) and (20.18) have the same form.

When all grains are identical spheres positioned on a lattice, we are able to sample all relevant molecular interactions within the half a *unit cell.* The half-unit cell is then the smallest REV we can choose. For better statistics in the computation, we may use a whole cell. Amorphous materials will have larger REVs.

20.2 Simulation Procedures

Molecular dynamics simulations were used to find an appropriate REV and the pressures of the REV. The single spherical grain and the fcc lattice were simulated to see this.

Molecular dynamics simulations are done using a three-dimensional box with side lengths L_x, L_y, L_z. The box corresponds to the picture in Fig. 20.1. The box can be elongated in the x-direction, so $L_x > L_y = L_z$. In our example, L_x was divided into m layers of thickness Δx, perpendicular to the transport direction, x. We obtained $\Delta x = L_x/m$. The volume of a layer was then $V_l = \Delta x L_y L_z$. The layer thickness was varied to order to find the proper REV.

The simulation box has two regions A and B, like in Cases I and II. Region A contains fluid (red) particles and grains (blue particles). Region B contains only fluid, see Fig. 20.1. Regions B=B_1+B_2 and A did not have the same size, but the layers had the same thickness, Δx. Lennard-Jones spline interaction potentials with a hard core were used here, see Appendix D.

In molecular simulations, periodic boundary conditions are used in all directions. In equilibrium simulations, all boundary conditions are the same. Away from equilibrium, reflecting particle boundary conditions [184] are used in the x-direction. Simulations in the canonical ensemble can be carried out with LAMMPS [166]. A Nosé–Hoover thermostat [185] was used in our case to keep the temperature constant and equal to $T^* = 2.0$ [41]. Superscript * indicates that a variable is given in reduced units, see Appendix D.

Case I. The single spherical grain has a radius $R = 5.5\sigma_0$, and four nearest neighbors in the periodic lattice. Here, σ_0 is the diameter of the fluid particles. The single spherical grain was placed in the center of the box. Periodic images of the spherical grain were created at distances L_y and L_z away in the y and z-directions, see Fig. 20.4(a). The surface-to-surface distance of the spherical grains was $d = L_y - 2R = L_z - 2R$. Two nearest neighbor-distances, $d = 4\sigma_0$ and $d = 11\sigma_0$, were considered.

Case II. The spherical grains were positioned in an fcc lattice with lattice constant a. In the reference system, $a = 20$. The value was later increased to 25, 30 and 40. Each grain had 12 nearest neighbors at a distance d_1. There are two shortest distances between two grain surfaces in the lattice, $d_1 = \frac{1}{2}(\sqrt{2}a - 4R)$ and $d_2 = a - 2R$, where $d_1 < d_2$, see the solid black lines in Fig. 20.1. Distances $d_1 = 4.14\sigma_0$ and $d_1 = 11.21\sigma_0$ were considered. These distances are almost the same as the distances considered in Case I. The corresponding other distances were $d_2 = 10.41\sigma_0$ and $d_2 = 20.41\sigma_0$.

The compressional energy of the fluid in one layer is $\widehat{p}_l^f V_l^f$. The integral pressure of the fluid in a layer of the present lattice did not depend on V_l^f, giving $\widehat{p}_l^f = p_l^f$. In both cases, we computed the volume of the grains V_l^r, the surface area Ω_l^{fr} and the compressional energy of layers l.

Particle interaction potential

The particles in the case studies interact with a Lennard-Jones/spline potential, $\phi_{\mathrm{LJ/s}}(r_{ij})$, which includes a hard core. Each particle type has a hard-core diameter R_{ii} and a soft-core diameter σ_{ii}. There were two types of particles, small particles with $\sigma_{ff} = \sigma_0$, $R_{ff} = 0$ and large particles with $\sigma_{rr} = 10\sigma_0$, $R_{rr} = 9\sigma_0$. The small particles are parts of the fluid (f), and the large particles are the grains (r). The hard-core and soft-core diameters for fluid-grain pairs are given by the Lorentz mixing rule

$$R_{fr} = \frac{1}{2}(R_{ff} + R_{rr}) \quad \text{and} \quad \sigma_{fr} = \frac{1}{2}(\sigma_{ff} + \sigma_{rr}) \tag{20.19}$$

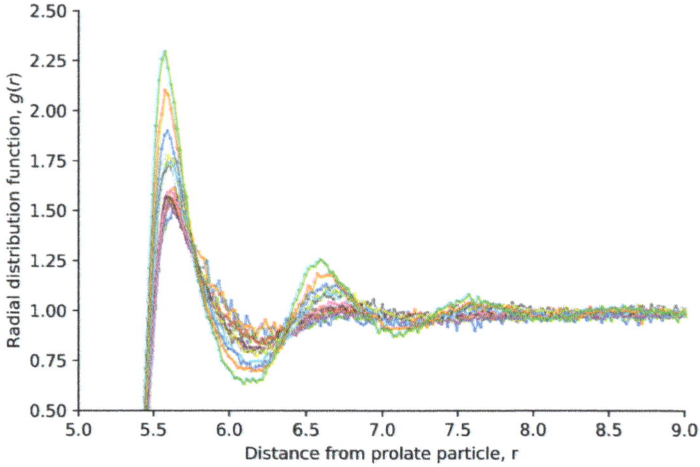

Figure 20.3. The radial distribution function of fluid particles around one grain in Case I. Results are shown for densities that vary between $\rho^* = 0.1$ and $\rho^* = 0.7$. Reproduced from [41] with permission from Frontiers.

We define the radius of the grain particles as $R \equiv (\sigma_{ff} + R_{rr})/2 = \sigma_{rr}/2 = 5\sigma_0$, which is the distance from the grain center to where the potential energy is zero. Fluid particles can occupy a position closer to the grain than this, this is illustrated in Fig. 20.3. The figure shows the radial distribution function, $g(r)$, of fluid particles around a single spherical grain. The density of fluid varied between $\rho^* = 0.1$ and $\rho^* = 0.7$. This shows that the average distance from the grain particle and the closest fluid particle is approximately $5.5\sigma_0$, but the fluid particles are able to occupy positions closer to the grain particle. The interaction strength ϵ_{ij} was set to ϵ_0 for all particle–particle pairs. The potential and its derivative (the force) are everywhere continuous.

Pressure computations

The pressure was computed as before (Chapter 18) according to Irving-Kirkwood [186]. This algorithm works for homogeneous systems. The contribution of the fluid to the compressional energy of layer l is [186]

$$p_l^f V_l^f = \frac{1}{3}\left\langle \sum_{i\in l} m_i(\mathbf{v}_i \cdot \mathbf{v}_i)\right\rangle - \frac{1}{6}\left\langle \sum_{i\in l}\sum_{j=1}^{N}(\mathbf{r}_{ij}\cdot \mathbf{f}_{ij})\right\rangle \tag{20.20}$$

where p_l^f is the contribution to the integral pressure due to the fluid particles, V_l^f, the fluid volume, m_i and $\mathbf{v}_i$ are the mass and velocity of fluid particle i. The first two sums are over all fluid particles i in layer l, while the second sum is over all the particles, j. Half of the virial contribution, the second term in Eq. (20.20), is assigned to particle i and the other half to particle j. Virial contributions were not assigned to the solid particles. The vector that connects particle i and j is $\mathbf{r}_{ij} \equiv \mathbf{r}_i - \mathbf{r}_j$, and $\mathbf{f}_{ij} = -\partial u_{ij}/\partial \mathbf{r}_{ij}$ is the force between them. The $\cdot$ means an inner product of the vectors.

The computation produced the integral pressure $\widehat{p}_l^f$ of the fluid. It was next confirmed that $\widehat{p}_l^f = p_l^f$ in layer l, see Figs. 20.4 and 20.6.

Computing the mass flux

A mass flux was generated across the system, using the *reflecting particle boundary method* developed by Li *et al.* [184]. Particles, which were moving from the right-hand side to the left-hand side, would pass the periodic boundaries at $x = 0$ and $x = L_x$ with probability $(1 - \alpha_p)$ and be reflected with probability α_p. Particles moving from left to right were passing freely through the boundary. A pressure difference across the system developed along the x-axis. The pressure difference increased with increasing α_p.

20.3 Integral and Differential Pressures

The equilibrium results of the molecular dynamics simulations are shown in Figs. 20.4–20.7. The contributions to the compressional energy in Eq. (20.6) were computed for Case I (single spherical grain) and Case II (fcc lattice of spherical grains). The layers are thin in region A and B in both cases compared to the unit cell (see Figs. 20.4 and 20.6).

The integral and the differential pressures in the bulk liquid region B are the same in all layers and equal to the fluid pressure, $\widehat{p}_l = p_l = p^f$. In region B of Fig. 20.1, $V_l^f = V_l$. The grain particles in region A are identical and the system is in equilibrium, so the integral pressure in all grains, $\widehat{p}^r$, is the same. Similarly, the surface tension γ^{fr} of all grains is the same. The integral fluid pressure between the grains $\widehat{p}_l^f$ is computed to be independent of V_l^f and to be equal to the fluid pressure in region B. This implies that $\widehat{p}_l^f = p_l^f = p^f$.

The integral pressure and the surface tension of the grains are not known. We know, however, that the properties are independent of the position along the axis. The integral grain pressure $\widehat{p}^r$ and surface tension γ^{fr} can then be obtained by fitting the function in Eq. (20.6) to known values of volume and area present in a layer, such that the integral pressure is the same in all layers. The layers used in the calculation are thin compared to the thickness of the REV. To obtain a constant REV-value, we average over five consecutive layers.

One spherical grain

The single spherical grain is illustrated in Fig. 20.4(a). Figures 20.4(b) and 20.4(c) present the volume of the grain in the layer, V_l^r, and the surface area between the grain and the fluid, Ω_l^{fr}, for the layers. Figure 20.4(b) gives the compressional energy for the layers along the x-axis of the simulation box. In order to be representative, the REV must include all of the solid sphere with sufficient boundaries left and right of the sphere. In order to find $p^{\text{REV}} V^{\text{REV}}$, we summed $\widehat{p}_l V_l$ over all five layers in the REV.

At equilibrium, $p^{\text{REV}} = p^f$. For the REV, we then have

$$p^f V^{\text{REV}} = \sum_{l\in\text{REV}} \widehat{p}_l V_l = p^f \sum_{l\in\text{REV}} V_l^f + \widehat{p}^r \sum_{l\in\text{REV}} V_l^r - \gamma^{fr} \sum_{l\in\text{REV}} \Omega_l^{fr} \tag{20.21}$$

We again used that $p_l^f = p^f$, $\widehat{p}_l^r = \widehat{p}^r$ and $\gamma_l^{fr} = \gamma^{fr}$. In this equation, we know all elements except $\widehat{p}^r$ and γ^{fr}.

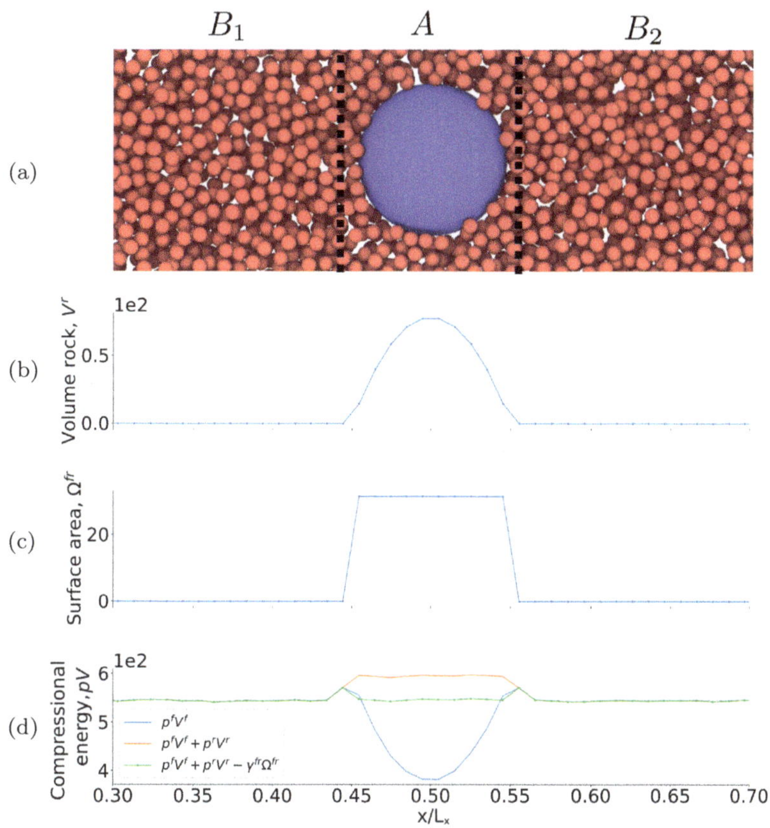

Figure 20.4. (a) One spherical grain, surrounded by a fluid phase, with characteristic length $d = 4\sigma_0$. (b) Grain volume in slit, V^r, (c) surface area Ω^{fr} and (d) compressional energy pV are shown as functions of position on the x-axis in the simulation box. Reproduced from [41] with permission from Frontiers.

The values of $\widehat{p}^r$ and γ^{fr} were obtained by fitting, such that the pressure, p^f in Eq. (20.21) is everywhere the same. Using these results, we calculated $p_l V_l$ of each layer from

$$\widehat{p}_l V_l = p^f V_l^f + \widehat{p}^r V_l^r - \gamma^{fr} \Omega_l^{fr} \tag{20.22}$$

The contributions to the compressional energy in Case I are shown in the bottom of Fig. 20.4(d). We see the contribution from the bulk fluid $p^f V_l^f$, the bulk fluid plus grain $p^f V_l^f + \widehat{p}^r V_l^r$ and the total compressional energy, $\widehat{p}_l V_l = p^f V_l^f + \widehat{p}^r V_l^r - \gamma^{fr}\Omega^{fr}$. The last expression gives the pressure of the REV when summed and divided with the volume of the REV.

Figure 20.4(d) shows that the fluid compressional energy is larger than any other contribution. The surface energy is not negligible, however. As the surface-to-volume ratio increases, the fluid contribution can become smaller than the surface contribution (not shown). In the present case, this will happen when the radius of the grains is $2.25\sigma_0$. For the grains used here, with $R = 5\sigma_0$, this does not happen.

The plots of $\widehat{p}^r$ and γ^{fr} as functions of p in region B are shown in Fig. 20.5. The values for $d = 4\sigma_0$ and $d = 11\sigma_0$ are given in the same plots. We see that the plots fall on top of each other. This shows that the integral pressure and the surface tension are independent of the distance d between the surfaces of the spheres, in the interval considered. If confinement effects were essential, we would expect that $\widehat{p}^r$ and γ^{fr} were functions of d. When d decreases below $4\sigma_0$, deviations may arise, possibly due to contributions from the disjoining pressure. Such a contribution is expected to vary with the surface area, and increase as the distance between interfaces becomes smaller. In plots like Fig. 20.5, we may see this as a reduction in the surface tension.

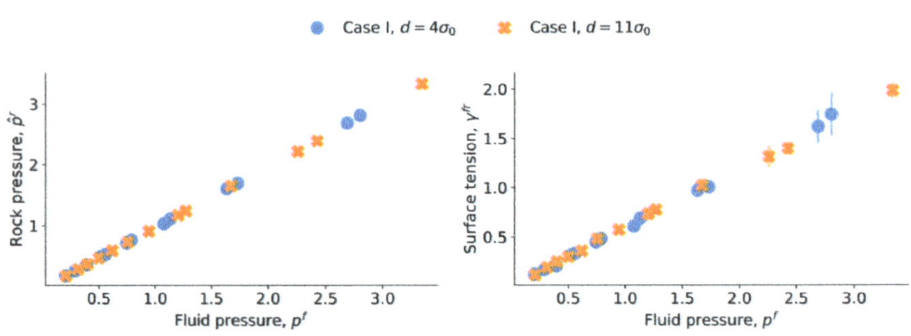

Figure 20.5. Fitted grain pressure $\widehat{p}^r$ and surface tension γ^{fr} as a function of pressure $p^f = p$ for a sphere (characteristic length $d = 4\sigma_0$ and $d = 11\sigma_0$). Reproduced from [41] with permission from Frontiers.

Lattice of spherical grains

Results for the lattice of spherical grains are illustrated in Figs. 20.6(a)–20.6(d). Figures 20.6(b) and 20.6(c) give the variation of the volume of the solid material V_l^r and of the surface area between the fluid and the solid, Ω^{fr}. Values are reported as a function of x. Figure 20.6(d) gives the compressional energy pV and its contributions as a function of the x-axis of the simulation box. When the REV of region A is properly chosen, $\widehat{p}^{\text{REV}} = p^{\text{REV}} = p$.

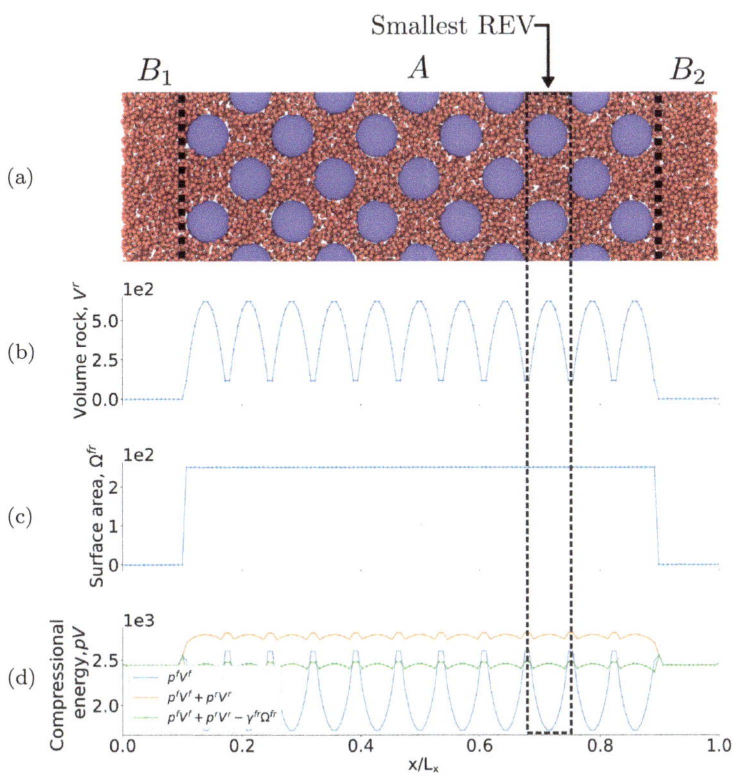

Figure 20.6. (a) The lattice of spherical grains surrounded by a fluid phase with $a = 20\sigma_0$, (b) volume of grain, V^r, (c) surface area Ω^{fr} and (d) compressional energy pV as a function of the x-axis of the simulation box. The smallest REV is a unit cell. Reproduced from [41] with permission from Frontiers.

In equilibrium, the pressure of the REV is constant in the bulk liquid phases, in regions B_1 and B_2, where p is the pressure of the fluid.

In order to obtain pV^{REV} in region A, we sum $p_l V_l$ over all the layers that make up the REV using Eq. (20.21). To proceed, we find first the values of all the elements in this equation, except $\widehat{p}^r$ and γ^{fr}. The values of $\widehat{p}^r$ and γ^{fr} are found by fitting, such that the pressure p is everywhere the same. Using these fitted values, it is possible to calculate $\widehat{p}_l V_l$ of each layer from Eq. (20.22). The three contributions to the compressional energy in this equation are shown in Fig. 20.6(d). They are as follows:

1. The bulk fluid contribution: $p_l^f V_l^f$.
2. The bulk fluid and grain contribution: $p_l^f V_l^f + \widehat{p}^r V_l^r$.
3. The total compressional energy: $\widehat{p}_l V_l = p_l^f V_l^f + \widehat{p}^r V_l^r - \gamma^{fr}\Omega^{fr}$.

Figure 20.6(d) shows that the bulk contribution is largest, but that the surface energy is significant.

It follows from Fig. 20.6(b) that a proper choice of the REV is a unit cell. With this choice all REV variables become identical (except the REVs at the boundaries). The integral over $\widehat{p}_l V_l$ in such a REV is everywhere the same and equal to pV^{REV}. At the outset the layers l were chosen to be smaller than the REV. Therefore, as a consequence $\widehat{p}_l V_l$ will vary from layer to layer, see Fig. 20.6(d).

The REV-values for $\widehat{p}^r$ and γ^{fr} are shown as functions of p^f in Fig. 20.7 for $d_1 = 4.14\sigma_0$ and $d_1 = 11.21\sigma_0$, corresponding to $a = 20$ and 30. We see that the value of d_1 has an impact on $\widehat{p}^r$ and γ^{fr}. Their values are larger, the smaller the distance is between the grains, as measured by d_1. (The lattice constant gives the interparticle distances in the fcc lattice, d_1 and d_2 by $d_1 = 0.5(\sqrt{2}a - 4R)$ and $d_2 = a - 2R$). The difference between the sets can be estimated. Consider then a difference in surface tension $\Delta\gamma^{fr}$. For the same fluid pressure in Eq. (20.13), we obtain a difference in integral pressure of $\Delta\widehat{p}^r = 3\Delta\gamma^{fr}/R$. This is nearly the value we find by comparing the lines in Fig. 20.6. One value can be predicted from the other using $R = 6\sigma_0$, while value in Fig. 20.3 is $R = 5\sigma_0$. The deviation may be due to the disjoining pressure. When the distance between the grain

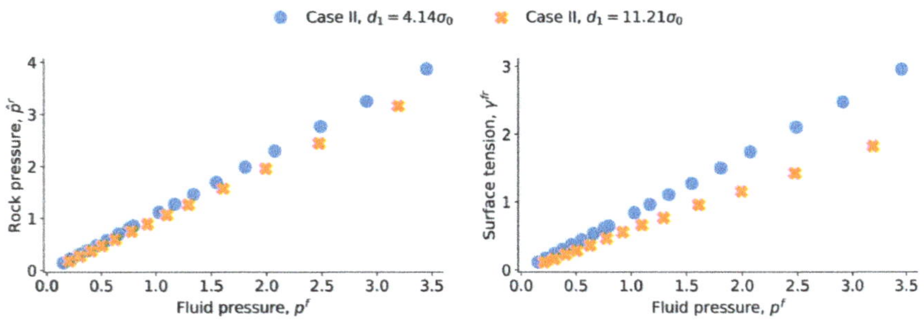

(a) (b)

Figure 20.7. Grain pressure $\widehat{p}^r$ and surface tension γ^{fr} as a function of pressure $p^f = p$ for an fcc lattice of spheres with characteristic length $d_1 = 4.14\sigma_0$ (a) and $d_1 = 11.21\sigma_0$ (b). Values are obtained keeping the equilibrium pressure constant. Reproduced from [41] with permission from Frontiers.

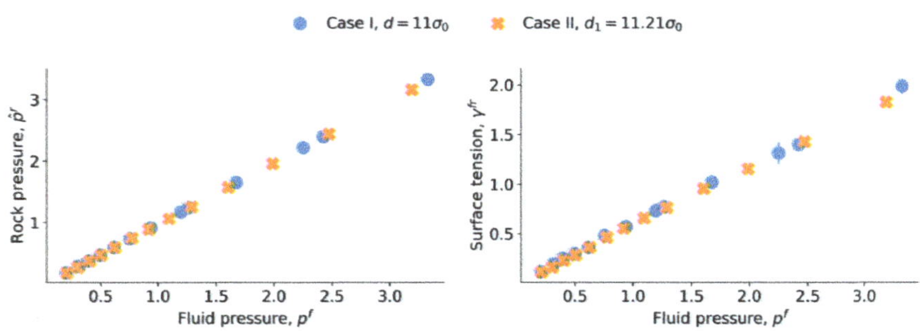

(a) (b)

Figure 20.8. Grain pressure $\widehat{p}^r$ and surface tension γ^{fr} as a function of pressure $p^f = p$ for a sphere (a, characteristic length $d = 11\sigma_0$) and an fcc lattice of spheres (b, characteristic length $d_1 = 11.21\sigma_0$). Values are obtained keeping the equilibrium pressure constant. Reproduced from [41] with permission from Frontiers.

surfaces increases, we expect that the dependence of confinement disappears. This thinking is supported by Fig. 20.8, where cases are shown with distances $d = 11\sigma_0$ (left) and $d_1 = 11.21\sigma_0$ (right), respectively. The curves for the single grain and fcc lattice of grains overlap.

We have seen in the present examples that the size of the REV includes the complete range of potential interactions available in the system. A larger REV is therefore not necessary. In order to determine a REV-property, we need to sample the whole space of possible interactions. We used layers thinner than the unit cell, to show this.

The state of equilibrium

To summarize, so far: We have seen above that the REV of a porous medium can be characterized by the compressional energy of the fluid and the solid phases, as well as by the surface between the fluid and the solid. When the characteristic length is measured in nanometer, the pressures and the surface tension depend on the size of the system. In this situation, there are two relevant pressures, rather than one, the integral and the differential pressure, respectively. The description is fully compatible with or equivalent to the Young–Laplace law.

In the examples taken here, the grain pressure and the surface tension were constant throughout the medium at equilibrium. Two porosities were studied, $\phi = 0.74$ and 0.92, respectively. For a spherical grain of radius R in an fcc lattice, the integral and the differential pressures are related by a scaling law, $\widehat{p}^r - p^r = \gamma/R$. A similar scaling law has been observed in other contexts (cf. Chapters 5 and 16) [40]. With knowledge of the various contributions to the compressional energy at equilibrium, we can find and construct a REV. This background is essential for understanding transport properties in porous media.

In the equilibrium study of the present cases, we have found that the pressure inside a grain in an fcc lattice and the corresponding surface tension, depend on the lattice constant a. A procedure was presented to find the pressure of a REV, from information of the equilibrium values of $\widehat{p}^r$ and γ^{fr} as a function of p^f. Results were presented for a particular porous medium, but they are likely to hold for other lattices, even amorphous materials if the REV can be defined with additive thermodynamic variables, as discussed in Chapter 17.

We have found that a REV can be defined for pores on the micro- as well as nanometer scale [24]. There is no conflict between the levels of description; they merge in the thermodynamic limit. The variables of the REV can be associated with an equation of state. This property allows us to develop a non-equilibrium thermodynamic theory for the nanoscale.

20.4 Transport at Varying Porosities

The driving force

A general thermodynamic theory of transport for flow in porous media was given in Chapter 19. When there is a volume flow, J_V, driven by an external pressure difference, Δp, we can write Darcy's law. The porous medium has the thickness Δx, and we have

$$J_V = -L_p \frac{\Delta p}{\Delta x} = -\frac{K}{\eta}\frac{\Delta p}{\Delta x} \tag{20.23}$$

The proportionality coefficient, L_p, or the hydraulic conductivity, is often expressed by the intrinsic permeability K and the bulk fluid viscosity η [143]. This equation applies when the medium is treated as a black box. As we have seen in the first part of this chapter, the relevant pressure of the REV is $\widehat{p}$, so the local driving force is $-d\widehat{p}/dx$. The local form of Eq. (20.23) is taken from Galteland [183], as follows:

$$J_V = -\frac{K}{\eta}\nabla\widehat{p} \tag{20.24}$$

A pressure gradient was generated for the porous fcc lattice, using the reflecting particle boundary method [41]. An overall pressure gradient between the external reservoirs B_1 and B_2 arose, of the order of magnitude 10^{12} bar $\cdot$ m^{-1}. The fluid on the left-hand side was then liquid-like, while the fluid on the right-hand side was vapor-like, cf. Fig. 20.9. The REV is also shown in this figure.

The REV fluid pressure p^f was first determined at global equilibrium from the sum over layers at equilibrium as described above. The corresponding local value of $\widehat{p}$ inside the lattice was next determined as a function of p^f as described above, from knowledge of the geometry of the lattice. We infer that this relationship applies

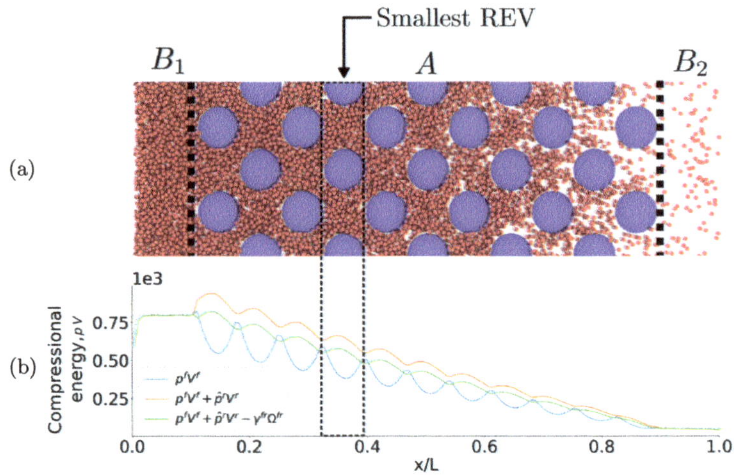

Figure 20.9. (a) Illustration of porous lattice in a pressure gradient, (b) compressional energy pV variation across the lattice. Reproduced from [41] with permission from Frontiers.

to any REV, also at local equilibrium. We can then use the relation to obtain $\widehat{p}$ once p^f is known. In our Case II, the lattice constant a varied between $20\sigma_0$ and $40\sigma_0$. The volume of the grains in a layer was V^r, and the surface area was Ω^{fr}. These parameters varied in the exact same way as in Figs. 20.6(b) and 20.6(c).

Figure 20.9(a) illustrates the system in a pressure gradient. Figure 20.9(b) shows the compressional energy, $\widehat{p}V$, of each layer along the x-axis. The dip in the pressure close to $x = 0$ is caused by the reflecting particle boundary. The reflecting particle boundary introduces a surface between the high pressure on the left-hand side and the low pressure on the right-hand side.

In order to compute a good REV variable away from equilibrium, we chose one layer as a reference point. We then computed the average of say five layers; two to the left, two to the right and the central layer. Moving one layer down the gradient, we repeated the procedure, and in this manner we obtained the property on a REV scale.

Such average values are plotted in Fig. 20.10. We see a smooth linear profile (central curve), supporting the idea of the procedure.

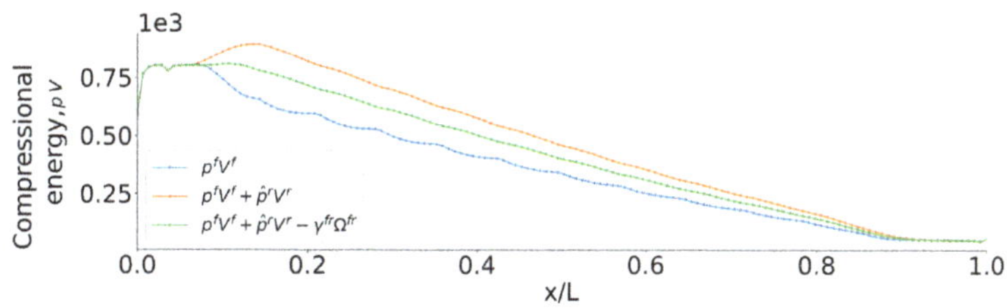

Figure 20.10. Compressional energy contributions for lattice parameter $a = 2d = 20\sigma_0$. Variations across the lattice, are smoothed over the representative elementary volume. Reproduced from [41] with permission from Frontiers.

Some traces of oscillation are still left in the separate contributions to the total compressional energy. The results of Fig. 20.10 apply to lattice constant $a = 20$. (The lattice constant gives the interparticle distances in the fcc lattice, d_1 and d_2 by $d_1 = 0.5(\sqrt{2}a - 4R)$ and $d_2 = a - 2R$). The integral as well as the differential pressures varied through the system. The relation between $\widehat{p} = p$ and $\widehat{p}^r$, γ^{fr} in a REV is the same as in Fig. 20.8, when we assume local equilibrium. Using this assumption we compute $\widehat{p}_l^r$ and γ_l^{fr} for each layer in each REV. By using the averaging procedure described above, the variation in the REV value becomes a smooth function. What is remarkable is that the total compressional energy (the central curve) does not jump at the boundaries, unlike the other curves. This follows, however, from the equilibrium condition used at this location.

The permeability

The intrinsic permeability of the lattice to the fluid, $L_p = K/\eta$, was computed from knowledge of the volume flux and the pressure gradient (the entropy production) [183]. We see from Fig. 20.10 that the variation in $\widehat{p}$ with x is essentially linear. In the simulations, the mass flux, J was kept constant (mass conservation). The flow is laminar, it has a Reynold's number of 0.72. The volume flux is related to the mass flux by the density ρ, $J_V = J/\rho$. This produces the relation $L_p = -(J/\rho)/(d\widehat{p}/dx)$. When the gradient in integral pressure is constant, the coefficient is inversely proportional to the density. The permeability may be expected to increase with increasing porosity. The possibility for systematic studies in this field seems large.

20.5 Concluding Remarks

In this chapter, we have shown how a thermodynamic driving force can be constructed for fluid transported through a porous medium. The pressure of the REV inside the medium was obtained from the REV compressional energy, with knowledge of the system geometry. We assumed equilibrium at the boundary to the external reservoir, and local equilibrium within each REV. The minimum size REV was half a unit cell, but one unit cell was used in practice to obtain smooth curves for the intergal pressure. A continuous description was obtained for transport across a series of REVs.

For the special case under study, the fcc lattice of varying porosity, we have described a procedure for determination of the intrinsic permeability as a function of porosity and fluid density. Only a one-component (one-phase) fluid was studied. The situation is increasingly more complicated with two-phase flow and an amorphous medium. The compressional energy may still prove useful as a starting point for calculations of driving forces and conductivities.

Chapter 21

Stretching of Small Polymers

Thermodynamic properties that characterize stretching of small polymers depend on the way that the stretching is controlled by the environment. We examine such stretching in two cases, when we control the length of the polymer or the applied force, i.e., when the thermodynamic ensembles of the small systems differ. As predicted, we find that the friction coefficient for isometric stretching differs from that for isotensional stretching. While the Legendre transform of one free energy into another does not work for small polymer stretching, the Legendre–Fenchel transform works.

The properties that describe biological and chemical processes are frequently size- and shape-dependent [40, 155, 187]. When the number of particles in a system goes down, the fluctuations obviously become relatively large. Consequently, it becomes rather challenging to compute average quantities precisely. While average properties can be well computed for large systems, we are often not able to properly describe statistical averages at the extreme end of the scale, for instance, for single polymers. Doubt has therefore been raised on the applicability of thermodynamic equations, for instance to describe stretching of small polymers [188]. We shall see that this scale can be perfectly well described using Hill's nanothermodynamics [4].

In order to do so we take as example the stretching of a small polymer. The small system is a polymer of varying lengths l, with a number of monomers, $n = 3, 7$ and 16. The last value turns out to be close to the thermodynamic limit value. From polymer chemistry we know that a bead on a chain for some purposes can be regarded as a particle [189, 190]. The polymer energy is extensive in n, provided that n is large enough. The environmental control variables are the temperature T and the polymer length (isometric stretching, one-dimensional canonical ensemble), or the external force applied to the polymer ends (isotensional stretching, one-dimensional isobaric ensemble). Each polymer property is an *ensemble average*.

It is already well known that the stretching depends on the set of variables that control the process, see, e.g., Süzen *et al.* [191] for a review. The purpose now is to show how the energy conversion can be understood in terms of nanothermodynamics [4]. We are building on the results of Bering and coworkers, for more details see [27, 54, 192, 193]. We shall see that the friction to stretching depends on the set of environmental control variables.

We saw in Chapter 3 that Legendre transforms can only be written for *ensembles of small systems*. Legendre–Fenchel transforms may then provide an alternative for small systems, see Section 4.9. We shall see here that this is indeed the case. We can successfully transform the Helmholtz energy into Gibbs energy, or, alternatively, the Gibbs- into Helmholtz energy using this transform. The polymer chosen to illustrate our points is poly-ethylene oxide (PEO). The polymer has been modeled by others using the united-atoms model for PEO monomers [194]. We shall vary the length of the polymer in terms of the number of monomers, and investigate the stretching properties using molecular dynamics simulations [27, 193] for just one polymer. The polymer exhibits all standard modes of movement under tension: torsion, bending and eventually bond breaking. Initially, there is an entropic regime, which obeys classical thermodynamics.

The stretching process is controlled by the environment in two ways; the isometric and the isotensional way. In the isometric case, the terminal positions of the polymer are fixed, giving it the length l. In the isotensional case, the endpoints of the hydrocarbon chain are controlled by a constant external force, f_{ext}. The schemes are illustrated in Figs. 21.1 and 21.2. The figures show polymers that are still in the entropic regime, not fully stretched.

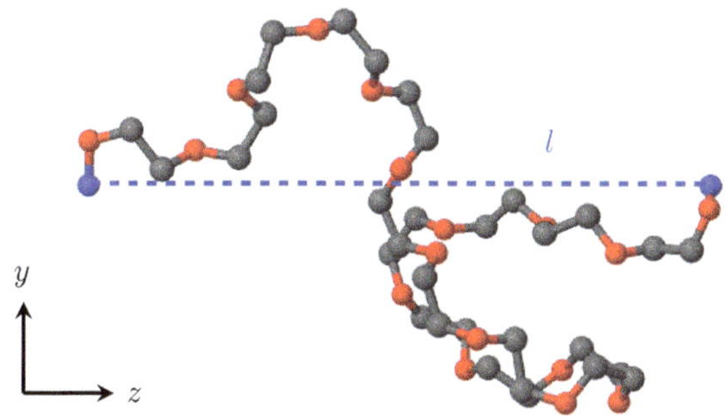

Figure 21.1. The isometric force mode: The coordinate system is chosen such that the z-axis is parallel to the end-to-end vector of the polymer, and the average force along $\hat{e}_z$ exerted on the two end-particles is sampled for fixed lengths. Reproduced from [27] with permission from ACS.

Figure 21.2. The isotensional force mode: A force of magnitude f is exerted on the end-particles along $\hat{\mathbf{e}}_\mathbf{z}$ on the left and right end-particles. Reproduced from [27] with permission from ACS.

Before we examine the numerical data from the simulation, we shall write the equations for the molecular stretching process, following Chapter 3. We shall next see that the thermodynamic equations, for equilibrium as well as nonequilibrium, capture the observations in a predictable way.

21.1 Theory

A theory for the response of a polymer to a constant applied force or to an internal force developed at constant length is of interest. The response depends on the magnitude of the perturbation. Typically, a small perturbation affects translational and torsional degrees of freedom. Bending requires forces that are larger. Only the largest forces will break a bond.

We shall describe the stretching using nanothermodynamics and see that there is a difference in the stretching dynamics for the two modes of force application when the polymer is small [192]. It is only for long or unstretched polymers that the rate laws for the two modes of stretching are the same [191].

The derivation of the rate law starts with the derivation of the entropy production. This follows the same systematic procedure for small as for large systems [25, 192]: We introduce conservation equations into Gibbs equation, and identify the entropy production and the entropy flow. The essential steps, which build upon ensemble expressions in Chapter 3, are given in what follows.

Isometric stretching

In isometric stretching, we control the temperature T and the length of the polymer, l, see Fig. 21.1. For the canonical ensemble of polymers, there is a change in the average internal energy of one polymer, $\bar{U}$, given by the first law of thermodynamics

$$d\bar{U} = dQ + \bar{f}_{\text{ext}} dl \tag{21.1}$$

The bar is used to denote the ensemble average of a property. The average internal energy can change by adding heat, dQ, and applying work, $\bar{f}_{\text{ext}} dl$, to the system. Here $\bar{f}_{\text{ext}}$ is the average external force working on the terminals of the polymer to change the length l. The heat exchange with a large reservoir at constant temperature introduces the entropy change in the system, $dS = dQ/T$.

On the other hand, we have the Gibbs equation for the system:

$$d\bar{U} = TdS + \bar{f} dl \tag{21.2}$$

Here $\bar{f}$ is the force delivered to the polymer terminals by movement of the monomers. By introducing the first law into Eq. (21.2), we obtain

$$dS = \frac{dQ}{T} + \frac{1}{T}(\bar{f}_{\text{ext}} - \bar{f}) dl \tag{21.3}$$

The first term on the right-hand side is minus the entropy change in the surroundings. The last term is the entropy change in the system and surroundings, or the total entropy change. Per unit of time, this gives the entropy production for one polymer:

$$\frac{dS_{\text{irr}}}{dt} = \frac{1}{T}(\bar{f}_{\text{ext}} - \bar{f})\frac{dl}{dt} \tag{21.4}$$

The stretching velocity has been introduced, $v \equiv dl/dt$. The difference in the external and internal force is $\Delta\bar{f} \equiv \bar{f}_{\text{ext}} - \bar{f}$.

In standard nonequilibrium thermodynamics [24], the force is a linear function of the velocity, giving

$$\Delta\bar{f} = \xi_l(l)v \tag{21.5}$$

where $\xi_l = \xi_l(l)$ is the friction coefficient for isometric stretching. This is one of the two coefficients that we want to find.

Once we know the friction coefficient, we can compute the entropy production from Eq. (21.4), giving $dS_{\text{irr}}/dt = v^2\xi_l(l)/T$. The entropy production is proportional to the square of the stretching velocity and to the friction coefficient of the length-controlled case. The entropy production is zero when the external force is balanced by the internal force, $\bar{f}_{\text{ext}} = \bar{f}$ (reversible conditions), just like in the situation when an external force on a piston, movable in a cylindric container, is balanced by the force exerted from the fluid inside the cylinder.

Isotensional stretching

In isotensional simulation experiments, we control the temperature T and the force on the ends of the polymer, f_{ext}. The first law becomes

$$d\bar{U} = dQ + f_{\text{ext}}d\bar{l} \tag{21.6}$$

The length of a single polymer is now fluctuating, and $\bar{l}$ indicates its average ensemble value. The expression for the added heat is the same as before, and the Gibbs equation for the average internal energy is

$$d\bar{U} = TdS + fd\bar{l} \tag{21.7}$$

By introducing the first law into Eq. (21.7), we can again identify the entropy flow into the surroundings and the entropy production per polymer

$$\frac{dS_{\text{irr}}}{dt} = \frac{1}{T}(f_{\text{ext}} - f)\frac{d\bar{l}}{dt} \tag{21.8}$$

The difference between the applied and internal force is expressed by $\Delta f = f_{\text{ext}} - f$, and the average stretching velocity is $\bar{v} = d\bar{l}/dt$. The rate-law in the force-controlled regime becomes

$$\Delta f = \xi_f(f)\bar{v} \tag{21.9}$$

where $\xi_f = \xi_f(f)$ is the friction coefficient for isotensional conditions. The entropy production follows as $dS_{\text{irr}}/dt = \bar{v}^2\xi_f(f)/T$. The entropy production is now proportional to the friction coefficient of the force-controlled case.

In the thermodynamic limit, the two friction coefficients are the same. Away from this limit, the rate laws depend on the set of environmental control variables. We shall see from stretching simulations of PEO that the friction coefficient for isometric stretching (Fig. 21.1) is about twice the value of the coefficient for isotensional stretching (Fig. 21.2) in the case of a polymer with 7 monomers.

The force in the entropic regime

Figures 21.1 and 21.2 picture polymers which are not fully stretched. These are natural shapes of the polymer for relatively small forces. In this regime, the polymer acts in a similar way as in the thermodynamic limit, because it has numerous degrees of freedom for movements. This is the entropic regime [189, 190].

It is common to model a polymer by a freely-jointed chain of monomers in the entropic regime [190]. The *effective* monomer length is b_{eff} and the effective number of monomers is N_{eff}, giving an unfolded length of the polymer equal to $l_{\text{unf}} = N_{\text{eff}}b_{\text{eff}}$ [189]. The statistics of a freely-jointed chain of monomers is the same as for a random walk of monomers and gives a radius of gyration, $R_{\text{g}} = l_{\text{unf}}/\sqrt{6N_{\text{eff}}}$, and the entropic force f_{S}.

$$f_{\text{S}} = \frac{3k_{\text{B}}Tl}{N_{\text{eff}}b_{\text{eff}}^2} \tag{21.10}$$

The length b_{eff} is expected to be close to the length of a monomer. The force is a linear function of l in the entropic regime. The slope can be used to find b_{eff} once l_{unf} is known, see Bering for a discussion [27]. We are interested in the regime beyond the entropic regime, where the movement is restricted, and the system must be considered as a small one.

When bead random walks become impossible, the force will start to limit the torsional degrees of freedom and next the bending degrees of freedom. Finally, a stretching of bonds will take place [195]. The force–length relationships are nonlinear in these regimes, as we shall see in what follows.

Helmholtz and Gibbs energies

Away from the entropic regime, we expect to be in the small system regime. In this regime, the thermodynamic properties that are normally extensive will show a non-trivial size dependency. The different ensembles have different fluctuations, and this leads to different size effects.

In the isometric experiment, there is a particular fluctuating force for each length. We write

$$\langle f(t)\rangle_l = \bar{f}(l) \tag{21.11}$$

and compute the corresponding Helmholtz energy from

$$F(l) = \int_{l_0}^{l} \bar{f}(l')\mathrm{d}l' \tag{21.12}$$

Examples of the integral along the length-axis of the force-elongation curve are shown in Fig. 21.4 for three polymers. The integral is the area below the curve in all cases.

In the isotensional experiment, there is a fluctuating length for each force. We write

$$\langle l(t)\rangle_f = \bar{l}(f) \tag{21.13}$$

and compute Gibbs energy from

$$G(f) = \int_{f_0}^{f} \bar{l}(f')\mathrm{d}f' \tag{21.14}$$

Examples of integrals along the force-axis of the force-elongation curve are also shown in Fig. 21.4. The integral is now illustrated by the area above the curves.

In the thermodynamic limit, we can relate F to G by a Legendre transformation. With $\Delta l = l - l_0$ and $\Delta \bar{f} = \bar{f}(l) - \bar{f}(l_0)$, we obtain for sufficiently large systems [192]

$$F(l) + G(f = \bar{f}(l)) = \Delta \bar{f} \Delta l \tag{21.15}$$

Legendre transforms like this do not in general apply to small systems. Equation (21.15) is valid in the special case when the force is linear in the elongation. This happens in the entropic regime.

The isometric and isotensional stretching modes differ in the work done to stretch the polymer. It is $\bar{f}l$ and $f\bar{l}$, respectively. Consider against this background a nonlinear force–elongation relationship $f = al + bl^2 + \cdots$, with constant parameters a and b. To linear order, the two applied works are then the same. With $f = al + bl^2$ the difference is $bl(\bar{l^2} - l\bar{l})$, which is different from zero. The presence of a nonlinear term breaks down the equality. It signals a failure to meet the thermodynamic limit condition.

In order to examine further the difference in entropy production, we can evaluate the expression $(\bar{f}dl/dt - fd\bar{l}/dt)$ from Eqs. (21.4) and (21.8). When b is positive, the entropy production is *larger* in the isometric case than in the isotensional case. We shall find a positive value of b here.

21.2 Model

In order to illustrate the equations above, we have chosen to study a chain of polyethylene oxide (PEO) of the form CH3–[O–CH2–CH2]$_n$–O–CH3 [27]. The PEO monomer consists of one oxygen and two carbons along with their hydrogen atoms. The polymer has been modeled with a united atoms model where each carbon is grouped with its bonded hydrogen atoms. With $n = 3, 7$ and 16, the number of united atoms are $N = 12, 24$ or 51. The endpoints of the chain are controlled by N, l, T (see Fig. 21.1) or by N, f_{ext}, T (see Fig. 21.2).

The LAMMPS [66] simulation package was used to simulate the interatomic force fields, see [27] for details. The potential energy has contributions from torsion, bending and stretching. The temperature was controlled to 300 K with a Langevin thermostat, which mimics the viscous aspect of a solvent. During sampling, the relaxation time was set to 1 ps, while the time step used in the simulations was 1 fs. All quantities presented were obtained as averages over 200 samples.

We investigated three molecular sizes, $N = 12, 24$ and 51. The forces varied from 0.01 up to 5 nN, or up to the failure limit of the polymer. The length-controlled simulations were sampled evenly in the length, while the force-controlled simulations were sampled evenly on a log scale in the force. This was done to distribute the data points evenly along the force–elongation curve. In order to ease the comparison between system sizes, the molecular length will be presented in units of the longitudinal length divided by the number of bonds $l_b \equiv l/(N-1)$ and $\bar{l}_b \equiv \bar{l}/(N-1)$.

21.3 The Stretching Regimes

In order to understand the response of the polymer to stretching, we consider the smallest radius R_c of a straight cylinder that can envelop the polymer completely at any length. Examples are shown for $N = 24$ in Fig. 21.3. Four snapshots illustrate the molecular conformation in various regimes. At small lengths, in the entropic regime, the radius R_c is 2.3 Å and is relatively constant. When the polymer is stretched to more than $l_b = 0.47$ Å, the torsional degree of freedom becomes gradually confined. This region is marked with orange background in the figure. At the end of this region, the polymer is unfolded to a planar configuration. Here, the C–O–C–C backbone changes from a trans-gauche (ttg) configuration to an all-trans configuration (ttt). The unbending and finally the bond stretching regions follow.

The force–elongation curves are now shown in Fig. 21.4 for $N = 12$ (a), $N = 24$ (b) and $N = 51$ (c). For each polymer, we identify different regimes. The polymer starts in an entropic regime. The range where torsion plays a role is shown with an orange background,

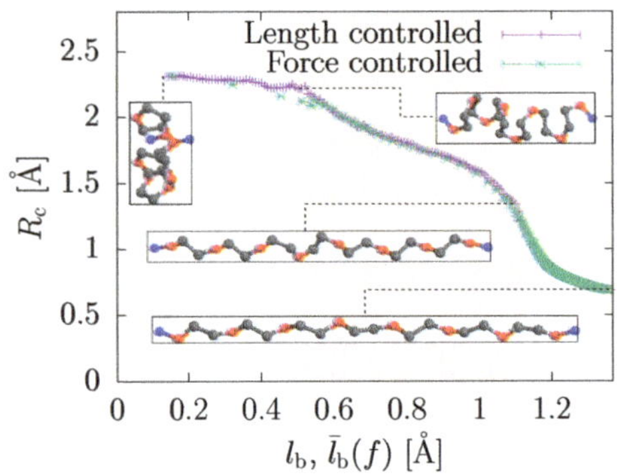

Figure 21.3. The radius R_c of the smallest straight cylinder that can envelop a polymer, as a function of polymer length. Four snapshots of the polymer are provided to illustrate the different stretching regimes for a polymer of length $N = 24$. The region for the torsional unfolding is marked with an orange background. The boundaries of this range are found from the inflection points of the curve. In the first two snapshots, the polymer attains a helical ttg order, while in the two last snapshots, the polymer is in a planar all-trans configuration. Reproduced from [27] with permission from ACS.

as in Fig. 21.3. A transition zone to the unbending, stretching and finally breaking regime follows. This is displayed in more detail in the inserts.

Entropic regime

The entropic regime is seen most clearly for $N = 51$ in Fig. 21.4(c). Lengths smaller than 0.05 Å are considered to be zero. There is a linear relation between the applied force and the molecular length of $0.05\,\text{Å} < l_b, \bar{l}_b < 0.47\,\text{Å}$, cf. Fig. 21.5 for an enlarged view.

The results for the entropic regime are according to expectations. In this regime, the polymer has so many degrees of freedom that it can be considered as large in a thermodynamic sense. This is confirmed by the fact that the length- and force-controlled cases appear identical. The curves for length- and force-controlled stretching fall on top of each other. From the slope of the curves and with $\bar{l}_{\text{unf}} = N_{\text{eff}}\bar{b}_{\text{eff}} = 0.47\,\text{Å}$, we find for the effective length $b_{\text{eff}} = 2.2\,\text{Å}$. This gives $N_{\text{eff}} = 11$ effective monomers [196]. The findings are in accordance with Eq. (21.15). So, within the accuracy

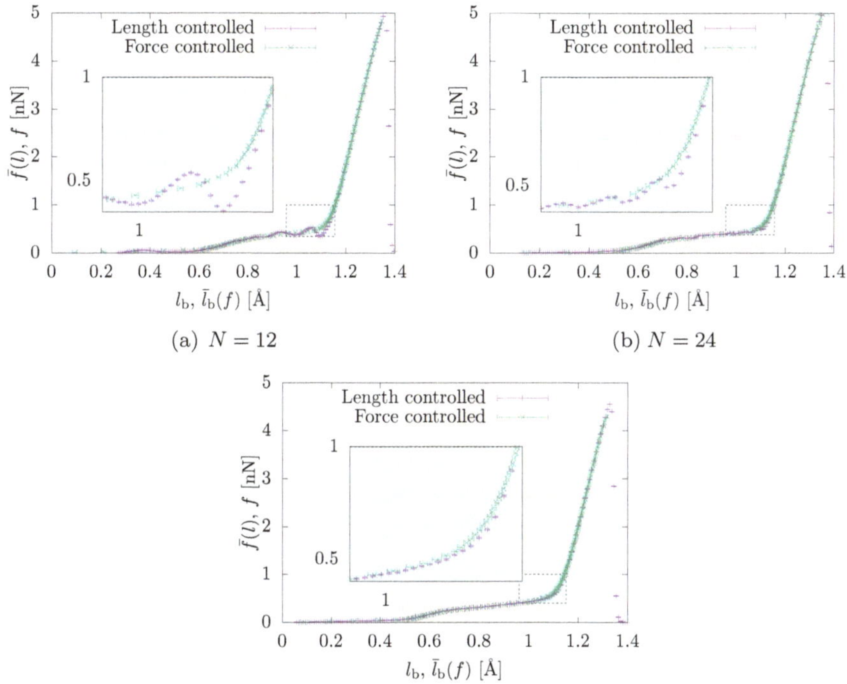

(a) $N = 12$

(b) $N = 24$

(c) $N = 51$

Figure 21.4. Force–elongation curves from the isometric and isotensional simulations for $N = 12$ (a), $N = 24$ (b) and $N = 51$ (c) as a function of length per bond. The region for the torsional unfolding is marked with an orange background. The transition region to the bond-stretching regime is shown more clearly in the insets. Reproduced from [27] with permission from ACS.

of the data presented in Fig. 21.5, the elongation behavior in the entropic regime is well described by an ideal freely jointed chain in the thermodynamic limit for forces up to about 0.05 nN.

Torsional regime

As the polymer is stretched beyond the entropic regime, the degrees of freedom are reduced, and the freely jointed chain model is no longer applicable. The torsional degrees of freedom will be gradually reduced. This occurs here when $0.47\,\text{Å} < l_b, \bar{l}_b < 1.1\,\text{Å}$. This regime is marked with an orange background in Figs. 21.3–21.4c. The start

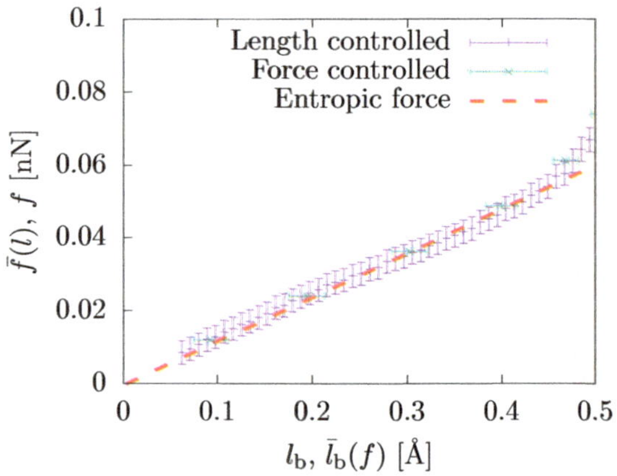

Figure 21.5. The entropic region below 0.4 Å, here shown for $N = 51$, is well described by a freely jointed chain with $N_{\text{eff}} = 10$ and $b_{\text{eff}} = 4$ (red stipled line). Reproduced from [27] with permission from ACS.

of the regime can be found by observing the deviation from linearity in Fig. 21.5. The end of the regime can be found from the inflection point of Fig. 21.3.

Strands of PEO are known to attain a helical shape in the crystalline state, in which the bonds of the C–O–C–C backbone are folded into a trans-gauche (ttg) order [197]. This is illustrated by the first two snapshots in Fig. 21.3. A further increase in the force gives rise to a transition from a helical ttg order to an elongated, planar all-trans configuration (ttt), as pictured in the last two snapshots in Fig. 21.3.

In Fig. 21.4, we see an increasing difference between the length- and force-controlled cases, increasing with decreasing molecular size. This behavior is emphasized in the inserts. Special for $N = 12$ are pronounced oscillations in the force–elongation curve. The oscillations are smaller for $N = 24$, and for $N = 51$ they are gone. *These oscillations in the length-controlled ensemble are finite-size effects that originate from local maxima in the potential of mean force associated with the unfolding of the polymer. The polymer is mechanically unstable, and all modes are not accessible in the force-controlled ensemble [187]. This is one of the reasons why the fluctuations differ between the two ensembles.*

Bending and bond-stretching regime

As the polymer is extended further, l_{b}, $\bar{l}_{\mathrm{b}}$ >1.1 Å, the monomers become elongated. The monomers are unbending and their potentials for bending and stretching give rise to a polymer-specific segment elasticity [195], which is increasingly dominated by the stretching of covalent bonds.

In this region, a small systematic difference appears in the force–elongation curves between the length- and force-controlled stretching experiments. This can be seen in the inserts of Fig. 21.4. The polymer is straightened out, as documented by the radius of the cylinder that envelops the polymer, cf. Fig. 21.3. The value shrinks to a value less than half the shortest bond length. The nonlinear contributions in the Morse potential for the bond stretching become increasingly prominent in this regime.

From the derivative of the force–elongation curve, shown in Fig. 21.6, we observe a maximum around $\bar{l}_b = 1.2$ Å. The probability for the bonds to rupture completely is now increasing. The force dropping to zero at the last points of the length-controlled simulations in Fig. 21.4 signifies rupture.

The nonlinearities in the force–elongation curve obtained from the stretching of the Morse potentials give rise to different fluctuations in the two ensembles, leading us to expect an effect of the small system size. The differences between the force–elongation curves in Fig. 21.4 are largest in the transition to the bond-stretching regime. The differences are not large, but they are finite and systematic.

Gibbs's and Helmholtz's energies

The difference in the Gibbs and Helmholtz energies, and the deviation from the Legendre transform in Eq. (21.15), are computed from the force–elongation curves in Fig. 21.4. The results are shown in Fig. 21.7. In order to compare results for different system sizes, we have divided by the work required to stretch the polymer completely.

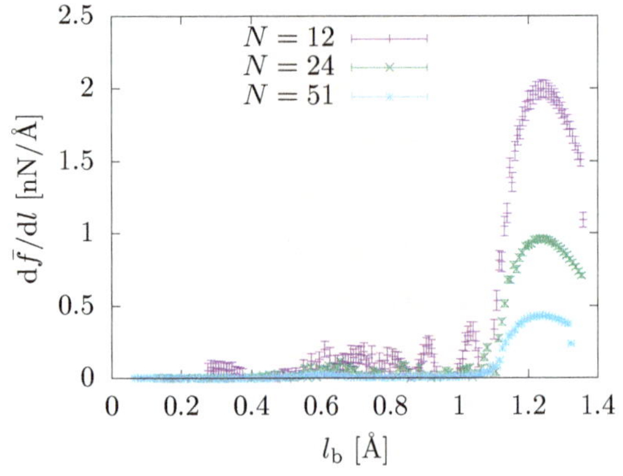

Figure 21.6. The derivative of the force–elongation curve from the length controlled simulations, $d\bar{f}/dl$ for $N = 12$, $N = 24$ and $N = 51$. The region for the torsional unfolding is marked with an orange background. We see that the maximum values coincide at about $l_b = 1.25$ Å. Reproduced from [27] with permission from American Chemical Society, ACS.

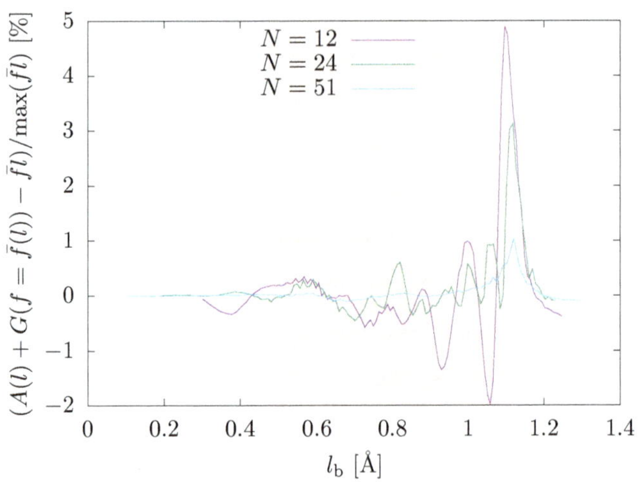

Figure 21.7. The percentage-wise difference in the Gibbs (G) and Helmholtz (A) energies for $N = 12$, $N = 24$ and $N = 51$ found by integration of the force–elongation curves in Fig. 21.4. We see a significant deviation from Eq. (21.15). The relative difference is largest for the smallest system. Reproduced from [27], with permission from ACS.

The largest free-energy difference is observed in the transition from the torsional regime to the bond-stretching regime, see also inserts of the force–elongation curves Fig. 21.4.

In the cases of $N = 12$ and $N = 24$, there is a clear correspondence between the deviation in the force–elongation curve in this region and the peak in the free-energy difference in Fig. 21.7. There is a significant deviation from Eq. (21.15), with the smallest system showing the largest deviation, as expected.

Legendre–Fenchel transforms

In Chapter 4, we presented a generalization of the Legendre transform, the Legendre–Fenchel transform; a transform which is well known in statistical physics [58, 198]. For polymer stretching, we have

$$-G_{\mathrm{LF}}\left(T, N, f(x_1)\right) = \max_{x}\left[f(x_1)x - F(T, N, x)\right] \tag{21.16}$$

where the function $F^*(T, N, f) = -G_{\mathrm{LF}}(T, N, f)$ is known as the Legendre–Fenchel transform [58, 198, 199] of $F(T, N, x)$ with respect to x at constant T and N. An important property of the Legendre-Fenchel transform is that it always yields convex functions. Therefore, the function $-G_{\mathrm{LF}}(f)$ is convex in f at constant T and N. Furthermore, if $F^*(f) = -G_{\mathrm{LF}}(f)$ is transformed again, we have

$$F^{**}(x) = \max_{f}[fx + G_{\mathrm{LF}}(f)] \tag{21.17}$$

Because $-G_{\mathrm{LF}}(f)$ is a convex function, the above transform, Eq. (21.17), reduces to the usual Legendre transform at points f for which $-G_{\mathrm{LF}}(f)$ is differentiable. This leads to

$$F^{**}(x) = f(x)x + G_{\mathrm{LF}}(f(x)) \tag{21.18}$$

where $f(x)$ is the unique solution to $dG_{\mathrm{LF}}(f)/df = -x$. Since $F^{**}(x)$ is simply the Legendre transform of $G_{\mathrm{LF}}(f)$, the former is the isotensional Helmholtz-free energy. Moreover, due to the properties of the Legendre-Fenchel transform, F^{**} is the *convex envelope* of the isometric-free energy F, namely, the largest convex function such that

$F^{**} \leq F$. This implies that the isometric and isotensional ensembles are not equivalent if F does not coincide with its convex envelope F^{**}.

In mathematical terms, the ensemble in-equivalence may arise because the Legendre–Fenchel transform is not necessarily self-dual (or involute), that is, $F^{**} \neq F$ when F is non-convex. In contrast, the convex envelope F^{**} of F has the same Legendre–Fenchel transform as F, meaning that $(F^{**})^* = F^*$ [58].

Results are given for $N = 12, 24, 51$ of $\bar{f}x - F, -G$, and $F^* = -G_{\mathrm{LF}}$ in Fig. 21.8(c)–21.8(c). The figures show clearly that the Legendre–Fenchel transform G_{LF} of F, in good approximation, is

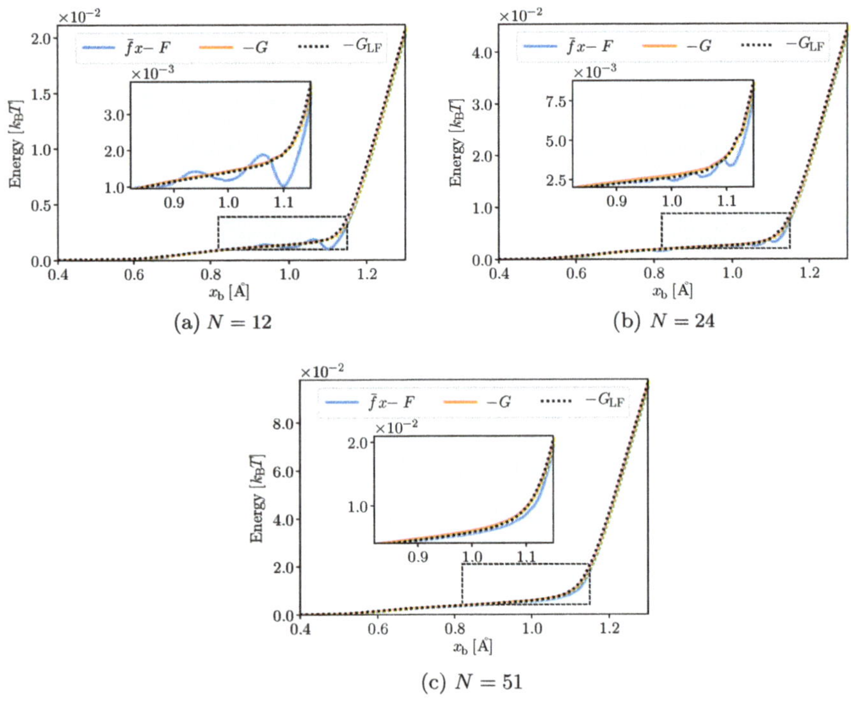

Figure 21.8. Energy as a function of length per bond of PEO-chains composed of $N = 12$, 24 and 51 united atoms. While the Legendre transform of the Helmholtz energy F is different from minus the Gibbs energy G, we see that the Legendre–Fenchel transform, G_{LF}, is an excellent approximation of F in all three cases. Reproduced from [27] with permission from Americal Chemical Society.

equal to G. Furthermore, they show that the Legendre transform $\overline{f}x - F$ of F is unequal to $-G$. In other words, we can use the Legendre–Fenchel transform, but not the Legendre transform for a small system.

21.4 Friction Coefficients

Force-controlled simulations

The friction coefficient $\xi_f = \xi_f(f)$ in Eq. (21.9) was computed for $N = 24$ and $N = 51$. The force was perturbed and the rate of change of the average length was determined in each case. Results are given for 200 independent tests, equilibrated for 150 different constant forces f_0 during 5 ns. At time $t = 0$, the force was increased. The length as a function of time before and after the increase in the force is shown in Fig. 21.9 for three force increments in the system with $N = 51$, averaged over 200 samples.

From these results, we find that the typical time scale for a linear force-response is ~0.5 ps for $N = 51$. Figure 21.9 shows that this conclusion does not depend on the magnitude of the force increment.

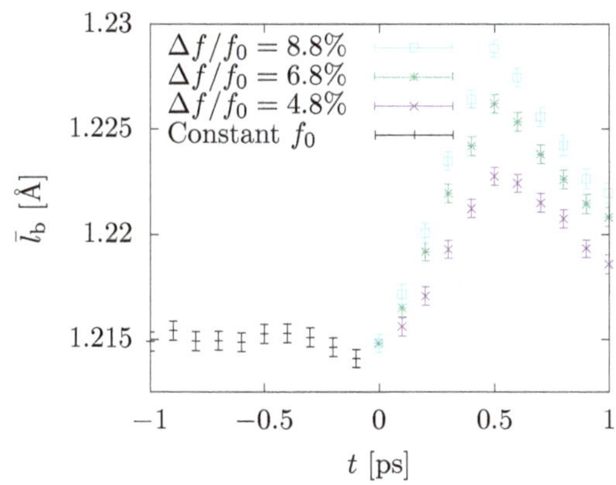

Figure 21.9. The length as a function of time for chains of length $N = 51$ before and after the force is increased by 4.8%, 6.8% and 8.8% from $f_0 = 2.3$ nN at $t = 0$. From this we conclude that the time scale for the linear response is ~0.5 ps for $N = 51$. Reproduced from [27] with permission from Americal Chemical Society.

The ratio of the force increment to the increase in the linear response is always the same, within the accuracy of the data points. Similar investigations of $N = 24$ gave a time scale of ∼0.2 ps.

The time scale for the linear regime is related to the relaxation time of the system, which depends on the length of the polymer. Other time scales in the range 0.1–1 ps were explored and found to give similar results. Within the chosen time scales, the linear response $d\bar{l}/dt$ was estimated for a range of force increments Δf, and plotted in Fig. 21.11 for $N = 51$. The polymer was equilibrated with $f_0 = 0.33$, 0.67 and 1.00 nN. The friction coefficient $\xi_f = \xi_f(f)$ was found from the slope of the force–velocity curves, cf. Eq. (21.9).

Unlike what is the case in the thermodynamic limit, the friction coefficient is largely dependent on the value of the force and the length of the polymer. This is the origin of the nonlinearity observed in $\overline{fl}$ and $f\bar{l}$.

Length-controlled simulations

In order to estimate the friction coefficient $\xi_l = \xi_l(l)$ in Eq. (21.5) for $N = 24$ and $N = 51$, the polymer was stretched at a range of velocities. The resulting increase in the force $\Delta\bar{f}$ associated with each stretching velocity was measured. A number of 200 independent samples were first equilibrated at 150 different constant lengths l_0 for 5 ns, and at time $t = 0$ the samples were stretched at 80 different constant velocities $v = dl/dt$ in the range 20 to 100 m/s for 1 ps. The force response from the polymer $\Delta\bar{f}$ for each stretching velocity was then averaged over the same time scale as used to estimate the linear response in the force-controlled simulations.

The resulting force–velocity curves for polymers with $N = 51$ and initial lengths $l_b = 0.824$ Å and $l_b = 1.192$ Å are shown in Fig. 21.11. Again, we found the friction coefficient $\xi_l = \xi_l(l)$ in Eq. (21.5) from the slope of the force–velocity curves.

The variation seen in the coefficient with the length of the polymer was similar to the results of the isotensional experiments, but the coefficients for force-controlled systems were systematically smaller than for the length-controlled systems. Because the fluctuations

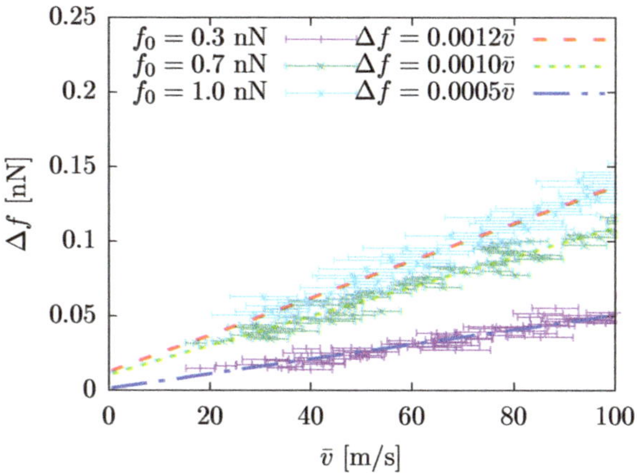

Figure 21.10. The estimated change in the velocity $\bar{v}$ from force-controlled stretching simulation of a polymer with $N = 24$ when the force is increased with Δf. Linear trends are observed, from which the friction coefficient is estimated. Reproduced from [27] with permission from Americal Chemical Society.

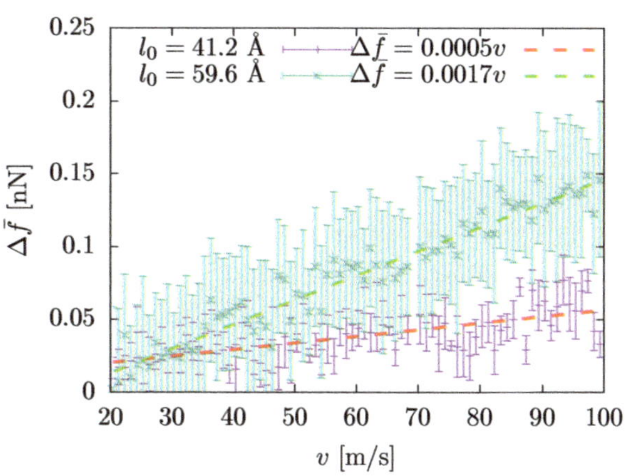

Figure 21.11. The estimated change in the force $\Delta \bar{f}$ from length-controlled stretching simulations of a polymer with $N = 51$, as the samples are stretched at a constant velocity v. Linear trends are observed, from which the friction coefficient is estimated. Reproduced from [27] with permission from Americal Chemical Society.

increased significantly at shorter lengths, we show only lengths for bonds larger than 0.4 Å. Both curves have a maximum near the relative length 1.2 Å per bond, which is where the Morse potential for bond-stretching is strongly nonlinear.

A difference in the friction coefficient can also be expected from a dynamical investigation of the system [200]. The length of the polymer acts as a slow variable, and the probability distribution of fast variables (the internal degrees of freedom of the polymer) are different for constant force and constant length.

21.5 Entropy Production

The force-controlled friction coefficient $\xi_f = \xi_f(f) = \xi_f(f(\bar{l}))$ and the length-controlled friction coefficient $\xi_l = \xi_l(l)$ found in the previous section are presented as a function of the length in Fig. 21.12 for polymers with $N = 24$ and $N = 51$. The difference between ξ_f and ξ_l is smaller for the largest polymer, as expected from Eq. (21.15).

The entropy production is found by multiplying the respective coefficient with the constant velocity squared over the temperature. The energy dissipation producing heat in the surroundings is the entropy production times the (constant) temperature. Apart from this trivial rescaling factor, the basic properties are considered to be temperature-independent under the assumption used.

For very short lengths, the entropy production should, by definition, go to zero. The uncertainty in the data in this region is rather high, but zero is within the margin of error. In the region of torsional unfolding, the ensemble difference is larger for the smaller system ($N = 24$) compared to the bigger system ($N = 51$). This is as expected from the discussion of the different stretching regimes. The entropy production reaches a maximum around 1.2 Å per bond for both system sizes. This is well into the bond-stretching regime. Again the ensemble difference is significantly larger for the smaller of the systems. This can now be explained by the nonlinearity of the Morse potential for bond-stretching, giving rise to different fluctuations in the two ensembles, and by the small-size effects in the isometric system.

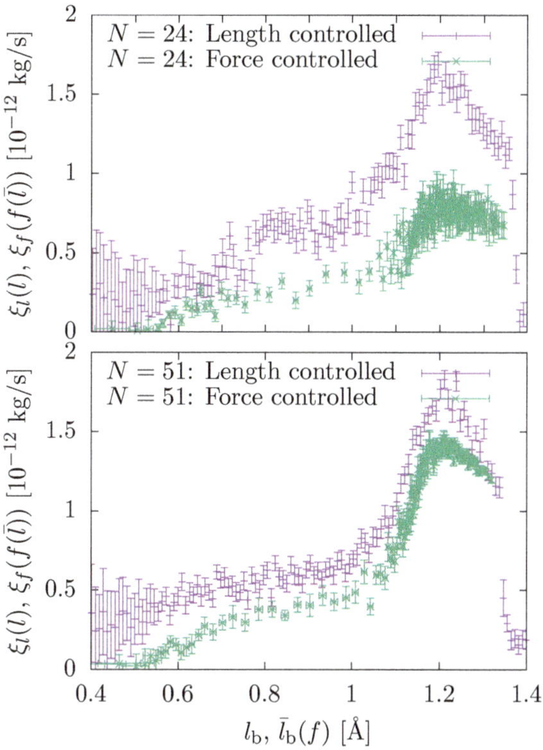

Figure 21.12. The estimated friction coefficient from the length-controlled and force-controlled simulations for $N = 24$ and $N = 51$. The ensemble deviation is most noticeable in the momomer-stretching regime, and most pronounced for the smallest system. Reproduced from [27] with permission from Americal Chemical Society.

By comparing the derivatives of the force–elongation curves of Fig. 21.6, we see that the position of the maxima appears to coincide. Coupling to low-frequency tangential phonons can very quickly dissipate energy in this regime.

We have seen above that the magnitude of the friction coefficient differs between the two stretching modes, with the length-controlled process having a higher friction coefficient than the force-controlled process. The force-controlled simulations display a significantly larger size-dependence than what is seen in the length-controlled simulations. It follows that the first process dissipates more energy regardless of the length of the polymer.

21.6 Concluding Remarks

In small-scale systems, away from the thermodynamic limit, we have seen that standard thermodynamics is no longer valid. Thermodynamic energies become non-extensive and statistical ensembles are not equivalent. Thermodynamic equations still apply, but now with the additional machinery offered by nanothermodynamics. We have also observed, in the lack of validity of Legendre transforms, that the Legendre–Fenchel transform may give a link between ensemble results (cf. Chapters 3 and 4).

The dissipation, which is generated at small scales, is sensitive to lack of equivalence between the statistical ensembles at these scales. The nature of the fluctuations depends on the way the system is controlled and gives rise to particular friction coefficients. We have seen that the friction coefficient, and thereby the entropy production, can more than double their value, from isometric stretching to isotensional stretching of small polymers. This is a good example of the fact that the properties of small systems depend on the way they are controlled.

The origin can be traced to the nonlinear parts of the force–elongation curve. The local minima and maxima in the potential of mean force are accessible only in the length-controlled ensemble. This was visible for $N = 24$, and even more so for $N = 12$, and can be traced back to polymer torsion.

The effects manifest themselves in a force–elongation relationship that becomes nonlinear for small systems. The regime is characterized by the system size (length) and by the number of degrees of freedom in the system, and is expressed in the friction coefficient. A difference in these will disappear when the length of the polymer is large enough, which in our case is above $N = 51$.

Through most of this book, we have investigated small systems in equilibrium. In this chapter, we have shown that non-equilibrium properties are also affected by the absence of the thermodynamic limit. The theoretical basis presented earlier could however well be applied to a systematic study of the irreversible processes that here take place at small scales. The study of dissipation at small scales is relevant for nanomachines. We expect that more examples will be demonstrated in the future.

Appendix A

Edge and Corner Effects: T, V, N

A small system may have significant edge and corner energies. We discuss these contributions here for the example given in Chapter 6.

The one-particle canonical partition function for component k in a small system with edges and corners follows from statistical mechanics [13]:

$$\begin{aligned} Q_{1,k}(T,V) &= \frac{V}{\Lambda_k^d} + \frac{O}{\Lambda_k^{d-1}} \exp\left(\beta U_k^s\right) + c_e \frac{L}{\Lambda_k^{d-2}} \exp\left(\beta U_k^e\right) \\ &\quad + c_c \exp\left(\beta U_k^c\right) = \frac{V}{\Lambda_k^d} \left[1 + c_s \frac{\Lambda_k}{L} \exp(\beta U_k^s) \right. \\ &\quad \left. + c_e \left(\frac{\Lambda_k}{L} \right)^2 \exp\left(\beta U_k^e\right) + c_c \left(\frac{\Lambda_k}{L} \right)^3 \exp\left(\beta U_k^c\right) \right] \end{aligned} \tag{A.1}$$

The **N**-particle canonical partition function is

$$Q(T,V,\mathbf{N}) = \prod_{k=1}^{n} Q_k(T,V,N_k) = \prod_{k=1}^{n} \frac{1}{N_k!} Q_{1,k}^{N_k}(T,V) \tag{A.2}$$

For the Helmholtz energy of a single replica, we have

$$\begin{aligned} F^{\mathrm{C}}(T,V,\mathbf{N}) &= -k_{\mathrm{B}} T \ln Q(T,V,\mathbf{N}) \\ &= -k_{\mathrm{B}} T \sum_{k=1}^{n} \ln Q_k(T,V,N_k) = \sum_{k=1}^{n} F_k^{\mathrm{C}}(T,V,N_k) \end{aligned} \tag{A.3}$$

With Stirling's formula for $N_k!$, we obtain

$$\begin{aligned} F_k^{\mathrm{C}}(T,V,N_k) &= k_{\mathrm{B}}T\ln N_k! - N_k k_{\mathrm{B}}T\ln Q_{1,k}(T,V) \\ &= N_k k_{\mathrm{B}}T\ln\left(\frac{N_k}{e}\right) + \frac{1}{2}k_{\mathrm{B}}T\ln(2\pi N_k) \\ &\quad - N_k k_{\mathrm{B}}T\ln Q_{1,k}(T,V) \end{aligned} \tag{A.4}$$

By introducing Eq. (A.1) into this, we obtain

$$\begin{aligned} F_k^{\mathrm{C}}(T,V,N_k) &= N_k k_{\mathrm{B}}T\ln\left(\frac{N_k\Lambda_k^d}{Ve}\right) + \frac{1}{2}k_{\mathrm{B}}T\ln(2\pi N_k) - N_k k_{\mathrm{B}}T \\ &\quad\times \ln\left[1 + c_s\frac{\Lambda_k}{L}\exp(\beta U_k^s) + c_e\left(\frac{\Lambda_k}{L}\right)^2\exp(\beta U_k^e)\right. \\ &\quad\left. + c_c\left(\frac{\Lambda_k}{L}\right)^3\exp(\beta U_k^c)\right] \end{aligned} \tag{A.5}$$

We can then derive p^{C}, s^{C}, μ_j^{C} from Eqs. (3.18)–(3.24)

$$\begin{aligned} p^{\mathrm{C}}(T,V,\mathbf{N}) &= -\left(\frac{\partial F^{\mathrm{C}}}{\partial V}\right)_{T,\mathbf{N}} = k_{\mathrm{B}}T\sum_{k=1}^{n} n_k \\ &\times\left(\frac{1 + \frac{d-1}{d}c_s\frac{\Lambda_k}{L}\exp(\beta U_k^s) + \frac{d-2}{d}c_e\left(\frac{\Lambda_k}{L}\right)^2\exp(\beta U_k^e)}{1 + c_s\frac{\Lambda_k}{L}\exp(\beta U_k^s) + c_e\left(\frac{\Lambda_k}{L}\right)^2\exp(\beta U_k^e) + c_c\left(\frac{\Lambda_k}{L}\right)^3\exp(\beta U_k^c)} + \mathbf{n}\cdot\frac{\partial}{\partial\mathbf{n}}\right. \\ &\times\left.\ln\left[1 + c_s\frac{\Lambda_k}{L}\exp(\beta U_k^s) + c_e\left(\frac{\Lambda_k}{L}\right)^2\exp(\beta U_k^e) + c_c\left(\frac{\Lambda_k}{L}\right)^3\exp(\beta U_k^c)\right]\right) \end{aligned} \tag{A.6}$$

where we used that U_k^s, U_k^e and U_k^c depend on the concentrations, but not on the volume. Furthermore, we note that $c_c(d-3) = 0$ for all d because $c_c = 0$ for $d = 1, 2$. The entropy density becomes

$$\begin{aligned} s^{\mathrm{C}}(T,V,\mathbf{N}) &= -\frac{1}{V}\left(\frac{\partial F^{\mathrm{C}}}{\partial T}\right)_{V,\mathbf{N}} \\ &= -k_{\mathrm{B}}\left\{\sum_{k=1}^{n}\left[n_k\ln(n_k\Lambda_k^d) - \frac{d+2}{2}n_k + \frac{1}{2V}\ln(2\pi N_k)\right] - \sum_{k=1}^{n} n_k\right. \end{aligned}$$

$$\times\left(\ln\left[1+c_s\frac{\Lambda_k}{L}\exp(\beta U_k^s)+c_e\left(\frac{\Lambda_k}{L}\right)^2\exp(\beta U_k^e)+c_c\left(\frac{\Lambda_k}{L}\right)^3\exp(\beta U_k^c)\right]\right.$$

$$-\frac{\frac{c_s}{2}\frac{\Lambda_k}{L}\exp(\beta U_k^s)+c_e\left(\frac{\Lambda_k}{L}\right)^2\exp(\beta U_k^e)+\frac{3c_c}{2}\left(\frac{\Lambda_k}{L}\right)^3\exp(\beta U_k^c)}{1+c_s\frac{\Lambda_k}{L}\exp(\beta U_k^s)+c_e\left(\frac{\Lambda_k}{L}\right)^2\exp(\beta U_k^e)+c_c\left(\frac{\Lambda_k}{L}\right)^3\exp(\beta U_k^c)}$$

$$\left.+\frac{c_s\frac{\Lambda_k}{L}\frac{\partial}{\partial T}\exp(\beta U_k^s)+c_e\left(\frac{\Lambda_k}{L}\right)^2\frac{\partial}{\partial T}\exp(\beta U_k^e)+c_c\left(\frac{\Lambda_k}{L}\right)^3\frac{\partial}{\partial T}\exp(\beta U_k^c)}{1+c_s\frac{\Lambda_k}{L}\exp(\beta U_k^s)+c_e\left(\frac{\Lambda_k}{L}\right)^2\exp(\beta U_k^e)+c_c\left(\frac{\Lambda_k}{L}\right)^3\exp(\beta U_k^c)}\right)\Bigg\} \tag{A.7}$$

The chemical potentials become

$$\mu_j^{\mathrm{C}}(T,V,\mathbf{N})=\left(\frac{\partial F^{\mathrm{C}}}{\partial N_j}\right)_{T,V,\mathbf{N}_j}=k_{\mathrm{B}}T\left\{\ln(n_j\Lambda_j^d)+\frac{1}{2N_j}\right.$$

$$-\ln\left[1+c_s\frac{\Lambda_j}{L}\exp(\beta U_j^s)+c_e\left(\frac{\Lambda_j}{L}\right)^2\exp(\beta U_j^e)+c_c\left(\frac{\Lambda_j}{L}\right)^3\exp(\beta U_j^c)\right]$$

$$-\sum_{k=1}^{n}n_k\frac{\partial}{\partial n_j}\ln\left[1+c_s\frac{\Lambda_k}{L}\exp(\beta U_k^s)+c_e\left(\frac{\Lambda_k}{L}\right)^2\exp(\beta U_k^e)\right.$$

$$\left.\left.+c_c\left(\frac{\Lambda_k}{L}\right)^3\exp(\beta U_k^c)\right]\right\} \tag{A.8}$$

The matrix of thermodynamic factors, Γ_{jk}^{C}, has the elements

$$\Gamma_{mj}^{\mathrm{C}}(T,V,\mathbf{N})=n_mA_{mj}^{\mathrm{C}}(T,V,\mathbf{N})=\frac{N_m}{k_{\mathrm{B}}T}\left(\frac{\partial\mu_j^{\mathrm{C}}}{\partial N_m}\right)_{T,V,\mathbf{N}_m}$$

$$=\frac{n_m}{k_{\mathrm{B}}T}\left(\frac{\partial\mu_j^{\mathrm{C}}}{\partial n_m}\right)_{T,V,\mathbf{N}_m}=\delta_{mj}\left(1-\frac{1}{2N_m}\right)$$

$$-n_m\left\{\frac{\partial}{\partial n_m}\ln\left[1+c_s\frac{\Lambda_j}{L}\exp\left(\beta U_j^s\right)+c_e\left(\frac{\Lambda_j}{L}\right)^2\exp(\beta U_j^e)\right.\right.$$

$$\left.+c_c\left(\frac{\Lambda_j}{L}\right)^3\exp(\beta U_j^c)\right]+\frac{\partial}{\partial n_j}\ln\left[1+c_s\frac{\Lambda_m}{L}\exp(\beta U_m^s)\right.$$

$$\left.\left.+c_e\left(\frac{\Lambda_m}{L}\right)^2\exp(\beta U_m^e)+c_c\left(\frac{\Lambda_m}{L}\right)^3\exp(\beta U_m^c)\right]\right\}$$

$$
- n_m \sum_{k=1}^{n} n_k \frac{\partial^2}{\partial n_m \partial n_j} \ln \left[1 + c_s \frac{\Lambda_k}{L} \exp(\beta U_k^s) \right.
$$

$$
\left. + c_e \left(\frac{\Lambda_k}{L} \right)^2 \exp(\beta U_k^e) + c_c \left(\frac{\Lambda_k}{L} \right)^3 \exp(\beta U_k^c) \right] \tag{A.9}
$$

We kept the last contribution as a differential. The contribution due to this differential contains terms proportional to the square of the first-order derivatives and terms proportional to the second-order derivatives of the surface, edge and corner energies. The symmetry relation given in Eq. (6.13) remains valid.

The above expressions show that Hadwiger's theorem is incorrect for the canonical ensemble when edge and corner contributions are taken into account.

Appendix B

Edge and Corner Effects: T, V, $\boldsymbol{\mu}$

A small system may have significant edge and corner energies. We discuss these contributions here for the example given in Chapter 6.

The one-particle grand canonical partition function in a small system with edges and corners follows from statistical mechanics [13]:

The grand canonical partition function equals

$$\Xi(T, V, \boldsymbol{\mu}) = \prod_{k=1}^{n} \sum_{N_k=0}^{\infty} \exp\left(\beta\mu_k N_k\right) Q_k(T, V, N_k)$$
$$= \exp\left(\sum_{k=1}^{n} \exp(\beta\mu_k) Q_{1,k}(T, V)\right) \tag{B.1}$$

where we used Eq. (A.2). By introducing the expressions above, thermodynamic properties can be derived in terms of T, V, $\boldsymbol{\mu}$, U_k^s, U_k^e and U_k^c. Using Eq. (4.56) we find for the integral pressure

$$\hat{p}(T, V, \boldsymbol{\mu}) = \frac{k_{\mathrm{B}}T}{V} \ln \Xi = \frac{k_{\mathrm{B}}T}{V} \sum_{k=1}^{n} \exp\left(\beta\mu_k\right) Q_{1,k}(T, V)$$
$$= k_{\mathrm{B}}T \sum_{k=1}^{n} \frac{\exp(\beta\mu_k)}{\Lambda_k^d} \left[1 + c_s \frac{\Lambda_k}{L} \exp(\beta U_k^s)\right.$$
$$\left. + c_e \left(\frac{\Lambda_k}{L}\right)^2 \exp(\beta U_k^e) + c_c \left(\frac{\Lambda_k}{L}\right)^3 \exp(\beta U_k^c)\right] \tag{B.2}$$

The differential pressure is given by

$$p^{\mathrm{GC}}(T,V,\boldsymbol{\mu}) = \left(\frac{\partial \hat{p}V}{\partial V}\right)_{T,\boldsymbol{\mu}} = k_{\mathrm{B}}T\sum_{k=1}^{n}\frac{\exp(\beta\mu_k)}{\Lambda_k^d}$$

$$\times\left[1 + c_s\frac{d-1}{d}\frac{\Lambda_k}{L}\exp\left(\beta U_k^s\right) + c_e\frac{d-2}{d}\left(\frac{\Lambda_k}{L}\right)^2\exp\left(\beta U_k^e\right)\right] \tag{B.3}$$

where we used that U_k^s, U_k^e and U_k^c do not depend on the volume. The entropy density is given by

$$s^{\mathrm{GC}}(T,V,\boldsymbol{\mu}) = \left(\frac{\partial \hat{p}}{\partial T}\right)_{V,\boldsymbol{\mu}} = k_{\mathrm{B}}\sum_{k=1}^{n}\frac{\exp(\beta\mu_k)}{\Lambda_k^d}\left\{\left[1 + \frac{d}{2} - \beta\mu_k\right]\right.$$

$$+\ c_s\frac{\Lambda_k}{L}\left[\frac{d+1}{2} - \beta(\mu_k + U_k^s) + \frac{\partial U_k^s}{k_{\mathrm{B}}\partial T}\right]\exp(\beta U_k^s)$$

$$+\ c_e\left(\frac{\Lambda_k}{L}\right)^2\left[\frac{d}{2} - \beta(\mu_k + U_k^e) + \frac{\partial U_k^e}{k_{\mathrm{B}}\partial T}\right]\exp(\beta U_k^e)$$

$$\left. +\ c_c\left(\frac{\Lambda_k}{L}\right)^3\left[1 - \beta\left(\mu_k + U_k^c\right) + \frac{\partial U_k^c}{k_{\mathrm{B}}\partial T}\right]\exp(\beta U_k^c)\right\} \tag{B.4}$$

The component densities are given by

$$n_j^{\mathrm{GC}}(T,V,\boldsymbol{\mu}) = \left(\frac{\partial \hat{p}}{\partial \mu_j}\right)_{T,V,\boldsymbol{\mu}_j} = \frac{\exp(\beta\mu_j)}{\Lambda_j^d}$$

$$\times\left[1 + c_s\frac{\Lambda_j}{L}\exp(\beta U_j^s) + c_e\left(\frac{\Lambda_j}{L}\right)^2\exp(\beta U_j^e) + c_c\left(\frac{\Lambda_j}{L}\right)^3\exp(\beta U_j^c)\right]$$

$$+\sum_{k=1}^{n}\frac{\exp(\beta\mu_k)}{\Lambda_k^d}\left[c_s\frac{\Lambda_k}{L}\frac{\partial U_k^s}{\partial \mu_j}\exp(\beta U_k^s) + c_e\left(\frac{\Lambda_k}{L}\right)^2\frac{\partial U_k^e}{\partial \mu_j}\exp\left(\beta U_k^e\right)\right.$$

$$\left. + c_c\left(\frac{\Lambda_k}{L}\right)^3\frac{\partial U_k^c}{\partial \mu_j}\exp(\beta U_k^c)\right] \tag{B.5}$$

For the inverse matrix of thermodynamic factors, we obtain

$$
\begin{aligned}
&\left(\boldsymbol{\Gamma}^{-1}\right)_{jm}^{\mathrm{GC}}(T,V,\boldsymbol{\mu})n_j^{\mathrm{GC}}(T,V,\boldsymbol{\mu}) = B_{jm}^{\mathrm{GC}}(T,V,\boldsymbol{\mu})\\
&= \frac{1}{\beta V}\left(\frac{\partial N_j^{\mathrm{GC}}}{\partial \mu_m}\right)_{T,V,\boldsymbol{\mu}_m} = \frac{1}{\beta}\left(\frac{\partial n_j^{\mathrm{GC}}}{\partial \mu_m}\right)_{T,V,\boldsymbol{\mu}_m} = \delta_{mj} n_j^{\mathrm{GC}}\\
&\times\left[1 + c_s\frac{\Lambda_j}{L}\exp(\beta U_j^s) + c_e\left(\frac{\Lambda_j}{L}\right)^2\exp(\beta U_j^e) + c_c\left(\frac{\Lambda_j}{L}\right)^3\exp(\beta U_j^c)\right]\\
&+\frac{\exp(\beta\mu_j)}{\Lambda_j^d}\left[c_s\frac{\Lambda_j}{L}\frac{\partial U_j^s}{\partial \mu_m}\exp(\beta U_j^s) + c_e\left(\frac{\Lambda_j}{L}\right)^2\frac{\partial U_j^e}{\partial \mu_m}\exp(\beta U_j^e)\right.\\
&\left.+c_c\left(\frac{\Lambda_j}{L}\right)^3\frac{\partial U_j^c}{\partial \mu_m}\exp(\beta U_j^c)\right]\\
&+\frac{\exp(\beta\mu_m)}{\Lambda_m^d}\left[c_s\frac{\Lambda_m}{L}\frac{\partial U_m^s}{\partial \mu_j}\exp(\beta U_m^s) + c_e\left(\frac{\Lambda_m}{L}\right)^2\frac{\partial U_m^e}{\partial \mu_j}\exp(\beta U_m^e)\right.\\
&\left.+c_c\left(\frac{\Lambda_m}{L}\right)^3\frac{\partial U_m^c}{\partial \mu_j}\exp(\beta U_m^c)\right]\\
&+\sum_{k=1}^{n}\frac{\exp(\beta\mu_k)}{\Lambda_k^d}\left[c_s\frac{\Lambda_k}{L}\left(\frac{\partial U_k^s}{\partial \mu_j}\frac{\partial U_k^s}{\partial \mu_m} + \frac{\partial^2 U_k^s}{\beta\partial\mu_j\partial\mu_m}\right)\exp(\beta U_k^s)\right.\\
&+c_e\left(\frac{\Lambda_k}{L}\right)^2\left(\frac{\partial U_k^e}{\partial \mu_j}\frac{\partial U_k^e}{\partial \mu_m} + \frac{\partial^2 U_k^e}{\beta\partial\mu_j\partial\mu_m}\right)\exp(\beta U_k^e)\\
&\left.+c_c\left(\frac{\Lambda_k}{L}\right)^3\left(\frac{\partial U_k^c}{\partial \mu_j}\frac{\partial U_k^c}{\partial \mu_m} + \frac{\partial^2 U_k^c}{\beta\partial\mu_j\partial\mu_m}\right)\exp(\beta U_k^c)\right] \qquad \text{(B.6)}
\end{aligned}
$$

The symmetry relation given in Eq. (6.21) remains valid.

The above expressions show that Hadwiger's theorem remains valid in the grand canonical ensemble when edge and corner contributions are taken into account.

Appendix C

Transformations in the Thermodynamic Limit

The T, p, N_j ensemble is convenient when a system is studied in a laboratory at constant temperature and pressure. From the small system method, however, we obtain results by extrapolation in the T, V, μ_j ensemble. A transformation between the different results is then needed. In this appendix, we show how the transformation can be done from the T, V, μ_j to the T, p, N_j ensemble, to give partial internal energies, partial enthalpies and partial compressional energies in the thermodynamic limit. The information can be found in the literature (see, e.g., [70, 71]), but we include it here for convenience.

Large system variables, like U, N_j, $p = \hat{p}$, do not depend on environmental control variables, so we remove superscript GC for these. We consider a binary mixture. The extension to multi-component mixtures is straightforward. In the first section, we explain the transformations for partial internal energies. In the second section, we extend the derivation to partial enthalpies and compressional energies.

C.1 Partial Internal Energies

Equation (7.6) gives the partial derivative of U with N_1 in the GC ensemble as follows:

$$\left(\frac{\partial U}{\partial N_1}\right)_{T,V,\mu_2} = \frac{\langle \mathcal{H} N_1 \rangle - U N_1}{\langle N_1^2 \rangle - N_1^2} \tag{C.1}$$

The brackets on the right-hand side indicate a GC average, but now for a large system. This implies, for instance, that $U = U(T, V, \boldsymbol{\mu})$ and $N_j = N_j(T, V, \boldsymbol{\mu})$.

We need to convert $(\partial U/\partial N_1)_{T,V,\mu_2}$ into $(\partial U/\partial N_1)_{T,p,N_2}$, and start with the total differential of U. The ensuing text follows the exposition of Schnell *et al.* [47], see also [70, 71].

$$dU = \left(\frac{\partial U}{\partial T}\right)_{V,\mathbf{N}} dT + \left(\frac{\partial U}{\partial V}\right)_{T,\mathbf{N}} dV + \left(\frac{\partial U}{\partial N_1}\right)_{T,V,N_2} dN_1 + \left(\frac{\partial U}{\partial N_2}\right)_{T,V,N_1} dN_2 \tag{C.2}$$

This implies that

$$\begin{aligned}\left(\frac{\partial U}{\partial N_1}\right)_{T,V,\mu_2} &= \left(\frac{\partial U}{\partial N_1}\right)_{T,V,N_2} + \left(\frac{\partial U}{\partial N_2}\right)_{T,V,N_1}\left(\frac{\partial N_2}{\partial N_1}\right)_{T,V,\mu_2}\\ \left(\frac{\partial U}{\partial N_2}\right)_{T,V,\mu_1} &= \left(\frac{\partial U}{\partial N_1}\right)_{T,V,N_2}\left(\frac{\partial N_1}{\partial N_2}\right)_{T,V,\mu_1} + \left(\frac{\partial U}{\partial N_2}\right)_{T,V,N_1}\end{aligned} \tag{C.3}$$

We use the identity (A.21) in Appendix A of Ben-Naim [71], which gives

$$\begin{aligned}\left(\frac{\partial N_2}{\partial N_1}\right)_{T,V,\mu_2} &= -\left(\frac{\partial \mu_2}{\partial N_2}\right)^{-1}_{T,V,N_1}\left(\frac{\partial \mu_2}{\partial N_1}\right)_{T,V,N_2}\\ &= -\left(\frac{\partial \mu_2}{\partial c_2}\right)^{-1}_{T,V,c_1}\left(\frac{\partial \mu_2}{\partial c_1}\right)_{T,V,c_2}\\ \left(\frac{\partial N_1}{\partial N_2}\right)_{T,V,\mu_1} &= -\left(\frac{\partial \mu_1}{\partial N_1}\right)^{-1}_{T,V,N_2}\left(\frac{\partial \mu_1}{\partial N_2}\right)_{T,V,N_1}\\ &= -\left(\frac{\partial \mu_1}{\partial c_1}\right)^{-1}_{T,V,c_2}\left(\frac{\partial \mu_1}{\partial c_2}\right)_{T,V,c_1}\end{aligned} \tag{C.4}$$

where $c_i \equiv N_i/V$. We now follow Kirkwood and Buff [70], as explained by Ben-Naim [71] (Eqs. (4.23) and (4.24)). This gives the particle correlation density

$$B_{ij} = B_{ji} = \frac{k_{\mathrm{B}}T}{V}\left(\frac{\partial N_j}{\partial \mu_i}\right)_{T,V,\boldsymbol{\mu}_i} = k_{\mathrm{B}}T\left(\frac{\partial c_j}{\partial \mu_i}\right)_{T,V,\boldsymbol{\mu}_i}$$
$$= c_i c_j G_{ij} + c_i \delta_{ij} = \frac{1}{V}\left(\langle N_i N_j\rangle - N_i N_j\right) \tag{C.5}$$

where G_{ij} is a Kirkwood–Buff integral. All averages refer to the GC ensemble. Furthermore:

$$A_{ij} = A_{ji} = \beta V\left(\frac{\partial \mu_j}{\partial N_i}\right)_{T,V,\mathbf{N}_i} = \beta\left(\frac{\partial \mu_j}{\partial c_i}\right)_{T,V,\mathbf{c}_i} \tag{C.6}$$

We note that the B- and A-matrices are each other's inverse:

$$\sum_j B_{ij}A_{jk} = \delta_{ik} \tag{C.7}$$

We introduce Eq. (C.6) into Eq. (C.4) and obtain

$$\left(\frac{\partial N_2}{\partial N_1}\right)_{T,V,\mu_2} = -\frac{A_{12}}{A_{22}} \quad \text{and} \quad \left(\frac{\partial N_1}{\partial N_2}\right)_{T,V,\mu_1} = -\frac{A_{12}}{A_{11}} \tag{C.8}$$

By using Eq. (C.8), we write Eq. (C.3) in the form

$$\left(\frac{\partial U}{\partial N_1}\right)_{T,V,\mu_2} = \left(\frac{\partial U}{\partial N_1}\right)_{T,V,N_2} - \left(\frac{\partial U}{\partial N_2}\right)_{T,V,N_1}\frac{A_{12}}{A_{22}}$$
$$\left(\frac{\partial U}{\partial N_2}\right)_{T,V,\mu_1} = -\left(\frac{\partial U}{\partial N_1}\right)_{T,V,N_2}\frac{A_{12}}{A_{11}} + \left(\frac{\partial U}{\partial N_2}\right)_{T,V,N_1} \tag{C.9}$$

The B-matrix is the inverse of the A-matrix. We invert the equations and obtain

$$\left(\frac{\partial U}{\partial N_1}\right)_{T,V,N_2} = B_{22}\left[A_{22}\left(\frac{\partial U}{\partial N_1}\right)_{T,V,\mu_2} + A_{12}\left(\frac{\partial U}{\partial N_2}\right)_{T,V,\mu_1}\right]$$
$$\left(\frac{\partial U}{\partial N_2}\right)_{T,V,N_1} = B_{11}\left[A_{11}\left(\frac{\partial U}{\partial N_2}\right)_{T,V,\mu_1} + A_{12}\left(\frac{\partial U}{\partial N_1}\right)_{T,V,\mu_2}\right] \tag{C.10}$$

The equations give $(\partial U/\partial N_1)_{T,V,N_2}$ and $(\partial U/\partial N_2)_{T,V,N_1}$.

We introduce the definition of the partial molar volumes (Eq. (4.30) of Ben-Naim [71])

$$V_j = \left(\frac{\partial V}{\partial N_j}\right)_{T,p,\mathbf{N}_j} = \left(\frac{\partial \mu_j}{\partial p}\right)_{T,\mathbf{N}} \tag{C.11}$$

and the identity (Eq. (A.21) of Ben-Naim [71])

$$\begin{aligned}&\left(\frac{\partial p}{\partial N_1}\right)_{T,V,N_2}\left(\frac{\partial N_1}{\partial V}\right)_{T,p,N_2}\left(\frac{\partial V}{\partial p}\right)_{T,\mathbf{N}}\\ &\quad = \left(\frac{\partial p}{\partial N_1}\right)_{T,V,N_2}\left(\frac{-V\kappa_T}{V_1}\right) = -1\end{aligned} \tag{C.12}$$

where κ_T is the isothermal compressibility

$$\kappa_T = -\frac{1}{V}\left(\frac{\partial V}{\partial p}\right)_{T,\mathbf{N}} \tag{C.13}$$

It follows from Eq. (C.12) that

$$\left(\frac{\partial p}{\partial N_1}\right)_{T,V,N_2} = \frac{V_1}{V\kappa_T} \tag{C.14}$$

κ_T and V_1 can be calculated from the fluctuations in the GC ensemble (see Eqs. (17) and (18) in Kirkwood and Buff [70]).

Euler's theorem gives the internal energy from Eq. (C.2):

$$U = N_1\left(\frac{\partial U}{\partial N_1}\right)_{T,V,N_2} + N_2\left(\frac{\partial U}{\partial N_2}\right)_{T,V,N_1} + V\left(\frac{\partial U}{\partial V}\right)_{T,\mathbf{N}} \tag{C.15}$$

It follows from Eq. (C.2) and Eq. (A.21) from Ben-Naim that

$$\left(\frac{\partial U}{\partial N_1}\right)_{T,p,N_2} = \left(\frac{\partial U}{\partial N_1}\right)_{T,V,N_2} - \left(\frac{\partial U}{\partial V}\right)_{T,\mathbf{N}}\left(\frac{\partial V}{\partial p}\right)_{T,\mathbf{N}}\left(\frac{\partial p}{\partial N_1}\right)_{T,V,N_2} \tag{C.16}$$

By using Eqs. (C.12) and (C.13), we find

$$\left(\frac{\partial U}{\partial N_1}\right)_{T,p,N_2} = \left(\frac{\partial U}{\partial N_1}\right)_{T,V,N_2} + \left(\frac{\partial U}{\partial V}\right)_{T,\mathbf{N}} V_1 \tag{C.17}$$

It follows from Eq. (C.15) that

$$\left(\frac{\partial U}{\partial V}\right)_{T,\mathbf{N}} = \frac{U}{V} - \frac{N_1}{V}\left(\frac{\partial U}{\partial N_1}\right)_{T,V,N_2} - \frac{N_2}{V}\left(\frac{\partial U}{\partial N_2}\right)_{T,V,N_1} \tag{C.18}$$

By introducing this into Eq. (C.17), the partial molar energy becomes

$$\left(\frac{\partial U}{\partial N_1}\right)_{T,p,N_2} = V_1\frac{U}{V} + \frac{N_2}{V}\left[V_2\left(\frac{\partial U}{\partial N_1}\right)_{T,V,N_2} - V_1\left(\frac{\partial U}{\partial N_2}\right)_{T,V,N_1}\right] \tag{C.19}$$

Equation (C.19), together with the analogous equation with the components interchanged, agrees with

$$U = N_1\left(\frac{\partial U}{\partial N_1}\right)_{T,p,N_2} + N_2\left(\frac{\partial U}{\partial N_2}\right)_{T,p,N_1} \tag{C.20}$$

This is the thermodynamic limit expression for the ensemble we are after, with expressions for partial molar properties given above. Partial molar properties of small systems do not obey this equality.

C.2 Partial Enthalpies and Compressional Energies

Equation (11.6) gives minus the partial compressional energy and Eq. (11.7), gives the partial molar enthalpy. In the first section, we showed how to convert $(\partial U/\partial N_1)_{T,V,\mu_2}$ into $(\partial U/\partial N_1)_{T,p,N_2}$. The same transformation can be used to convert $(\partial(pV)/\partial N_i)_{T,V,\boldsymbol{\mu}_i}$ into $(\partial(pV)/\partial N_i)_{T,p,\mathbf{N}_i}$, and $(\partial H/\partial N_i)_{T,V,\boldsymbol{\mu}_i}$ into $(\partial H/\partial N_i)_{T,p,\mathbf{N}_i}$.

We follow the procedure above and use the matrices in Eqs. (C.5) and (C.6). In analogy with Eq. (C.10), we obtain

$$\begin{aligned}
\left(\frac{\partial pV}{\partial N_1}\right)_{T,V,N_2} &= B_{22}\left[A_{22}\left(\frac{\partial pV}{\partial N_1}\right)_{T,V,\mu_2} + A_{12}\left(\frac{\partial pV}{\partial N_2}\right)_{T,V,\mu_1}\right] \\
\left(\frac{\partial pV}{\partial N_2}\right)_{T,V,N_1} &= B_{11}\left[A_{11}\left(\frac{\partial pV}{\partial N_2}\right)_{T,V,\mu_1} + A_{12}\left(\frac{\partial pV}{\partial N_1}\right)_{T,V,\mu_2}\right]
\end{aligned} \tag{C.21}$$

and

$$\left(\frac{\partial H}{\partial N_1}\right)_{T,V,N_2} = B_{22}\left[A_{22}\left(\frac{\partial H}{\partial N_1}\right)_{T,V,\mu_2} + A_{12}\left(\frac{\partial H}{\partial N_2}\right)_{T,V,\mu_1}\right]$$
$$\left(\frac{\partial H}{\partial N_2}\right)_{T,V,N_1} = B_{11}\left[A_{11}\left(\frac{\partial H}{\partial N_2}\right)_{T,V,\mu_1} + A_{12}\left(\frac{\partial H}{\partial N_1}\right)_{T,V,\mu_2}\right] \tag{C.22}$$

The partial molar volumes were defined in Eq. (C.11). We use them and find

$$\left(\frac{\partial pV}{\partial N_1}\right)_{T,p,N_2} = pV_1 + \frac{N_2}{V}\left[V_2\left(\frac{\partial pV}{\partial N_1}\right)_{T,V,N_2} - V_1\left(\frac{\partial pV}{\partial N_2}\right)_{T,V,N_1}\right] \tag{C.23}$$

and

$$\left(\frac{\partial H}{\partial N_1}\right)_{T,p,N_2} = H\frac{V_1}{V} + \frac{N_2}{V}\left[V_2\left(\frac{\partial H}{\partial N_1}\right)_{T,V,N_2} - V_1\left(\frac{\partial H}{\partial N_2}\right)_{T,V,N_1}\right] \tag{C.24}$$

These equations, together with the analogous equations with the components interchanged, agree with

$$pV = N_1\left(\frac{\partial pV}{\partial N_1}\right)_{T,p,N_2} + N_2\left(\frac{\partial pV}{\partial N_2}\right)_{T,p,N_1} \tag{C.25}$$

and

$$H = N_1\left(\frac{\partial H}{\partial N_1}\right)_{T,p,N_2} + N_2\left(\frac{\partial H}{\partial N_2}\right)_{T,p,N_1} \tag{C.26}$$

These are thermodynamic limit expressions for the ensemble that we are seeking, with expressions for partial molar properties above. Partial molar properties of small systems will in general not agree with these equalities.

Appendix D

Interaction Potentials in Atomistic Simulations

In this book, we make extensive use of atomistic simulations in order to compute thermodynamic properties, verify relations and prove assumptions. Several interaction potentials are available for this purpose. State-of-the-art interaction potentials are often trained by applications to artificial neural networks, and can include bond formation as well as different oxidation states. They have an accuracy comparable to that of density functional theory (DFT), but this is not needed when the purpose is to verify theoretical relations.

Classical interaction potentials rely on functional forms of the interaction potential that allow for fast evaluation on regular computer architectures. A common assumption is that non-bonded interactions, can be reduced to pair interactions. This means that the total non-bonded energy of the system can be calculated by consideration of interactions of all pairs in the system as follows:

$$U_{\text{pot}} = \sum_{i} \sum_{j>i} \phi_{\text{pair}}(r_{ij}) \tag{D.1}$$

where ϕ_{pair} is the pair interaction potential.

For the purpose of theory development, and shifted to use one of the four potentials: Lennard-Jones (LJ), cutoff and shifted Lennard-Jones (cut), Weeks–Chandler–Anderson (WCA) and Lennard-Jones/Spline (LJ/s). We present their functional form, and briefly discuss the shape of the potential, and the system-specific parameters that go into these equations.

The LJ potential is the most commonly used interaction potential, with the following functional form:

$$\phi_{\mathrm{LJ}}(r_{ij}) = 4\varepsilon_{ij}\left[\left(\frac{\sigma_{ij}}{r_{ij}}\right)^{12} - \left(\frac{\sigma_{ij}}{r_{ij}}\right)^{6}\right] \tag{D.2}$$

Here ϕ_{LJ} is the particle–particle interaction energy and r_{ij} is the particle–particle distance. There are two type-specific parameters: ε and σ; where the former determines the depth of the potential well, i.e., the lowest value of the interaction energy, and σ is the distance where the interaction energy between the particles is zero. The parameters ε and σ can have various sizes, but in this book, most of the simulations have been run with reduced unit sizes of one, see Table D.1. The LJ-potential is plotted for reduced units, $\varepsilon = \sigma = 1$, in Fig. D.1.

Table D.1 presents the set of reduced units used in the simulations. The particle diameter σ_{11} gives the unit of length, while ε_{11} is the unit of energy. In the case of a mixture, the reduced units are given by those of the first component. In reduced units, the particle diameter, as well as the minimum LJ-potential energy of the particles of the first component are always unity. For more details, see [55].

Table D.1. Reduced units. The basic unit for length is σ_{11}, for energy, ε_{11}, and for mass, m_1.

Dimension	Real units	Reduced units
Length	r	$r^* = r/\sigma_{11}$
Time	t	$t^* = (t/\sigma_{11})\sqrt{\varepsilon_{11}/m_1}$
Mass	M	$m^* = M/m_1$
Temperature	T	$T^* = k_B T/\varepsilon_{11}$
Energy	u	$u^* = u/\varepsilon_{11}$
Pressure	P	$P^* = P\sigma_{11}^3/\varepsilon_{11}$
Density	ρ	$\rho^* = \rho\sigma_{11}^3$

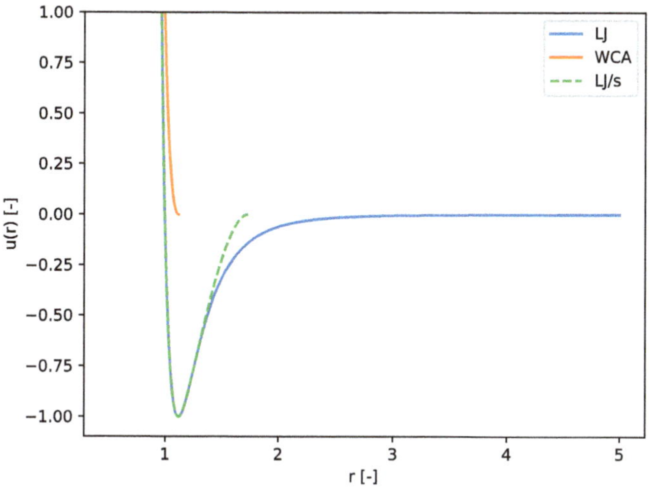

Figure D.1. Lennard-Jones (LJ), Weeks–Chandler–Anderson (WCA) and LJ/s interaction potentials. We show the particle–particle distance, r, on the x-axis, and the pair interaction energy on the y-axis.

The cutoff and shifted LJ potential is given by

$$\phi_{\text{cut}}(r_{ij}) = 4\varepsilon_{ij}\left\{\left[\left(\frac{\sigma_{ij}}{r_{ij}}\right)^{12} - \left(\frac{\sigma_{ij}}{r_{ij}}\right)^{6}\right] - \left[\left(\frac{\sigma_{ij}}{r_{\text{cut}}}\right)^{12} - \left(\frac{\sigma_{ij}}{r_{\text{cut}}}\right)^{6}\right]\right\}\theta(r_{\text{cut}} - r_{ij}) \tag{D.3}$$

where θ is the Heaviside function, which is zero for a negative argument and one for a positive argument, $\phi_{\text{cut}}(r_{ij})$ is zero for $r_{ij} > r_{\text{cut}}$. Common choices for the cutoff are $r_{\text{cut}} = 2.5\sigma$ and 4σ. Because of the finite cutoff distance, it is computationally less demanding than the full LJ-potential.

The Weeks–Chandler–Anderson (WCA) potential is a special case of the cutoff and shifted LJ-potential for which one chooses the cutoff at the minimum $r_{\text{WCA}} = \sigma_{ij}\ (2^{1/6})$ of the LJ-potential.

$$\phi_{\text{WCA}}(r_{ij}) = \varepsilon_{ij}\left\{4\left[\left(\frac{\sigma_{ij}}{r_{ij}}\right)^{12} - \left(\frac{\sigma_{ij}}{r_{ij}}\right)^{6}\right] + 1\right\}\theta(r_{\text{WCA}} - r_{ij}) \tag{D.4}$$

The WCA potential is a purely repulsive potential. In Fig. D.1, it is shown for $\sigma = \varepsilon = 1$. The WCA potential is always positive or zero, and the force between the particles will only be repulsive. It is short-ranged compared with the LJ-potential, as it converges to zero after the cutoff distance. It is used to study the nature of repulsive interactions. Because of the shorter cutoff distance, it is computationally less demanding than the cutoff and shifted LJ-potential for a larger cutoff.

The particles in the case studies interact with a LJ/s potential, which includes a hard core

$$\phi_{\mathrm{LJ/s}}(r_{ij}) = \begin{cases} \infty & \text{if} \quad r < R_{ij} \\ 4\epsilon_{ij}\left[\left(\frac{\sigma_{ij}-R_{ij}}{r-R_{ij}}\right)^{12} - \alpha_{ij}\left(\frac{\sigma_{ij}-R_{ij}}{r-R_{ij}}\right)^{6}\right] & \text{if } R_{ij} < r < r_{s,ij} \\ a_{ij}(r-r_{c,ij})^2 + b_{ij}(r-r_{c,ij})^3 & \text{if} \quad r_s < r < r_{c,ij} \\ 0 & \text{if} \quad r > r_{c,ij} \end{cases} \tag{D.5}$$

Each pair of particles has a hard-core diameter R_{ij} and a soft-core diameter σ_{ij}. The potential and its derivative are continuous in $r = r_{c,ij}$. The parameters a_{ij}, b_{ij} and $r_{s,ij}$ are determined so that the potential and its derivative (the force) are continuous at $r = r_{s,ij}$. The second derivative is continuous and zero at $r = r_{s,ij}$. On the basis of these conditions, we obtain

$$r_{s,ij} = \left(\frac{26}{7\alpha_{ij}}\right)^{1/6}(\sigma_{ij} - R_{ij}) + R_{ij} \tag{D.6}$$

$$r_{c,ij} = \frac{67}{48}\left(\frac{26}{7\alpha_{ij}}\right)^{1/6}(\sigma_{ij} - R_{ij}) + R_{ij} \tag{D.7}$$

$$a_{ij} = -\frac{24192}{3211}\frac{\epsilon_{ij}\alpha_{ij}^2}{(r_{s,ij} - R_{ij})^2} \tag{D.8}$$

$$b_{ij} = -\frac{387072}{61009}\frac{\epsilon_{ij}\alpha_{ij}^2}{(r_{s,ij} - R_{ij})^2} \tag{D.9}$$

The parameter α_{ij} is an interaction parameter used to control the attractive interaction between particles. It varies between zero and one. For small values, the interaction is strongly repulsive.

References

[1] S.K. Schnell, T.J.H. Vlugt, J.-M. Simon, D. Bedeaux, and S. Kjelstrup. "Thermodynamics of a small system in a μT reservoir". *Chem. Phys. Letters* 504 (2011), pp. 199–201. DOI: 10.1016/j.cplett.2011.01.080.

[2] S.K. Schnell, T.J.H. Vlugt, J.-M. Simon, D. Bedeaux, and S. Kjelstrup. "Thermodynamics of small systems embedded in a reservoir: A detailed analysis of finite size effects". *Mol. Phys.* 110 (2012), pp. 1069–1079. DOI: 10.1080/00268976.2011.637524.

[3] T.L. Hill. "Thermodynamics of small systems". *J. Chem. Phys.* 36 (1962), pp. 3182–3197. DOI: 10.1063/1.1732447.

[4] T.L. Hill. *Thermodynamics of Small Systems. Part 1 and 2.* Benjamin, New York, 1963.

[5] N.F.A. van der Vegt, K. Haldrup, S. Roke, J. Zheng, M. Lund, and H.J. Bakker. "Water-mediated ion pairing: Occurrence and relevance". *Chem. Rev.* 116 (2016), pp. 7626–7641. DOI: org/10.1021/acs.chemrev.5b00742.

[6] S.H. Jamali, L. Wolff, T.M. Becker, A. Bardow, T.J.H. Vlugt, and O.A. Moultos. "Finite-size effects of binary mutual diffusion coefficients from molecular dynamics". *J. Chem. Theory Comput.* 14 (2018), pp. 2667–2677. DOI: 10.1021/acs.jctc.8b00170.

[7] J. Milzetti, D. Nayar, and N.F.A van der Vegt. "Convergence of Kirkwood–Buff integrals of ideal and nonideal aqueous solutions using molecular dynamics simulations". *J. Phys. Chem. B* 122 (2018), pp. 5515–5526. DOI: 10.1021/acs.jpcb.7b11831.

[8] R. Cortes-Huerto, K. Kremer, and R. Potestio. "Communication: Kirkwood-Buff integrals in the thermodynamic limit from small-sized molecular dynamics simulations". *J. Chem. Phys.* 145 (2016), p. 141103. DOI: org/10.1063/1.4964779.

[9] H.S. Ashbaugh, K. Weiss, S.M. Williams, B. Meng, and L.N. Surampudi. "Temperature and pressure dependence of methane correlations and osmotic second virial coefficients in water". *J. Phys. Chem. B* 119 (2015), pp. 6280–6294. DOI: 10.1021/acs.jpcb.5b02056.

[10] H.B. Callen. *Thermodynamics and an Introduction to Thermostatistics.* McGraw-Hill, New York, 1960.

[11] J. Smith, H.C. Van Ness, M.M. Abbott, and M. Swihart. *Introduction to Chemical Engineering Thermodynamics.* 8th edn. John Wiley, New York, 2018.

[12] B. Widom. *Statistical Mechanics: A Concise Introduction for Chemists.* Cambridge University Press, 2002.

[13] D.A. McQuarrie. *Statistical Mechanics.* University Science Books, Sausalito, California, 2000.

[14] T.L. Hill. *Thermodynamics of Small Systems.* Dover, New York, 1994.

[15] D. Bedeaux, S. Kjelstrup, and S.K. Schnell. *Nanothermodynamics. General Theory.* Norwegian University of Science and Technology, PoreLab, Trondheim, 2020.

[16] S. Kjelstrup, ed. "Special issue on nanoscale thermodynamics". *Nanomaterials* (2021), p. 584. DOI: 10.3390/books978-3-0365-1169-6.

[17] D. Bedeaux and S. Kjelstrup. "Fluctuation-dissipation theorems for multiphase flow in porous media". *Entropy* 24 (2022), p. 46. DOI: org/10.3390/e24010046.

[18] D. Frenkel and B. Smit. *Understanding Molecular Simulation: From Algorithms to Applications*, Vol. 1. Elsevier, 2001.

[19] R. Clausius. *The Mechanical Theory of Heat.* Taylor and Francis, London, 1867.

[20] J.W. Gibbs. *The Scientific Papers of J.W. Gibbs.* Dover, New York, 1961.

[21] T.L. Hill. "Perspective: Nanothermodynamics". *Nano Letters* 1 (2001), pp. 111–112. DOI: 10.1021/nl010010d.

[22] T.L. Hill. "A different approach to nanothermodynamics". *Nano Letters* 1 (2001), pp. 273–275. DOI: 10.1021/nl010027w.

[23] S.R. de Groot and P. Mazur. *Non-Equilibrium Thermodynamics.* Dover, New York, 1984.

[24] S. Kjelstrup, D. Bedeaux, E. Johannessen, and J. Gross. *Non-Equilibrium Thermodynamics for Engineers.* 2nd edn. World Scientific, Singapore, 2017.

[25] S. Kjelstrup and D. Bedeaux. *Non-Equilibrium Thermodynamics for Heterogeneous Systems.* Series on Advances in Statistical Mechanics. Vol. 20, 2nd edn. World Scientific, Singapore, 2020.

[26] J.M. Rubi, D. Bedeaux, and S. Kjelstrup. "Unifying thermodynamics and kinetic descriptions of single-molecule processes: RNA unfolding under tension". *J. Phys. Chem. B* 111 (2007), pp. 9598–9602. DOI: 10.1021/jp073413w.

[27] E. Bering, S. Kjelstrup, D. Bedeaux, J.M. Rubi, and A.S. de Wijn. "Entropy production beyond the thermodynamic limit from single-molecule stretching simulations". *J. Phys. Chem. B* 124 (2020), pp. 8909–8917. DOI: 10.1021/acs.jpcb.0c05963.

[28] I. Latella, A. Perez-Madrid, A. Campa, L. Casati, and S. Ruffo. "Thermodynamics of non-additive systems". *Phys. Rev. Letters* 114 (2015), p. 230601. DOI: 10.1103/PhysRevLett.114.230601.

[29] D.P. Landau and K. Binder. *A Guide to Monte Carlo Simulations in Statistical Physics.* Cambridge, 2000.

[30] J.I. Siepmann, I.R. McDonald, and D. Frenkel. "Finite-size corrections to the chemical potential". *J. Phys.: Condens. Matt.* 4 (1992), pp. 679–691. DOI: 10.1088/0953-8984/4/3/009.

[31] X. Liu, A. Bardow, and T.J.H. Vlugt. "Multicomponent Maxwell-Stefan diffusivities at infinite dilution". *Ind. Eng. Chem. Res.* 50 (2011), pp. 4776–4782. DOI: 10.1021/ie102515w.

[32] X. Liu, A. Martin-Calvo, E.S. McGarrity, S.K. Schnell, S. Calero, J.-M. Simon, D. Bedeaux, S. Kjelstrup, A. Bardow, and T.J.H. Vlugt. "Fick diffusion coefficients in ternary liquid systems from equilibrium molecular dynamics simulations". *Ind. Eng. Chem. Res.* 51 (2012), pp. 10247–10258. DOI: 10.1021/ie301009v.

[33] P. Krüger, S.K. Schnell, D. Bedeaux, S. Kjelstrup, T.J.H. Vlugt, and J.-M. Simon. "Kirkwood-Buff integrals for finite volumes". *J. Phys. Chem. Letters* 4 (2013), pp. 235–238. DOI: 10.1021/jz301992u.

[34] P. Krüger and T.J.H. Vlugt. "Size and shape dependence of finite volume Kirkwood-Buff integrals". *Fluid Phase Equilibria* 97 (2018), p. 051301. DOI: 10.1103/PhysRevE.97.051301.

[35] N. Dawass, P. Krüger, S.K. Schnell, J.-M. Simon, and T.J.H. Vlugt. "Kirkwood-Buff integrals from molecular simulations". *Fluid Phase Equilibria* 486 (2019), pp. 21–36. DOI: 10.1016/j.fluid.2018.12.027.

[36] D. Bedeaux and S. Kjelstrup. "Hill's nano-thermodynamics is equivalent with Gibbs' thermodynamics for surfaces of constant curvatures". *Chem. Phys. Letters* 707 (2018), pp. 40–43. DOI: 10.1016/j. cplett.2018.07.031.

[37] T.T. Trinh, D. Bedeaux, J.-M. Simon, and S. Kjelstrup. "Calculation of the chemical potential and the activity coefficient of two layers of CO_2 adsorbed on a graphite surface". *Phys. Chem. Chem. Phys.* 17 (2015), pp. 1226–1233. DOI: 10.1039/c4cp03782k.

[38] B.A. Strøm, J. He, S. Kjelstrup, and D. Bedeaux. "When thermodynamic properties of adsorbed films depend on size: Fundamental theory and case study". *Nanomaterials* 19 (2020), pp. 1691–1711. DOI: 10.3390/nano10091691.

[39] B.A. Strøm, D. Bedeaux, and S.K. Schnell. "Adsorption of an ideal gas on a small spherical adsorbent". *Nanomaterials* 19 (2021), pp. 431–441. DOI: 10.3390/nano11020431.

[40] M.T. Rauter, O. Galteland, M. Erdös, O.A. Moultos, T.J.H. Vlugt, S.K. Schnell, D. Bedeaux, and S. Kjelstrup. "Two-phase equilibrium conditions in nanopores". *Nanomaterials* 10 (2020), pp. 1–18. DOI: 10.3390/nano10040608.

[41] O. Galteland, S. Kjelstrup, D. Bedeaux, and B. Hafskjold. "Pressures inside a nano-porous medium. The case of a single phase fluid". *Frontiers Phys.* 7 (2019), pp. 60. DOI: 10.3389/fphy.2019.00060.

[42] S. Kjelstrup, D. Bedeaux, A. Hansen, B. Hafskjold, and O. Galteland. "Non-isothermal transport of multi-phase fluids in porous media. Constitutive equations". *Frontiers Phys.* 6 (2019), p. 150. ISSN: 2296-424X. DOI: 10.3389/fphy.2018.00150.

[43] J. Milzetti, D. Nayar, and N.F.A. van der Vegt. "Convergence of Kirkwood-Buff integrals of ideal and nonideal aqueous solutions using molecular dynamics simulations". *J. Phys. Chem. B* 122 (2018), pp. 5515–5526. DOI: 10.1021/acs.jpcb.7b11831.

[44] P. Ganguly and N.F.A. van der Vegt. "Convergence of sampling Kirkwood-Buff integrals of aqueous solutions with molecular dynamics simulations". *J. Chem. Theory Comput.* 9 (2013), pp. 1347–1355. DOI: 10.1021/ct301017q.

[45] N. Dawass, P. Krüger, S.K. Schnell, D. Bedeaux, S. Kjelstrup, J.-M. Simon, and T.J.H. Vlugt. "Finite-size effects of Kirkwood-Buff integrals from molecular simulations". *Molecular Simulations* 44 (2018), pp. 599–612. DOI: 10.1080/08927022.2017.1416114.

[46] R. Skorpa, J.-M. Simon, D. Bedeaux, and S. Kjelstrup. "The reaction enthalpy of hydrogen dissociation calculated with the Small System Method from the simulation of molecular fluctuations". *Phys. Chem. Chem. Phys.* 16 (2014), pp. 19681–19693. DOI: 10.1039/C4CP02768J.

[47] S.K. Schnell, R. Skorpa, D. Bedeaux, S. Kjelstrup, T.J.H. Vlugt, and J.-M. Simon. "Partial molar enthalpies and reaction enthalpies from equilibrium molecular dynamics simulation". *J. Chem. Phys.* 141 (2014), p. 144501. DOI: 10.1063/1.4896939.

[48] B.A. Strøm, J.-M. Simon, S.K. Schnell, S. Kjelstrup, J. He, and D. Bedeaux. "Size and shape effects on the thermodynamic properties of nanoscale volumes of water". *Phys. Chem. Chem. Phys.* 19 (2017), pp. 9016–9027. DOI: 10.1039/C7CP00874K.

[49] M. Erdös, O. Galteland, D. Bedeaux, S. Kjelstrup, O.A. Moultos, and T.J.H. Vlugt. "Gibbs ensemble Monte Carlo simulation of fluids in confinement: Relation between the differential and integral pressures". *Nanomaterials* 10 (2020), pp. 293–778. DOI: 10.3390/nano10020293.

[50] W. Dong. "Thermodynamics of interfaces extended to nanoscales by introducing integral and differential surface tensions". *PNAS* 118 (2021), e2019873118. DOI: 10.1073/pnas.2019873118.

[51] O. Galteland, D. Bedeaux, and S. Kjelstrup. "Nanothermodynamic description and molecular simulation of a single-phase fluid in a slit pore". *Nanomaterials* 11 (2021), pp. 165–185. DOI: org/10.3390/nano11010165.

[52] R.C. Tolman. "The effect of droplet size on surface tension". *J. Chem. Phys.* 17 (1949), pp. 333–337. DOI: 10.1063/1.1747247.

[53] W. Helfrich. "Elastic properties of lipid bilayers: theory and possible experiments". *Z. Naturforsch C.* 28c (1973), pp. 693–703. DOI: 10.1515/znc-1973-11-1209.

[54] E. Bering, D. Bedeaux, S. Kjelstrup, A.S. de Wijn, I. Latella, and J.M. Rubi. "A Legendre–Fenchel transform for molecular stretching energies". *Nanomaterials* 10 (2020), p. 2355. DOI: 10.3390/nano10122355.

[55] D. Frenkel and B. Smit. *Understanding Molecular Simulations.* Academic Press, 2002.

[56] R. Skorpa, J.-M. Simon, D. Bedeaux, and S. Kjelstrup. "Equilibrium properties of the reaction $H_2 \rightleftharpoons 2H$". *Phys. Chem. Chem. Phys.* 6 (2014), pp. 1227–1237. DOI: 10.1039/C3CP54149E.

[57] G. Ciccotti and J.P. Ryckaert. "Molecular dynamics simulation of rigid molecules". *Computer Physics Reports* 4 (1986), pp. 345–392. DOI: 10.1016/0167-7977(86)90022-5.

[58] H. Touchette. "The large deviation approach to statistical mechanics". *Phys. Rep.* 478 (2009), pp. 1–69. DOI: 10.1016/j.physrep.2009.05.002.

[59] O. Galteland, E. Bering, K.R. Kristiansen, D. Bedeaux, and S. Kjelstrup. "Legendre-Fenchel transform capture layering transitions in porous media". *Nanoscale Advanced* (2022), pp. 165–185. DOI: 10.1039/D1NA00846C.

[60] D. Bedeaux and P. Mazur. "Mesoscopic non-equilibrium thermodynamics in quantum systems". *Physica A* 298 (2001), pp. 81–100. DOI: 10.1016/S0378-4371(01)00223-0.

[61] C.Z. Qiao, S.L. Zhao, H.L. Liu, and W. Dong. "Connect the thermodynamics of bulk and confined fluids: Confinement-adsorption scaling". *Langmuir* 35 (2019), pp. 3840–3847. DOI: 10.1021/acs.langmuir.8b03126.

[62] V. Bråten, Ø. Wilhelmsen, and S.K. Schnell. "Chemical potential differences in the macroscopic limit from fluctuations in small systems". *J. Chem. Inf. Model.* 61 (2021), pp. 840–855. DOI: org/10.1021/acs.jcim.0c01367.
[63] H. Hadwiger. *Vorlesungen uber Inhalt, Oberflache und Isoperimentrie.* Springer, Berlin, 1957.
[64] E.M. Blokhuis. "Existence of a bending rigidity for a hard-sphere liquid near a curved hard wall: Validity of the Hadwiger theorem". *Phys. Rev. E* 87 (2013), p. 022401. DOI: 10.1103/Phys RevE.87.022401.
[65] J.D. Weeks, D. Chandler, and H.C. Andersen. "Role of repulsive forces in determining the equilibrium structure of simple liquids". *J. Chem. Phys.* 54 (1971), pp. 5237–5247. DOI: 10.1063/1.1674820.
[66] S. Plimpton. "Fast parallel algorithms for short-range molecular dynamics". *J.Comp. Phys.* 117 (1995), pp. 1–19. DOI: 10.1006/jcph.1995.1039.
[67] S.K. Schnell, X. Liu, J.-M. Simon, A. Bardow, D. Bedeaux, T.J.H. Vlugt, and S. Kjelstrup. "Calculating thermodynamic properties from fluctuations at small scales". *J. Phys. Chem. B* 115 (2011), pp. 10911–10918. DOI: 10.1021/jp204347p.
[68] X. Liu, S.K. Schnell, J.-M. Simon, D. Bedeaux, S. Kjelstrup, A. Bardow, and T.J.H. Vlugt. "Fick diffusion coefficients of liquid mixtures directly obtained from equilibrium molecular dynamics". *J. Phys. Chem. B* 115 (2011), pp. 12921–12929. DOI: 10.1021/jp208360s.
[69] V. Bråten, D. Bedeaux, Ø. Wilhelmsen, and S.K. Schnell. "Small size effects in open and closed systems: What can we learn from ideal gases about systems with interacting particles?" *J. Chem. Phys.* 155 (2021), pp. 244504–17. DOI: org/10.1063/5.0076684.
[70] J.G. Kirkwood and F.P. Buff. "The statistical mechanical theory of solutions. I". *J. Chem. Phys.* 19 (1951), pp. 774–778. DOI: 10.1063/1.1748352.
[71] A. Ben-Naim. *A Molecular Theory of Solutions.* Oxford University Press, 2006.
[72] X. Liu, S.K. Schnell, J.-M. Simon, P. Krüger, D. Bedeaux, S. Kjelstrup, A. Bardow, and T.J.H. Vlugt. "Diffusion coefficients from molecular dynamics simulations in binary and ternary mixtures". *Int. J. Thermophys.* 34 (2013), pp. 1169–1196. DOI: 10.1007/s10765-013-1482-3.
[73] I. Oppenheim and P. Mazur. "Virial expansion for finite closed systems; canonical ensembles". *Physica* XXIII (1957), pp. 197–215.

[74] S.H. Jamali, L. Wolff, T.M. Becker, M. de Groen, M. Ramdin, R. Hartkamp, A. Bardow, T.J.H. Vlugt, and O.A. Moultos. "OCTP: A tool for on-the-fly calculation of transport properties of fluids with the order-n algorithm in LAMMPS". *J. Chem. Inf. Model.* 59 (2019), pp. 1290–1294. DOI: 10.1021/acs.jcim.8b00939.

[75] M. Heidari, K. Kremer, R. Potestio, and R. Cortes-Huerto. "Fluctuations, finite-size effects and the thermodynamic limit in computer simulations: Revisiting the spatial block analysis method". *Entropy* 20 (2018), pp. 222–16. DOI: 10.3390/e20040222.

[76] N. Dawass, P. Krüger, S.K. Schnell, O.A. Moultos, I.G. Economou, T.J.H. Vlugt, and J.-M. Simon. "Kirkwood-Buff integrals using molecular simulation: Estimation of surface effects". *Nanomaterials* 10 (2020), pp. 771–788. DOI: 10.3390/nano10040771.

[77] X. Liu, T.J.H. Vlugt, and A. Bardow. "Maxwell-Stefan diffusivities in liquid mixtures: Using molecular dynamics for testing model predictions". *Fluid Phase Equilibria* 301 (2011), pp. 110–117. DOI: 10.1016/j.fluid.2010.11.019.

[78] A. Anbari, H.-T. Chien, S.S. Datta, W. Deng, D.A. Weitz, and J. Fan. "Microfluidic model porous media: Fabrication and applications". *Small* 14 (2018), pp. 1703575–15. DOI: 10.1002/smll.201703575.

[79] G.D.C. Kuiken. *Thermodynamics of Irreversible Processes: Applications to Diffusion and Rheology.* Wiley, New York, 1994.

[80] R. Taylor and R. Krishna. *Multicomponent Mass Transfer.* Wiley, New York, 1993.

[81] E.L. Cussler. *Diffusion, Mass Transfer in Fluid Systems*, 2nd edn. Cambridge University Press, 1997.

[82] X. Liu, T.J.H. Vlugt, and A. Bardow. "Maxwell-Stefan diffusivities in binary mixtures of ionic liquids with dimethyl sulfoxide (DMSO) and H_2O". *J. Phys. Chem. B* 115 (2011), pp. 8506–8517. DOI: 10.1021/jp203026c.

[83] R. Krishna and J.M. van Baten. "The darken relation for multicomponent diffusion in liquid mixtures of linear alkanes: An investigation using molecular dynamics (MD) simulations". *Ind. Eng. Chem. Res.* 44 (2005), p. 6939. DOI: 10.1021/ie050146c.

[84] L. Onsager. "Reciprocal relations in irreversible processes. I". *Phys. Rev.* 37 (1931), pp. 405–426. DOI: 10.1103/PhysRev.37.405.

[85] L. Onsager. "Reciprocal relations in irreversible processes. II". *Phys. Rev.* 38 (1931), pp. 2265–2279. DOI: 10.1103/PhysRev.38.2265.

[86] B. Poling, J. Prausnitz, and J. O'Connell. *The Properties of Gases and Liquids.* McGraw-Hill, New York, 2004.

[87] J.M. Ortiz de Zarate and J.V. Sengers. "Frame-invariant Fick diffusion matrices of multicomponent fluid mixtures". *Phys. Chem. Chem. Phys.* 22 (2020), pp. 17597–17604. DOI: 10.1039/d0cp01110j.
[88] R. Krishna and J.A. Wesselingh. "The Maxwell-Stefan approach to mass transfer". *Chem. Eng. Sci.* 52 (1997), pp. 861–911. DOI: 10.1016/S0009-2509(96)00458-7.
[89] R. Krishna and J.M. van Baten. "Diffusion of alkane mixtures in zeolites: Validating the Maxwell-Stefan formulation using MD simulations". *J. Phys. Chem. B* 109 (2005), pp. 6386–6396. DOI: 10.1021/jp044257.
[90] A. Bardow, E. Kriesten, M.A. Voda, F. Casanova, B. Blümich, and W. Marquardt. "Prediction of multicomponent mutual diffusion in liquids: Model discrimination using NMR data". *Fluid Phase Equilib.* 278 (2009), pp. 27–35. DOI: 10.1016/j.fluid.2009.01.001.
[91] A.P. Hardt, D.K. Anderson, R. Rathbun, B.W. Mar, and A.L.J. Babb. "Self-diffusion in liquids. II. Comparison between mutual and self- diffusion coefficients". *Phys. Chem.* 63 (1959), pp. 2059–2061. DOI: 10.1021/j150582a021.
[92] A. Perera, L. Zoranic, F. Sokolic, and R.J. Mazighi. "A comparative Molecular Dynamics study of water–methanol and acetone–methanol mixtures". *J. Molecular Liquids* 159 (2011), pp. 52–59. DOI: 10.1016/j.molliq.2010.05.006.
[93] J. Gmehling and U. Onken. *Vapor-Liquid Equilibrium Data Collection.* DECHEMA, Frankfurt, Germany, 2005.
[94] C.S. Schacht, T.J.H. Vlugt, and J. Gross. "Using an analytic equation of state to obtain quantitative solubilities of CO_2 by molecular simulation". *J. Phys. Chem. Lett.* 2 (2011), pp. 393–396. DOI: 10.1021/jz101688h.
[95] B. Hafskjold and S. K. Ratkje. "Criteria for local equilibrium in a system with transport of heat and mass". *J. Stat. Phys.* 78 (1995), pp. 463–497. DOI: 10.1007/BF02183360.
[96] S.K. Schnell, T.J.H. Vlugt, J.-M. Simon, D. Bedeaux, and S. Kjelstrup. "Thermodynamics of a small system in a μ, T reservoir". *Chem. Phys. Lett.* 504 (2011), pp. 199–201. DOI: 10.1016/j.cplett.2011.01.080.
[97] A. Saul and W. Wagner. "A fundamental equation for water covering the range from the melting line to 1273 K at pressures up to 25 000 MPa". *J. Phys. Chem. Ref. Data* 18 (Oct. 1989), pp. 1537–1564. DOI: 10.1063/1.555836.
[98] K. Sawada and T. Fujimoto. "Effective ionization and dissociation rate coefficients of molecular hydrogen in plasma". *J. Appl. Phys.* 78 (1995), pp. 2913–2924. DOI: http://dx.doi.org/10.1063/1.360037.

[99] W.R. Magro, D.M. Ceperley, C. Pierleoni, and B. Bernu. "Molecular dissociation in hot, dense hydrogen". *Phys. Rev. Lett.* 76 (1996), pp. 1240–1243.

[100] T.L. Ward and T. Dao. "Model of hydrogen permeation behavior in palladium membranes". *J. Mem. Sci.* 153 (1999), pp. 211–231.

[101] S. Yun and S.T. Oyama. "Correlations in palladium membranes for hydrogen separation: A review". *J. Mem. Sci.* 375 (2011), pp. 28–45. DOI: 10.1016/j.memsci.2011.03.057.

[102] R. Skorpa, M. Voldsund, M. Takla, S.K. Schnell, D. Bedeaux, and S. Kjelstrup. "Assessing the coupled heat and mass transport of hydrogen through a palladium membrane". *J. Mem. Sci.* 394–395 (2012), pp. 131–139. DOI: 10.1016/j.memsci.2011.12.033.

[103] T.P. Perng and J.K. Wu. "A brief review note on mechanisms of hydrogen entry into metals". *Materials Letters* 57 (2003), pp. 3437–3438. DOI: 10.1016/S0167-577X(03)00095-8.

[104] D. Kohen, J.C. Tully, and F.H. Stillinger. "Modeling the interaction of hydrogen with silicon surfaces". *Surface Science* 397 (1998), pp. 225–236. DOI: 10.1016/S0039-6028(97)00739-5.

[105] F.H. Stillinger and T.A. Weber. "Molecular dynamics simulation for chemically reactive substances. Fluorine". *J. Chem. Phys.* 8 (1988), pp. 5123–5133. DOI: 10.1063/1.454665.

[106] J. Xu, S. Kjelstrup, and D. Bedeaux. "Molecular dynamics simulations of a chemical reaction; conditions for local equilibrium in a temperature gradient". *Phys. Chem. Chem. Phys.* 8 (2006), pp. 2017– 2027.

[107] J. Xu, S. Kjelstrup, D. Bedeaux, and J.-M. Simon. "The heat of transfer in a chemical reaction at equilibrium". *J. Non-Equilibr. Thermodyn.* 32 (2007), pp. 341–349. DOI: 10.1515/JNETDY.2007.026.

[108] D.L. Diedrich and J.B. Anderson. "An accurate quantum Monte Carlo calculation of the barrier height for the reaction $H + H_2 = H_2 + H$". *Science* 258 (1992), pp. 786–788.

[109] D.L. Diedrich and J.B. Anderson. "Exact quantum Monte Carlo calculations of the potential energy surface for the reaction $H + H_2 \rightarrow H2 + H$". *J. Chem. Phys.* 100 (1994), pp. 8089–8095.

[110] D.A. Reed and G. Erlich. "Surface diffusion, atomic jump rates and thermodynamics". *Surface Science* 102 (1981), pp. 588–609. ISSN: 0039–6028. DOI: 10.1016/0039-6028(81)90048-0.

[111] K.S. Shing and S.-T. Chung. "Calculation of infinite-dilution partial molar properties by computer simulation". *AIChE Journal* 34 (1988), pp. 1973–1980.

[112] A.C.T. van Duin, S. Dasgupta, F. Lorant, and W.A. Goddard III. "ReaxFF: A reactive force field for hydrocarbons". *J. Phys. Chem. A* 105 (2001), pp. 9396–9409. DOI: 10.1021/jp004368u.

[113] L.P.H. Jeurgens, Z.M. Wang, and E.J. Mittemeijer. "Thermodynamics of reactions and phase transformations at interfaces and surfaces". *Int. J. Mater. Res.* 100 (2009), pp. 1281–1307. DOI: 10.3139/146.110204.

[114] R. Bradley Shumbera, H.H. Kan, and J.F. Weaver. "Oxidation of Pt(1 0 0)-hex-R0.7° by gas-phase oxygen atoms". *Surf. Sci.* 601 (2007), pp. 235–246. DOI: 10.1016/j.susc.2006.09.040.

[115] F. Costache, S. Kouteva-Arguirova, and J. Reif. "Sub-damage-threshold femtosecond laser ablation from crystalline Si: surface nanostructures and phase transformation". *Applied Physics A* 79 (2004), pp. 1429– 1432. DOI: 10.1007/S00339-004-2803-Y.

[116] M.P. Cox, G. Ertl, R. Imbihl, and J. Rustig. "Catalytic CO oxidation on Pt surfaces: A lattice-gas cellular automaton model". *Surf. Sci.* 134 (1983), pp. L517–L523. DOI: 10.1016/0378-4371(92)90274-T.

[117] I. Inzoli, J.-M. Simon, S. Kjelstrup, and D. Bedeaux. "Thermal diffusion and partial molar enthalpy variations of n-butane in silicalite-1". *J. Phys. Chem. B* 112 (2008), pp. 14937–14951. DOI: 10.1021/jp804778u.

[118] I. Inzoli, S. Kjelstrup, D. Bedeaux, and J.-M. Simon. "Transport coefficients of n-butane into and through the surface of silicalite-1 from non-equilibrium molecular dynamics study". *Microporous Mesoporous Materials* 125 (2009), pp. 112–125. DOI: 10.1016/j.micromeso.2008.12.030.

[119] X. He, J. Arvid Lie, E. Sheridan, and M.-B. Hägg. "CO_2 capture by hollow fibre carbon membranes: Experiments and process simulations". *Energy Procedia* 1 (2009), pp. 261–268. DOI: 10.1016/j.egypro.2009.01.037.

[120] J. Collell and G. Galliero. "Determination of the thermodynamic correction factor of fluids confined in nano-metric slit pores from molecular simulation". *J. Chem. Phys.* 140 (2014), p. 194702. DOI: 0.1063/1.4875703.

[121] T.T. Trinh, T.J.H. Vlugt, M.B. Hägg, D. Bedeaux, and S. Kjelstrup. "Selectivity and self diffusion of carbon dioxide and hydrogen in a mixture on a graphite surface". *Front. Chem.* 1 (2013), p. 38. DOI: 10.3389/fchem.2013.00038.

[122] T.S. van Erp, T.T. Trinh, S. Kjelstrup, and K.S. Glavatskiy. "On the relation between the Langmuir and thermodynamic flux equations". *Front. Phys.* 1 (2009), p. 30. DOI: 10.3389/fphy.2013.00036.

[123] J.J. Potoff and J.I. Siepmann. "Vapor–liquid equilibria of mixtures containing alkanes, carbon dioxide, and nitrogen". *Am. I. Chem. Eng.* 47 (2001), pp. 1676–1682. DOI: 10.1002/aic.690470719.

[124] M.P. Allen and D.J. Tildesley. *Computer Simulation of Liquids.* Oxford University Press, New York, 1987.

[125] O. Boucher, D. Randall, P. Artaxo, C. Bretherton, G. Feingold, P. Forster, V.-M. Kerminen, Y. Kondo, H. Liao, P. Lohmann, P. Rasch, S.K. Satheesh, S. Sherwood, B. Stevens, and X.Y. Zhang. *Climate Change 2013: The Physical Science Basis. Contribution of Working Group I to the Fifth Assessment Report of the Intergovernmental Panel on Climate Change. Chapter 7–Clouds and Areosols.* Cambridge University Press, Cambridge, UK, 2013, pp. 571–657. DOI: 10.1017/CBO9781107415324.016.

[126] A.I. Rusanov. "The thermodynamics of processes of new phase formation". *Russ. Chem. Rev.* 33 (1964), pp. 385–399. DOI: 10.1070/RC1964v033n07ABEH001432.

[127] J.W.P. Schmelzer, I. Gutzow, and J. Schmelzer. "Curvature-dependent surface tension and nucleation theory". *J. Colloid Interface Sci.* 178 (1996), pp. 657–665. DOI: 10.1006/jcis.1996.0163.

[128] D. Reguera and H. Reiss. "Nucleation in confined ideal binary mixtures: The Renninger-Wilemski problem revisited". *J. Chem. Phys.* 119 (2003), pp. 1533–1546. DOI: 10.1063/1.1579685.

[129] J.W.P. Schmelzer, G.S. Boltachev, and V.G. Baidakov. "Classical and generalized Gibbs' approaches and the work of critical cluster formation in nucleation theory". *J. Chem. Phys.* 124 (2006), p. 194503. DOI: 10.1063/1.2196412.

[130] V.G. Dubrovskii. "Refinement of nucleation theory for vapor-liquid-solid nanowires". *Cryst. Growth Des.* 17 (2017), pp. 2589–2593. DOI: 10.1021/acs.cgd.7b00124.

[131] A. Aasen, E.M. Blokhuis, and Ø. Wilhelmsen. "Tolman lengths and rigidity constants of multicomponent fluids: Fundamental theory and numerical examples". *J. Chem. Phys.* 148 (2018), p. 204702. DOI: 10.1063/1.5026747.

[132] R.H. Fowler and E.A. Guggenheim. *Statistical Thermodynamics: A Version of Statistical Mechanics for Students of Physics and Chemistry.* New York, Macmillan, 1939.

[133] A.L. Myers and J.M. Prausnitz. "Thermodynamics of mixed-gas adsorption". *AIChE J.* 11 (1965), pp. 121–127. DOI: 10.1002/aic.690110125.

[134] T.L. Hill. "Statistical mechanics of adsoption. IX. Adsorption thermodynamics and solution thermodynamics". *J. Chem. Phys.* 18 (1950), pp. 246–256. DOI: 10.1063/1.1747615.

[135] G. Mie. "Zur kinetischen Theorie der einatomigen Körper". *Ann. Phys.* 316 (1903), pp. 657–697. DOI: 10.1002/andp.19033160802.

[136] C. Avendano, T. Lafitte, A. Galindo, C.S. Adjiman, G. Jackson, and E.A. Müller. "SAFT-γ force field for the simulation of molecular fluids. 1. A single-site coarse grained model of carbon dioxide". *J. Phys. Chem. B* 115 (2011), pp. 11154–11169. DOI: 10.1021/jp204908d.

[137] S.C. Graham, J.B. Homer, and J.L.J. Rosenfeld. "The formation and coagulation of soot aerosols generated by the pyrolysis of aromatic hydrocarbons". *Proc. R. Soc. London. Ser. A Math. Phys. Sci.* 344 (1975), pp. 259–285. DOI: 10.1098/rspa.1975.0101.

[138] P. Virtanen, R. Gommers, T.E. Oliphant, M. Haberland, T. Reddy, D. Cournapeau, E. Burovski, P. Peterson, W. Weckesser, J. Bright, S.J. van der Walt, B. Brett, J. Wilson, K.J. Millman, N. Mayorov, A.R.J. Nelson, E. Jones, R. Kern, E. Larson, C.J. Carey, I. Polat, Y. Feng, E.W. Moore, J. VanderPlas, D. Laxalde, J. Perktold, R. Cimrman, I. Henriksen, E.A. Quintero, C.R. Harris, A.M. Archibald, A.H. Ribeiro, F. Pedregosa, and P. van Mulbregt. "SciPy 1.0: Fundamental algorithms for scientific computing in python". *Nature Methods* 17 (2020), pp. 261–272. DOI: 10.1038/s41592-019-0686-2.

[139] J. Bear. *Dynamics of Fluids in Porous Media.* Dover, New York, 1988.

[140] S.M. Hassanizadeh and W.G. Gray. "Pressure tensor for inhomogeneous fluids". *Adv. Water Resour.* 13 (1990), pp. 169–186. DOI: 10.1016/0309-1708(90)90040-B.

[141] W.G. Gray and S.M. Hassanizadeh. "Macroscale continuum mechanics for multiphase porous-media flow including phases, interfaces". *Adv. Water Resour.* 21 (1998), pp. 261–281. DOI: 10.1016/S0309-1708(96)00063-2.

[142] R. Hilfer. "Macroscopic equations of motion for two phase flow in porous media". *Phys. Rev. E* 58 (1998), pp. 2090–2096. DOI: 10.1103/PhysRevE.58.2090.

[143] R. Helmig. *Multiphase Flow and Transport Processes in the Subsurface.* Springer, Berlin, 1997, p. 367.

[144] S. Kjelstrup, D. Bedeaux, A. Hansen, B. Hafskjold, and O. Galteland. "Non-isothermal transport of multi-phase fluids in porous media. The entropy production". *Front. Phys.* 6 (2018), p. 126. DOI: 10.3389/fphy.2018.00126.

[145] H.H. Khanamiri, C.F. Berg, P.A. Slotte, S. Schlüter, and O. Torsæter. "Description of free energy for immiscible two-fluid flow in porous media by integral geometry and thermodynamics". *Water Resource Res.* 48 (2018), pp. 9045–9059. DOI: 10.1029/2018WR023619.

[146] J. McClure, T. Ramstad, Z. Li, R.T. Armstrong, and S. Berg. "Modeling geometric state for fluids in porous media: Evolution of the Euler characteristic". In: *Transp. Porous Media* 133 (2020), pp. 229–250. DOI: 10.1007/s11242-020-01420-1.
[147] S. Sinha, A. Hansen, D. Bedeaux, and S. Kjelstrup. "Effective rheology of bubbles moving in a capillary tube". *Phys. Rev. E* 78 (2013), pp. 025001–5. DOI: 10.1103/PhysRevE.87.025001.
[148] I. Savani, S. Sinha, A. Hansen, D. Bedeaux, S. Kjelstrup, and M. Vassvik. "A Monte Carlo algorithm for immiscible two-phase flow in porous media". *Transp. Porous Media* 116 (2017), pp. 869–888. DOI: 10.1007/s11242-016-0804-x.
[149] M. Winkler, M.Aa. Gjennestad, D. Bedeaux, S. Kjelstrup, R. Cabriolu, and A. Hansen. "Onsager-symmetry obeyed in athermal mesoscopic systems: Two-phase flow in porous media". *Front. Phys.* 8 (2020), pp. 60–68. DOI: 10.3389/fphy.2020.00060.
[150] S. Kjelstrup, D. Bedeaux, I. Inzoli, and J.-M. Simon. "Criteria for validity of thermodynamic equations from non-equilibrium molecular dynamics simulations". *Energy* 33 (2008), pp. 1185–1190. DOI: 10.1016/j.energy.2008.04.005.
[151] X. Dong, H. Liu, J. Hou, K. Wu, and Z. Chen. "Phase equilibria of confined fluids in nanopores of tight and shale rocks considering the effect of capillary pressure and adsorption film". *Ind. Eng. Chem. Res.* 55 (2016), pp. 798–811. DOI: 10.1021/acs.iecr.5b04276.
[152] K.E. Gubbins, Y. Long, and M. Śliwinske-Bartkowiak. "Thermodynamics of confined nano-phases". *J. Chem. Thermodyn.* 74 (2014), pp. 169–183. DOI: 10.1016/j.jct.2014.01.024.
[153] N. Giovambattista, P.J. Rossky, and P.G. Debenedetti. "Phase transitions induced by nanoconfinement in liquid water". *Phys. Rev. Lett.* 102 (2009), p. 050603. DOI: 10.1103/PhysRevLett.102.050603.
[154] E. Keshavarzi and M. Kamalvand. "Energy effects on the structure and thermodynamic properties of nanoconfined fluids (a density functional theory study)". *J. Phys. Chem. B* 113 (2009), pp. 5493–5499. DOI: 10.1021/jp808466p.
[155] E. Braun, J.J. Chen, S.K. Schnell, L.C. Lin, J.A. Reimer, and B. Smit. "Nanoporous materials can tune the critical point of a pure substance". *Angew. Chemie — Int. Ed.* 54 (2015), pp. 14349–14352. DOI: 10.1002/anie.201506865.
[156] J.G. Pharoah, K. Karan, and W. Sun. "On effective transport coefficients in PEM fuel cell electrodes: Anisotropy of the porous trans- port layers". *J. Power Sources* 161 (2006), pp. 214–224. DOI: 10.1016/j.jpowsour.2006.03.093.

[157] J. Newman and W. Tiedemann. "Porous-electrode theory with battery applications". *AIChE J.* 21 (1975), pp. 25–41. DOI: 10.1002/aic.690210103.

[158] H. Guo, Y. Wyart, J. Perot, F. Nauleau, and P. Moulin. "Low-pressure membrane integrity tests for drinking water treatment: A review". *Water Res.* 44 (2010), pp. 41–57. DOI: 10.1016/j.watres.2009.09.032.

[159] A.M. Alklaibi and N. Lior. "Transport analysis of air-gap membrane distillation". *J. Membr. Sci.* 255 (2005), pp. 239–253. DOI: 10.1016/j.memsci.2005.01.038.

[160] N. Kuipers, J.H. Hanemaaijer, H. Brouwer, J. van Medevoort, A. Jansen, F. Altena, P. van der Vleuten, and H. Bak. "Simultaneous production of high-quality water and electrical power from aqueous feedstocks and waste heat by high-pressure membrane distillation". *Desalin. Water Treat.* 55 (2015), pp. 2766–2776. DOI: 10.1080/19443994.2014.946724.

[161] H. Eslami and N. Mehdipour. "Local chemical potential and pressure tensor in inhomogeneous nanoconfined fluids". *J. Chem. Phys.* 137 (2012), p. 144702. DOI: 10.1063/1.4757016.

[162] L.S. Bennethum and T. Weinstein. "Three pressures in porous media". *Transp. Porous Media* 54 (2004), pp. 1–34. DOI: 10.1023/A:1025701922798.

[163] D.H. Tsai. "The virial theorem and stress calculation in molecular dynamics". *J. Chem. Phys.* 70 (1979), pp. 1375–1382. DOI: 10.1063/1.437577.

[164] J.G. Kirkwood and F.P. Buff. "The statistical mechanical theory of surface tension". *J. Chem. Phys.* 17 (1949), pp. 338–347. DOI: 10.1063/1.1747248.

[165] B. Hafskjold, K.P. Travis, A.B. Hass, M. Hammer, A. Aasen, and Ø. Wilhelmsen. "Thermodynamic properties of the 3D Lennard-Jones/spline model". *Mol. Phys.* 117 (2019), pp. 3754–3769. DOI: 10.1080/00268976.2019.1664780.

[166] S. Plimpton. "Fast parallel algorithms for short — range molecular dynamics". *J. Comput. Phys.* 117 (1995), pp. 1–19. DOI: 10.1006/jcph.1995.1039.

[167] C. Braga and K.P. Travis. "A configurational temperature Nosé-Hoover thermostat". *J. Chem. Phys.* 123 (2005), p. 134101. DOI: 10.1063/1.2013227.

[168] B. Hafskjold and T. Ikeshoji. "Microscopic pressure tensor for hardsphere fluids". *Phys. Rev. E* 66 (2002), p. 011203. DOI: 10.1103/PhysRevE.66.011203.

[169] T. Ikeshoji, B. Hafskjold, and H. Furuholt. "Molecular-level calculation scheme for pressure in inhomogeneous systems of flat and spherical layers". *Mol. Sim.* 29 (2003), pp. 101–109. DOI: 10.1080/102866202100002518a.

[170] B. Todd, D.J. Evans, and P.J. Daivis. "Pressure tensor for inhomogeneous fluids". *Phys. Rev. E* 52 (1995), p. 1627. DOI: 10.1103/PhysRevE.52.1627.

[171] B.D. Todd and P.J. Daivis. *Nonequilibrium Molecular Dynamics: Theory, Algorithms and Applications.* Cambridge University Press, Cambridge, 2017.

[172] T. Holt. "Transport and equilibrium properties of a cation exchange membrane". PhD thesis. Institute of Physical Chemistry, Norwegian Institute of Technology, University of Trondheim, 1983.

[173] K.S. Førland, T. Førland, and S. Kjelstrup Ratkje. *Irreversible Thermodynamics. Theory and Application.* Wiley, Chichester, W1988.

[174] H.B. Callen and T.A. Welton. "Irreversibility and generalized noise". *Phys. Rev.* 83 (1951), p. 34. DOI: 10.1103/PhysRev.83.34.

[175] M.S. Green. "Markoff random processes and the statistical mechanics of time-dependent phenomena. II. Irreversible processes in fluids". *J. Chem. Phys.* 22 (1954), p. 398. DOI: 10.1063/1.1740082.

[176] R. Kubo. "The fluctuation-dissipation theorem". *Rep. Prog. Phys.* 29 (1966), p. 255. DOI: 10.1088/0034-4885/29/1/306.

[177] J.M. Ortiz de Zarate and J.V. Sengers. *Hydrodynamic Fluctuations in Fluids and Fluid Mixture.* Elsevier, Amsterdam, 2006, p. 309.

[178] A. Hansen and T. Ramstad. "Towards a thermodynamics of immiscible two-phase steady-state flow in porous media". *Comput. Geosci.* 13 (2009), pp. 227–234. DOI: 10.1007/s10596-008-9109-7.

[179] M. Erpelding, S. Sinha, K.T. Tallakstad, A. Hansen, E.G. Flekkøy, and K.J. Måløy. "History independence of steady state in simultaneous two-phase flow through two-dimensional porous media". *Phys. Rev. E* 88 (2013), p. 053004. DOI: 10.1007/s11242-018-1139-6.

[180] A. Hansen, S. Sinha, D. Bedeaux, S. Kjelstrup, M.Aa. Gjennestad, and M. Vassvik. "Relations between seepage velocities in immiscible, incompressible two-phase flow in porous media". *Transport in Porous Media* 125 (2018), pp. 565–587. DOI: 10.1007/s11242-018-1139-6.

[181] W.G. Gray and C.T. Miller. "Thermodynamically constrained averaging theory approach for modeling flow and transport phenomena in porous medium systems: 8. Interface and common curve dynamics". *Adv. Water Resour.* 33 (2010), pp. 1427–1443. DOI: 10.1016/j.advwatres.2010.07.002.

[182] J. Magda, M. Tirrell, and H. Davis. "Molecular dynamics of narrow, liquid-filled pores". *J. Chem. Phys.* 83 (1985), pp. 1888–1901. DOI: 10.1063/1.449375.

[183] O. Galteland. *Nanothermodynamics and Molecular Simulations of Fluids in Porous Media.* PhD thesis, Department of Chemistry, Norwegian University of Science and Technology, NTNU, Trondheim, Norway, 2022.

[184] J. Li, D. Liao, and S. Yip. "Coupling continuum to molecular-dynamics simulation: reflecting particle method and the field estimator". *Phys. Rev. E* 57 (1998), pp. 7259–7267. DOI: 10.1103/PhysRevE.75.7259.

[185] W.G. Hoover and B.L. Holian. "Kinetic moments method for the canonical ensemble distribution". *Phys. Lett. Sect. A* 211 (1996), pp. 253–257. DOI: 10.1016/0375-9601(95)00973-6.

[186] J.H. Irving and J.G. Kirkwood. "The statistical mechanical theory of transport processes. IV. The equations of hydrodynamics". *J. Chem. Phys.* 18 (1950), pp. 817–829. DOI: 10.1006/jcph.1995.1039.

[187] I. Franco, G.C. Schatz, and M.A. Ratner. "Single-molecule pulling and the folding of donor-acceptor oligorotaxanes: Phenomenology and interpretation". *J. Chem. Phys.* 131 (2009), p. 124902. DOI: 10.1063/1.3223729.

[188] D. Keller, D. Swigon, and C. Bustamante. "Relating single-molecule measurements to thermodynamics". *Biophys. J.* 84 (2003), pp. 733–738. DOI: 10.1016/S0006-3495(03)74892-9.

[189] P.-G. de Gennes. *Scaling Concepts in Polymer Physics.* Cornell University Press, Ithaca N.Y., 1979. DOI: 10.1163/_q3_SIM_00374.

[190] M. Doi and S.F. Edwards. *The Theory of Polymer Dynamics.* Clarendon Press, 1986, p. 391.

[191] M. Süzen, M. Sega, and C. Holm. "Ensemble inequivalence in single-molecule experiments". *Phys. Rev. E* 79 (2009), p. 051118. DOI: 10.1103/PhysRevE.79.051118.

[192] J.M. Rubi, D. Bedeaux, and S. Kjelstrup. "Thermodynamics for single-molecule stretching experiments". *J. Phys. Chem. B* 110 (2006), pp. 12733–12737. DOI: 10.1021/jp061840o.

[193] E. Bering. *Stretching, Breaking, and Dissolution of Polymeric Nanofibres by Computer Experiments. PhD.thesis.* PoreLab, Department of Physics, Trondheim: Norwegian University of Science and Technology, 2021.

[194] C. Chen, P. Depa, V.G. Sakai, J.K. Maranas, J.W. Lynn, I. Peral, and J.R.D. Copley. "A comparison of united atom, explicit atom, and coarse-grained simulation models for poly(ethylene oxide)". *J. Chem. Phys.* 124 (2006), 234901(11). DOI: 10.1063/1.2204035.

[195] F. Oesterhelt, M. Rief, and H.E. Gaub. "Single molecule force spectroscopy by AFM indicates helical structure of poly(ethylene-glycol) in water". *New J. Phys.* 1 (1999), pp. 6.1–6.11. DOI: 10.1088/1367-2630/1/1/006.

[196] J.E. Mark and P.J. Flory. "The configuration of the polyoxyethylene chain". *J. Am. Chem. Soc.* 87 (1965), pp. 1415–1423. DOI: 10.1021/ja01085a001.

[197] H. Tadokoro. *Structure of Crystalline Polymers.* Krieger, Florida, 1990, p. 465.

[198] A. Campa, T. Dauxois, D. Fanelli, and S. Ruffo. *Physics of Long-Range Interacting Systems.* Oxford University Press, 2014.

[199] R. Ellis, K. Haven, and B. Turkington. "Large deviation principles and complete equivalence and nonequivalence results for pure and mixed ensembles". *J. Stat. Phys.* 101 (2000), pp. 999–1064. DOI: 10.1023/A:1026446225804.

[200] W. Just, K. Gelfert, N. Baba, A. Riegert, and H. Kantz. "Elimination of fast chaotic degrees of freedom: On the accuracy of the born approximation". *J. Stat. Phys.* 112 (2003), pp. 277–292. DOI: 10.1023/A:1023635805818.

Index

Aasen, A., 234, 286
Abbott, M.M., 3, 11
Adjiman, C.S., 245
adsorption, 7, 219
algorithm, 95
Alklaibi, M.A., 275
Allen, M.P., 220
Altena, F., 275
Anbari, A., 163
Anderson, D.K., 190
Anderson, H.C., 88
Anderson, J.B., 196
Archibald, A.M., 250
Armstrong, R.T., 259
Artaxo, P., 233
Ashbaugh, H.S., 1
Avendano, C., 245

Baba, N., 352
Babb, A.L.J., 190
Baidakov, V.G., 234
Bak, H., 275
Bakker, H.J., 1
Bardow, A., 1, 10, 149, 161, 170, 186, 190, 195
Bear, A., 256, 259, 311
Becker, T.M., 1, 161, 186
Ben-Naim, A., 149, 170, 203, 366
bending regime, 345
Bennethum, L.S., 275
Berendsen's theorem, 57
Berg, C.F., 256, 329
Berg, S., 259
Bering, E., 8, 22, 37, 46, 334
Bernu, B., 195
Binder, K., 6
Blümich, B., 190
Blokhuis, E., 87, 234
Boltachev, G.S., 234
bond-stretching regime, 345
Boucher, O., 233
Braun, E., 275, 333
Bretherton, C., 233
Brett, B., 250
Bright, J., 250
Brouwer, H., 275
Buff, F.P., 149, 170, 203, 287, 366
Burovski, E., 250
Bustamante, C., 333

Cabriolu, R., 256, 266
Calero, S., 10, 161, 186
Callen, H.B., 3, 11, 305
Campa, A., 6, 347
canonical ensemble, 334
Carey, C.J., 250
Casanova, F., 190
Casati, L., 6
Ceperley, C., 195
Chamberlin, R.V., 5
Chandler, D., 88
characteristic size, 85
chemical potential, 4, 45
  integral, 144
  integral $\widehat{\mu}$, 45
Chen, C., 334

Chen, J.J., 275, 333
Chen, Z., 275
Chien, H.-T., 163
Ciccotti, G., 57
Cimrman, R., 250
classical mechanics, 56
Clausius, R., 3, 11
Collell, J., 219, 232
compressional energy, 41
contact line
  three-phase, 260, 270
Copley, J.R.D., 334
correlation length, 128, 151
correlations
  cross, 305
  self, 305
  particle density, 91
Cortes-Huerto, R., 1, 10
Costache, F., 219
coupling coefficient, 305
Cournapeau, D., 250
Cox, M.P., 219
curvature
  Gaussian, 87
Cussler, E.L., 185

Daivis, P.J., 287
Dao, T., 195
Darcy scale, 298
Darcy's law, 329
Dasgupta, S., 217
Datta, S.S., 163
Dauxois, T., 347
Dawass, N., 10, 158, 161
de Broglie wavelength, 104
de Gennes, P.-G., 333, 334, 338
de Groen, M., 161
de Groot, S.R., 5, 298
de Wijn, A.S., 8, 334
Debenedetti, P.G., 275
degrees of freedom, 57
Deng, W., 163
Depa, P., 334
Diedrich, D.L., 196
diffusion coefficients
  Fick, 150
  Maxwell-Stefan, 150
disjoining pressure, 184
Doi, M., 333, 334
Dong, W., 85, 103, 178, 184
Dong, X., 275
Dubrovski, V.G., 234

Economou, I.G., 161
Edwards, S.F., 333
egde- and corner energy, 357, 361
Ellis, R., 347
energy
  compressional, 38, 165, 281, 313, 321
  corner, 85
  internal, 3
  line, 23, 85
  potential, 56
  replica, 4
  surface, 23, 85
ensemble
  canonical, 14, 124, 334
  grand canonical, 14, 124, 163, 263
  isobaric, 14, 124, 334
  microcanonical, 14, 124
  nanocanonical, 14, 124
  of replicas, 6, 16, 234
  of small systems, 12
  osmotic, 14, 124
  independence, 151
ensemble distribution
  normalization of, 124
enthalpy
  differential, 164
  integral, 164
  molar, 169
  partial molar, 169
entropic regime, 334, 342
  of polymers, 338
entropy, 3, 12, 17, 27, 58, 135
  configurational, 261, 266
entropy production, 33, 37, 41, 46, 49, 332, 336–338, 340, 352

entropy-hat, 28
environmental variables, 7, 13, 24, 103
equation of state, 14, 15, 165
  for small system, 110
  integral chemical potential, 144
  single particle, 146
  single replica, 135
  single replica integral pressure, 139
equilibrium, 328
  conditions for, 51
equilibrium time correlations, 186
Erdös, M., 10, 236, 276, 289, 333
ergodic hypothesis, 19, 89
Erpelding, M., 306
Ertl, G., 219
Eslami, H., 275
Euler equation, 18
Euler homogeneous function, 16, 18, 234, 368
  of degree one, 20, 272
  of degree zero, 20
Euler, L., 4
Evans, D.J., 287
expansion
  in inverse length, 106
extensive, 4, 123

Fan, J., 163
Fanelli, D., 347
Feingold, G., 233
Feng, Y., 250
Fick's diffusivities, 185
Flekkøy, E.G., 306
Flory, P.J., 342
fluctuation
  internal energy, 133
  particle number, 133, 166
  thermodynamic properties, 147
  volume, 133
fluctuation-dissipation theorem, 298, 303
fluctuations, 303
force-elongation curve, 342, 345
Forster, P., 233
Fowler, E.A., 240
Franco, I., 333
Frenkel, D., 3, 6, 220, 286
Fujimoto, T., 195
Furuholt, H., 287

Galindo, A., 245
Galliero, G., 219, 232
Galteland, O., 8, 10, 18, 236, 256, 266, 276, 279, 289, 312, 328, 329, 333
ganglia, 274, 306
Ganguly, P., 10, 151
Gaub, H.E., 339
Gelfert, K., 352
Gibbs energy, 35, 53, 339, 345
Gibbs energy-hat, 36
Gibbs equation, 18, 33
Gibbs, J.W., 3, 11, 19, 178, 248
Gibbs-Duhem equation, 18
Giovambattista, N., 275
Gjennestad, M.Aa., 256
Gmehling, J., 192
Goddard, W.A., 217
Gommers, R., 250
Graham, S.C., 247
grand potential, 40, 68, 258, 263, 281
  statistical mechanics, 269
Gray, W.G., 256, 298, 311
Green, M.S., 305
Gross, J., 149, 192
Gubbins, K.E., 275, 286
Guggenheim, R.H., 240
Guo, H., 275
Gutzow, I., 234

Haberland, M., 250
Hadwiger's theorem, 87, 106, 112, 115, 116
Hadwiger, H., 87, 116
Hafskjold, B., 8, 195, 276, 279, 286, 287, 311
Haldrup, K., 1
Hamaker constant, 247

Hamiltonian, 123, 136, 138
  of small systems, 55
Hammer, M., 286
Hanemaaijer, J.H., 275
Hansen, A., 8, 256, 266, 279, 306, 311
Hardt, A.P., 190
Harris, C.R., 250
Hartkamp, R., 161
Hass, A.B., 286
Hassanizadeh, S.M., 256, 298, 311
Haven, K., 347
He, J., 10, 234, 275
He, X., 219
heat capacity, 137
  small system, 88, 142
Helfrich energy, 87
Helfrich, W., 19, 178
Helmholtz energy, 31, 53, 106, 339, 345
Helmholtz energy-hat, 31
Helmig, R., 256, 329
Henriksen, I., 250
Hilfer, R., 256
Hill's postulate, 15
  limitations, 6
Hill–Gibbs equation, 6, 25, 26, 42
Hill–Gibbs–Duhem equation, 18, 22, 35, 40, 272
Holm, C., 334
Holt, T., 297
Homer, J.B., 247
Hou, J., 275
Hunter, J.D., 250
hydrogen dissociation, 196
Hägg, M.-B., 219, 220

Ikeshoji, T., 195, 287
Imbihl, R., 219
integral relation, 166
  internal energy, 117, 120, 123
  pressure, 175
interaction potential
  three-particle, 196
internal energy, 12, 136
  coarse grained, 262
  configurational contribution, 262
  fluctuation, 133
Inzoli, I., 219, 266
isobaric ensemble, 334
isometric stretching, 8
isomorphic, 278
isotensional stretching, 8
isotherm
  Hill–Gibbs, 244
isothermal compressibility, 138, 142

Jackson, G., 245
Jamali, S.H., 1, 161, 186
Jansen, A., 275
Jeurgens, L.P.H., 219
Johannessen, E., 149
Jones, E., 250
Just, W., 352

Kamalvand, M., 275
Kantz, H., 352
Karan, K., 275
Keller, D., 333
Kerminen, V.-M., 233
Kern, R., 250
Keshavarzi, E., 275
Khanamiri, H.H., 256
kinetic energy, 56
Kirkwood, J.G., 149, 170, 203, 284, 287, 366
Kirkwood–Buff integrals, 7, 8, 112, 128, 131, 147, 149, 203
Kohen, D., 196
Kondo, Y., 233
Kouteva-Arguirova, S., 219
Krüger, P., 10, 150, 158, 161, 186, 203
Kremer, K., 1, 10
Kriesten, E., 190
Krishna, R., 185, 190
Kubo, R., 305
Kuiken, G.D.C., 185
Kuipers, N., 275

Lafitte, T., 245
LAMMPS, 83, 89, 340
Landau, D.P., 6
Langevin thermostat, 340
Larson, E., 250
Latella, I., 6, 334
Laxalde, D., 250
Legendre transforms, 13, 19, 73, 120, 123, 156
Legendre-Fenchel, 21, 22
Legendre-Fenchel transforms, 13, 55, 347, 349
Lennard-Jones cut-off potential, 372
Lennard-Jones potential, 88, 372, 374
Lennard-Jones spline potential, 286, 320, 372, 374
Liao, H., 233
Lie, A., 219
Lin, L.C., 275, 333
Lindeberg, E., 297
Lior, N., 275
Liouville equation, 55, 80
Liouville-von Neumann equation, 55, 80
Liu, H., 275
Liu, H.L., 85, 103
Liu, X., 149, 161, 170, 186
Lohmann, P., 233
Long, Y., 275, 286
long-range potential, 6
Lorant, F., 217
Lund, M., 1
Lynn, J.W., 334

Måløy, K.J., 306
Magro, W.R., 195
Mar, B.W., 190
Maranas, J.K., 334
Mark, J.E., 342
Marquardt, W., 190
Martin-Calvo, A., 10, 161, 186
mass density
 coarse grained, 261
Maxwell relations, 21, 32, 40, 45, 109, 112, 184
Maxwell velocity distribution, 124
Maxwell–Stefan diffusivities, 185
Mayorov, N., 250
Mazighi, R.J., 192
Mazur, P., 5, 81, 150, 298
McClure, J.E., 259
McDonald, I.R., 6, 220
McGarrity, E.S., 10, 161, 186
McQuarrie, D.A., 3, 11
mean field theory, 274
Mehdipour, N., 275
Meng, B., 1
Mie, G., 245
Millman, K.J., 250
Milzetti, J., 1, 10, 160
Minkowski functionals, 87
Mittemeijer, E.J., 219
molecular dynamics simulations, 3, 89
molecular stretching experiments, 8
Monte Carlo simulations, 89
Moore, E.W., 250
Morse potential, 345
Moulin, P., 275
Moultos, O.A., 1, 10, 161, 236, 276, 289, 333
multi-phase flow, 299
Myers, A.L., 244

nanocanonical, 46, 73
nanothermodynamics, 5, 11, 13
Nauleau, F., 275
Nayar, D., 1, 10, 151
Nelson, A.R.J., 250
Newman, J., 275
noise
 Gaussian white, 308
non-extensive, 11

O'Connel, J., 188
Oesterhelt, F., 339
Oliphant, T.E., 250
Onken, U., 192
Onsager coefficients, 305
Oppenheim, I., 150

Ortiz de Zarate, J.M., 188, 305
Oyama, S.T., 195

p-hat, $\widehat{p}$, 39, 82, 110
pair correlation function, 117, 122, 123, 128, 151
  for small system, 168
pair-potential, 167, 374
palladium membrane, 196
partial molar energy
  in small system, 119
partial molar properties
  of small systems, 175, 370
particle correlation, 141, 367
particle correlation density, 91
particle number, 4, 12
  average, 144
  upper limit, 15
  fluctuation, 133
partition function, 55
  canonical ensemble, 135
  grand canonical ensemble, 139, 263
  isobaric ensemble, 137
  microcanonical ensemble, 134
  nanocanonical ensemble, 145
  osmotic ensemble, 143
Pedregosa, F., 250
Peral, I., 334
Perera, A., 192
Perez-Madrid, A., 6
Perktold, J., 250
Perng, T., 195
Perot, J., 275
Peterson, P., 250
Pharoah, J., 275
phase space, 124
Pierloni, C., 195
plasma modeling, 196
Plimpton, S., 286
Polat, I., 250
Poling, B., 188
pore
  vapor pressure, 275
porous media, 178
porous media transport, 8
Potestio, R., 1, 10
Potoff, J.J., 220
Prausnitz, J.M., 188
pressure, 3
  differential, 38, 82, 109, 139, 163, 166, 276, 278, 283, 314, 316, 321
  integral, 38, 68, 82, 109, 139, 163, 276, 278, 281, 313, 316, 321
  mechanical, 287
pressure p-hat, $\widehat{p}$, 163
pressure tensor, 287
probability distribution, 17, 55, 59, 124
  in phase space, 269

Qiao, C.Z., 85, 103
quantum mechanics, 56
Quintero, E.A., 250

Ramdin, M., 161
Ramstad, T., 259, 306
Randall, D., 233
Rasch, P., 233
Rathbun, R., 190
Ratner, M.A., 333
Rauter, M.T., 10, 18, 236, 276, 287, 289, 328, 333
Reddy, T., 250
reduced units, 89, 372
Reguera, D., 234
Reif, J., 219
Reimer, J.A., 275, 333
Reiss, H., 234
replica energy, 4, 12, 23, 31, 42, 46, 113
  determination of, 25
  grand canonical ensemble, 263
replica ensemble, 4
replica entropy, 27
representative elementary volume
  REV, 255, 266, 299, 312
Ribeiro, A.H., 250

Rief, M., 339
Riegert, A., 352
rigidities, 85, 108
Roke, S., 1
Rosenfeld, J.L.J., 247
Rossky, P.J., 275
Rubi, J.M., 5, 334
Ruffo, S., 6, 347
Rusanov, A.I., 234
Rustig, J., 219
Ryckaert, J.P, 57

Süzen, M., 334
Sakai, V.G., 334
sampling routine, 96
Satheesh, S.K., 233
saturation
    non-wetting fluid, 306
    wetting fluid, 306
Saul, A., 195
Savani, I., 266, 306, 312
Sawada, K., 195
scaling law, 40, 85, 102, 316, 317
Schacht, C., 192
Schatz, G.C., 333
Schmelzer, J.W., 234
seepage velocity, 306
Sega, M., 334
Sengers, J.V., 188, 305
shape, 12, 15, 83
    of small system, 15
    as variable, 24, 93
shape-dependence, 43
Sheridan, E., 219
Sherwood, S., 233
Siepman, J.I., 220
Siepmann, J.I., 6
Simon, J.-M., 10, 57, 83, 149, 158, 161, 168, 170, 186, 195, 219, 236, 266, 275, 366
simply connected system, 6
simulation box, 96
single polymer
    isometric stretching, 46
    isotensional stretching, 37
Sinha, S., 266, 306, 312
size, 12, 15
size-dependence, 6, 11, 23, 83, 103
Skorpa, R., 10, 57, 168, 195, 218, 366
slit pore
    with droplet, 276
Sliwinska-Bartkowiak, M., 275, 286
Slotte, P.A., 256
Small System Method, 7, 8, 83, 85, 95
small system scaling, 85
Smit, B., 3, 275, 286, 333
Smith, J., 3, 11, 117
Sokolic, F., 192
spreading pressure, 240
Stevens, B., 233
Stillinger, F.H., 196
Stirling's approximation, 61, 267
Strøm, B.A., 10, 51, 234, 275, 279
sub-sampling, 89
subdivision potential, 4, 12, 17, 40, 46, 110, 114, 146, 235, 244
subsystems
    weakly coupled, 269
Sun, W., 275
Surampudi, L.N., 1
surface
    curvature, 266
    dividing, 248, 260
    autonomous nature, 270
    energy, 57, 85
surface tension, 184
    differential, 184, 315
    integral, 184, 314
Swigon, D., 333
Swihart, M., 3, 11
symmetry relation, 360
system
    simply connected, 15

Tadokoro, H., 344
Takla, M., 195
Tallakstad, K.T., 306
Taylor, R., 185, 190
temperature, 3

thermodynamic factor, 108, 109, 111, 140, 156, 359
thermodynamic limit, 24, 156, 165, 338
three-particle interactions, 57
Tiedemann W., 275
Tildesley, D.J., 220
time-scale, 299
Todd, B.D., 287
Tolman length, 85, 108, 234
Tolman, R.C., 19, 178, 234
Torsaeter, O., 256
torsional regime, 344
Touchette, H., 347
Travis, K.P., 286
Trinh, T.T., 8, 219, 220, 275
Tsai, D.H., 276
Tully, J.C., 196
Turkington, B., 347

van Baten, J.M., 161, 186
van der Vegt, N.F.A., 1, 10, 160
van der Vleuten, P., 275
van der Walt, S.J., 250
van Duin, A.C.T., 217
van Erp, T., 220
van Medevoort, J., 275
van Mulbregt, P., 250
van Ness, H.C., 3, 11, 117
VanderPlas, J., 250
variables
  additive, 258, 260
  coarse-grained, 258, 298
  configurational, 258, 267
Vassvik, M., 266, 312
Virtanen, P., 250
Vlugt, T.J.H., 1, 10, 83, 150, 158, 161, 170, 186, 195, 220, 236, 275, 289, 333
Voda, M.A., 190
Voldsund, M., 195
volume, 3, 12
  fluctuation, 133

Wagner, W., 195
Wang, Z.M., 219
Ward, T.L., 195
WCA potential, 88, 372
Weber, T.A., 196
Weckesser, W., 250
Weeks, J.D., 88
Weeks-Chandler-Anderson potential, 372, 374
weight function, 158
Weinstein, T., 275
Weiss, K., 1
Weitz, D.A., 163
Welton, T.A., 305
Wesselingh, J.A., 188
Widom, B., 3, 11
Wilhelmsen, Ø., 234, 286
Williams, S.M., 1
Wilson, J., 250
Winkler, M., 256, 266
Wolff, L., 1, 161, 186
work
  by the differential pressure, 164
  by the integral pressure, 164
Wu, J.K., 195
Wu, K., 275
Wyart, Y., 275

Xu, J., 196

Young equation, 282
Young's law, 276
Young-Laplace equation, 282
Young-Laplace law, 276, 316, 317, 328
Yun, S., 195

Zhang, X.Y., 233
Zhao, S.L., 85, 103
Zheng, J., 1
Zoranic, L., 192

www.ingramcontent.com/pod-product-compliance
Ingram Content Group UK Ltd.
Pitfield, Milton Keynes, MK11 3LW, UK
UKHW022105170926
8624IPUK00024B/125

9 789811 274992